LONDON MATHEMATICAL SOCIETY LECTURE NOTE SERIES

Managing Editor: Professor J.W.S. Cassels, Department of Pure Mathematics and Mathematical Statistics, University of Cambridge, 16 Mill Lane, Cambridge CB2 1SB, England

The books in the series listed below are available from booksellers, or, in case of difficulty, from Cambridge University Press.

London Mathematical Society Lecture Note Series, 177

Applications of Categories in Computer Science

Proceedings of the LMS Symposium, Durham 1991

Edited by
M. P. Fourman
University of Edinburgh
P. T. Johnstone
University of Cambridge
A. M. Pitts
University of Cambridge

CAMBRIDGE
UNIVERSITY PRESS

Published by the Press Syndicate of the University of Cambridge
The Pitt Building, Trumpington Street, Cambridge CB2 1RP
40 West 20th Street, New York, NY 10011-4211, USA
10 Stamford Road, Oakleigh, Victoria 3166, Australia

www.cambridge.org
Information on this title: www.cambridge.org/9780521427265

First published 1992

Library of Congress cataloguing in publication data available

British Library cataloguing in publication data available

ISBN-13 978-0-521-42726-5 paperback
ISBN-10 0-521-42726-6 paperback

Transferred to digital printing 2005

Contents

Preface

The London Mathematical Society Symposium on Applications of Categories in Computer Science took place in the Department of Mathematical Sciences at the University of Durham from 20 to 30 July 1991. Although the interaction between the mathematical theory of categories and theoretical computer science is by no means a recent phenomenon, the last few years have seen a marked upsurge in activity in this area. Consequently this was a very well-attended and lively Symposium. There were 100 participants, 73 receiving partial financial support from the Science and Engineering Research Council. The scientific aspects of the meeting were organized by Michael Fourman (Edinburgh), Peter Johnstone (Cambridge) and Andrew Pitts (Cambridge). A programme committee consisting of the three organizers together with Samson Abramsky (Imperial College), Pierre-Louis Curien (ENS, Paris) and Glynn Winskel (Aarhus) decided the final details of the scientific programme. There were 62 talks, eight of which were by the four key speakers who were Pierre-Louis Curien, Peter Freyd (Pennsylvania), John Reynolds (Carnegie-Mellon) and Glynn Winskel.

The papers in this volume represent final versions of a selection of the material presented at the Symposium, or in one case (the paper which stands last in the volume) of a development arising out of discussions which took place at the Symposium. We hope that they collectively present a balanced overview of the current state of research on the intersection between categories and computer science. All the papers have been refereed; we regret that pressure of space obliged us to exclude one or two papers that received favourable referee's reports. We should like to thank all those who submitted papers and all who acted as referees, in particular those (the majority) who made efforts to conform to the tight timetable we set. Their co-operation is much appreciated.

Michael Fourman

February 1992

Peter Johnstone

Andrew Pitts

L.M.S. DURHAM SYMPOSIUM ON
APPLICATIONS OF CATEGORIES IN COMPUTER SCIENCE
20-30 July 1991

1. Professor J.L. MacDonald	30. Dr. P. Taylor	59. Professor E.G. Wagner
2. Professor G. Winskel	31. Dr. D. Rydeheard	60. Dr. C.-H.L. Ong
3. Professor C. Gunter	32. Dr. A. Obtulowicz	61. Mr. H. Yan
4. Dr. E. Moggi	33. Mr. A. Hutchinson	62. Dr. R.F.C. Walters
5. Dr. P.-L. Curien	34. Mr. I. Stark	63. Mr. N. Ghani
6. Dr. P.T. Johnstone	35. Dr. G.C.Wraith	64. Professor M. Barr
7. Dr. A.M. Pitts	36. B. Carpenter	65. Professor S. Abramsky
8. Professor M.P. Fourman	37. Mr. O. de Moor	66. Professor P.J. Scott
9. Professor P.J. Freyd	38. Dr. R. Dyckhoff	67. Professor J.W. Gray
10. Professor J.C. Reynolds	39. Mr. A. Laakkonen	68. Mr. Y. Fu
11. Dr. J. Meseguer	40. Mr. B. Hilken	69. Dr. S. Vickers
12. Professor G.M. Kelly	41. Mr. R. Blute	70. Dr. D.T. Sannella
13. Dr. B. Jacobs	42. Mr. E. Ritter	71. Mr. R. Diaconescu
14. Mr. S. Ambler	43. Dr. F.J. Oles	72. Mr. S. Carmody
15. Mr. A. Lent	44. Dr. H. Simmons	73. Dr. T. Streicher
16. Mr. J. McKinna	45. Dr. C.B. Jay	74. Dr. J. Launchbury
17. Mr. W. Phoa	46. Professor S. Brookes	75. Professor D. Schmidt
18. Dr. D. Murphy	47. Dr. A. Stoughton	76. Dr. F.-J. de Vries
19. Dr. Y. Lafont	48. Professor M. Mislove	77. Mr. M.J. Cole
20. Ms S. Finkelstein	49. Dr. P. Dybjer	78. Dr. J.G. Stell
21. Dr. V.C.V. de Paiva	50. Mr. J.T. Wilson	79. Dr. R. Cockett
22. Dr. P. Wadler	51. Dr. D. Gurr	80. Professor R.A.G. Seely
23. Dr. M. Johnson	52. Professor D.B. Benson	81. Dr. A. Carboni
24. Dr. C. Brown	53. Dr. M. Droste	82. Professor C. Wells
25. Ms P. Gardner	54. Dr. G. Rosolini	83. Dr. P. Mendler
26. Dr. C.J. Mulvey	55. Mr. T. Fukushima	84. Mr. A. Simpson
27. Dr. F. Lamarche	56. Dr. S. Kasangian	85. Mr. M. Fiore
28. Professor P. Mulry	57. Mr. R.L. Crole	
29. Dr. R. Wachter	58. Professor R. Tennent	

Computational Comonads
and Intensional Semantics

Stephen Brookes[*]   *Shai Geva*[*]

Abstract

We explore some foundational issues in the development of
a theory of intensional semantics, in which program denotations
may convey information about computation strategy in addition
to the usual extensional information. Beginning with an "extensional" category $\mathcal{C}$, whose morphisms we can think of as functions
of some kind, we model a notion of computation using a comonad
with certain extra structure and we regard the Kleisli category of
the comonad as an intensional category. An intensional morphism,
or algorithm, can be thought of as a function from computations
to values, or as a function from values to values equipped with a
computation strategy. Under certain rather general assumptions
the underlying category $\mathcal{C}$ can be recovered from the Kleisli category by taking a quotient, derived from a congruence relation
that we call extensional equivalence. We then focus on the case
where the underlying category is cartesian closed. Under further
assumptions the Kleisli category satisfies a weak form of cartesian
closure: application morphisms exist, currying and uncurrying of
morphisms make sense, and the diagram for exponentiation commutes up to extensional equivalence. When the underlying category is an ordered category we identify conditions under which
the exponentiation diagram commutes up to an inequality. We
illustrate these ideas and results by introducing some notions of
computation on domains and by discussing the properties of the
corresponding categories of algorithms on domains.

[*]School of Computer Science, Carnegie Mellon University, Pittsburgh, Pa 15213, USA.
This research was supported in part by National Science Foundation grant CCR-9006064
and in part by DARPA/NSF grant CCR-8906483.

1   Introduction

Most existing denotational semantic treatments of programming languages
are extensional, in that they abstract away from computational details and
ascribe essentially extensional meanings to programs. For instance, in the
standard denotational treatment of imperative while-programs the meaning
of a program is taken to be a partial function from states to states; and in
the standard denotational model of the simply typed λ-calculus, the meaning
of a term of function type is taken to be a continuous function. Extensional
models are appropriate if one wants to reason only about extensional properties of programs, such as partial correctness of while-programs. However,
such models give no insight into questions concerning essentially intensional
aspects of program behavior, such as efficiency or complexity. For instance,
in a typical extensional model all sorting programs denote the same function,
regardless of their computation strategy, and therefore regardless of their
worst- or average-case behavior. It is desirable to have a semantic model in
which sensible comparisons can be made between programs with the same
extensional behavior, on the basis of their computation strategy.

We emphasize that we regard intensionality as a relative term; given a programming language we might wish to provide an extensional semantics and
also an intensional semantics that contains more computational information
and is thus at a lower level of abstraction. We would like to be able to extract
extensional meanings from intensional meanings, and to show that the intensional semantics "fits properly" on top of the extensional semantics. Suppose
that we have an extensional semantics provided in a category whose objects
represent sets of data values and whose morphisms are functions of some
kind; and that we have an intensional semantics in a category with the same
objects but with morphisms that we regard as algorithms, which correspond
to functions equipped with a computation strategy. It seems reasonable that
we should be able to define an equivalence relation on algorithms (in the
same hom-set) that identifies all pairs of algorithms with the same "extensional part"; that composition of algorithms should respect this equivalence;
and that quotienting the algorithms from A to B by this equivalence relation
should yield precisely the extensional morphisms.

In this paper we set out a basis for a category-theoretic approach to intensional semantics, motivated by the following intuition. If the extensional
meaning of a program may be modelled as some kind of function from data
values to data values, then we can obtain an intensional semantics by introducing a notion of computation and defining an intensional meaning to be a
function from computations to values. This accords with an intuitive operational semantics for programs in which the execution of a program proceeds
lazily in a demand-driven, coroutine-like manner [10]: the program responds

to requests for output (say, from a user) by performing input computation until it has sufficient information to supply an output value. We formalize what we mean by a notion of computation in abstract terms as follows. Suppose that extensional meanings are given in some category $\mathcal{C}$. Then, for each object A, we specify an object TA of computations over A and we specify how to lift a morphism f from A to B into a morphism Tf from TA to TB; we require that T be a functor on $\mathcal{C}$. We specify, for each object A, a morphism $\epsilon_A : TA \rightarrow A$ from computations to values and a morphism $\delta_A : TA \rightarrow T^2A$ that maps a computation over A to a computation over TA. Intuitively, ϵ maps a computation to the value it computes, and δ shows how a computation may itself be computed. We require that (T, ϵ, δ) be a comonad over $\mathcal{C}$. Then we regard the Kleisli category of this comonad as an intensional category; it has the same objects as $\mathcal{C}$, and an intensional morphism from A to B is just a morphism in $\mathcal{C}$ from TA to B.

We say that a comonad is computational if there is a natural way to convert a data value into a degenerate computation returning that value. This enables us to extract from an algorithm a function from values to values, and we obtain an extensional equivalence relation on algorithms by identifying all pairs of algorithms that determine the same function. We show that if the comonad is computational then the Kleisli category collapses onto $\mathcal{C}$ under extensional equivalence. This means that an intensional semantics based on a computational comonad can be viewed as a conservative extension of a corresponding extensional semantics based on the underlying category. We may therefore use this framework for reasoning at different levels of abstraction, while being assured that the addition of intensional information does not interfere with purely extensional aspects of program behavior.

We then show that, assuming certain further conditions concerning products, the Kleisli category satisfies an intensional analogue of the cartesian closedness property. This generalizes from the known result that if the underlying category is cartesian closed and the (functor part of the) comonad preserves products then the Kleisli category is also cartesian closed. When the underlying category $\mathcal{C}$ is an ordered category, we identify conditions under which the Kleisli construction preserves certain lax forms of cartesian closedness.

Throughout the paper we motivate our definitions and results by means of notions of computation on domains. We focus primarily on three forms of computation at differing levels of abstraction. At the end of the paper we discuss briefly some further examples that indicate the broader applicability of our ideas. We also mention the relationship between our work and the sequential algorithms model of Berry and Curien [6].

We assume familiarity with elementary category theory and domain theory.

We refer the reader to [11] and [1] for categorical background and to [8] for the relevant domain theory.

2   Computations, Comonads and Algorithms

Let $\mathcal{C}$ be a category that we regard as providing an extensional framework. We wish to encapsulate in abstract terms what a notion of computation over $\mathcal{C}$ is, and to build an "intensional" category whose morphisms can be thought of as extensional morphisms equipped with additional computational information. We model a notion of computation over $\mathcal{C}$ using a comonad over $\mathcal{C}$, the functor part of which maps an object A to an object TA representing computations over A. The two other components of the comonad describe how to extract a value from a computation, and how a computation is built up from its sub-computations. We then take an intensional morphism from A to B to be an extensional morphism from TA to B, essentially a morphism from *input computations* over A to *output values* in B. This leads us to use for our intensional category the Kleisli category $\mathcal{C}_T$ [11], which has the same objects as $\mathcal{C}$ and in which the morphisms from A to B are exactly the $\mathcal{C}$-morphisms from TA to B. Typically $\mathcal{C}$ is a category in which morphisms are functions of some kind, and we will refer to intensional morphisms in $\mathcal{C}_T$ as *algorithms* to emphasize their computational content. In case we need to compare Kleisli categories for different comonads over the same underlying category we will use the term T-algorithm, indicating the comonad explicitly.

2.1   Comonads and the Kleisli category

Definition 2.1 A *comonad* over a category $\mathcal{C}$ is a triple (T, ϵ, δ) where $T : \mathcal{C} \to \mathcal{C}$ is a functor, $\epsilon : T \to I_{\mathcal{C}}$ is a natural transformation from T to the identity functor, and $\delta : T \to T^2$ is a natural transformation from T to T^2, such that the following associativity and identity conditions hold, for every object A:

$$T(\delta_A) \circ \delta_A = \delta_{TA} \circ \delta_A$$
$$\epsilon_{TA} \circ \delta_A = T(\epsilon_A) \circ \delta_A = \mathrm{id}_{TA}.$$

Figures 1 and 2 express these requirements in diagrammatic form.   •

Definition 2.2 Given a comonad (T, ϵ, δ) over $\mathcal{C}$, the *Kleisli category* $\mathcal{C}_T$ is defined by:

- The objects of $\mathcal{C}_T$ are the objects of $\mathcal{C}$.

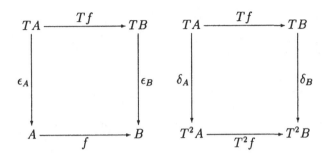

Figure 1. Naturality of ϵ and δ in a comonad: these diagrams commute, for all A, B, $f : A \to^{\mathcal{C}} B$.

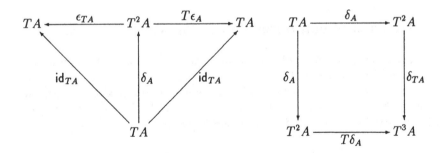

Figure 2. Identity and associativity laws of a comonad: these diagrams commute, for all A.

- The morphisms from A to B in $\mathcal{C}_T$ are the morphisms from TA to B in $\mathcal{C}$.

- The identity morphism $\widehat{\mathrm{id}}_A$ on A in $\mathcal{C}_T$ is $\epsilon_A : TA \to^{\mathcal{C}} A$.

- The composition in $\mathcal{C}_T$ of $a : A \to^{\mathcal{C}_T} B$ and $a' : B \to^{\mathcal{C}_T} C$, denoted $a' \bar{\circ} a$, is the composition in $\mathcal{C}$ of $\delta_A : TA \to^{\mathcal{C}} T^2 A$, $Ta : T^2 A \to^{\mathcal{C}} TB$ and $a' : TB \to^{\mathcal{C}} C$, i.e.,

$$a' \bar{\circ} a = a' \circ Ta \circ \delta_A.$$

The associativity and identity laws of the comonad ensure that $\mathcal{C}_T$ is a category [11].   •

This use of morphisms from TA to B to model algorithms from A to B fits well with an intuitive operational semantics based on the coroutine mechanism [10]. A program responds to requests for output by requiring some computation on its input (typically, to evaluate some portion of the input) until it has enough information to determine what output value to produce. Execution is lazy, in that computation is demand-driven. The operational behavior of algorithm composition can be described as follows. Let $a : A \to^{\mathcal{C}_T} B$ and $a' : B \to^{\mathcal{C}_T} C$. Then $a' \bar{\circ} a$ responds to a request for output (in C) by performing an input computation t over A, transforming this into a computation t' over B by applying δ and then Ta to t, and supplying t' as argument to a'. For further details concerning operational semantics we refer to [5].

3   Notions of computation on domains

Our main examples will be based on a category of domains and continuous functions. To avoid repetition and to be precise, let us remark that by a domain we mean a directed-complete, bounded-complete, algebraic partial order with a least element. That is, a domain $(D, \sqsubseteq)$ is a set D equipped with a partial order $\sqsubseteq$ satisfying the following conditions:

- D has a least element, denoted $\perp_D$.

- Every non-empty directed subset $X \subseteq D$ has a least upper bound $\bigsqcup X$.

- Every non-empty bounded subset $X \subseteq D$ has a least upper bound $\bigsqcup X$.

- Every element of D is the least upper bound of its (directed set of) finite approximations.

A set X is directed iff for all $x, y \in X$ there is a $z \in X$ such that $x \sqsubseteq z$ and $y \sqsubseteq z$. A set X is bounded (or consistent) iff there is a $z \in D$ such that $x \sqsubseteq z$ for all $x \in X$. An element $e \in D$ is finite (or isolated) iff, for all directed subsets $X \subseteq D$, if $e \sqsubseteq \bigsqcup X$ then $e \sqsubseteq x$ for some $x \in X$.

We remark that none of our example comonads really requires that the underlying domain be algebraic, and nor does the presence of a least element play any prominent rôle (except, of course, in justifying the existence of least fixed points). We could just as well work in the category of directed-complete, bounded complete partial orders and continuous functions. Nevertheless, the property of algebraicity is very natural in the computational setting and all of our example functors on domains preserve algebraicity. At the end of the paper we will discuss further examples based on different categories and different types of domain.

3.1 Increasing paths

The first notion of computation that we introduce models in abstract terms a sequence of time steps in which some incremental evaluation is being performed. For example, a program with two inputs may need to evaluate one or more of its input arguments and it may attempt to perform evaluations in parallel or in sequence; moreover, it may only require partial information about its arguments, as is typically the case, say, when an argument is a function and the program needs to apply that argument to an already known parameter. One natural way to formalize this form of computation is as an increasing sequence of values drawn from a domain, whose partial order reflects the amount of information inherent in a value.

We define the comonad T_1 of "increasing paths" as follows[1].

- For a given domain $(D, \sqsubseteq)$, let $T_1 D$ be the set of infinite increasing sequences over D, ordered componentwise. Thus, the elements of $T_1 D$ have form $\langle d_n \rangle_{n=0}^{\infty}$, where for each $n \geq 0$, $d_n \sqsubseteq d_{n+1}$; and we define $\langle d_n \rangle_{n=0}^{\infty} \sqsubseteq_{T_1 D} \langle d'_n \rangle_{n=0}^{\infty}$ iff for all $n \geq 0$, $d_n \sqsubseteq_D d'_n$.

- For a continuous function $f : D \to D'$, let $T_1 f : T_1 D \to T_1 D'$ be the function that applies f componentwise. That is, $(T_1 f)\langle d_n \rangle_{n=0}^{\infty} = \langle f d_n \rangle_{n=0}^{\infty}$.

- For $t \in T_1 D$ let $\epsilon_D t$ be the least upper bound of t. That is, $\epsilon \langle d_n \rangle_{n=0}^{\infty} = \bigsqcup_{n=0}^{\infty} d_n$.

---

[1]This comonad, adapted for Scott domains, was first introduced in [4].

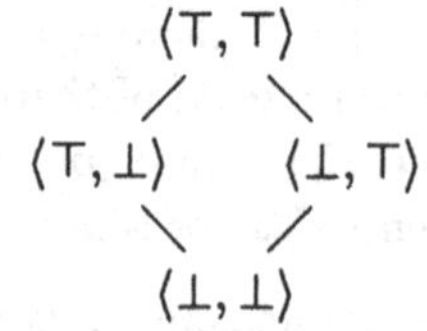

$$\langle \top, \top \rangle$$

$$\langle \top, \bot \rangle \qquad \langle \bot, \top \rangle$$

$$\langle \bot, \bot \rangle$$

Figure 3. The domain Two $\times$ Two.

- For $t \in T_1 D$ let $\delta_D t$ be the sequence of prefixes of t, each represented as an infinite sequence. That is, if $t = \langle d_n \rangle_{n=0}^{\infty}$, then for each $n \geq 0$, $(\delta t)_n = d_0 \ldots d_n d_n^{\omega}$.

Intuitively, a computation may be viewed as a (time-indexed) sequence of increments in the amount of information known about a data value, and the value "computed" by such a computation is its least upper bound; each computation is itself built up progressively from its prefixes. Equivalently, we can regard a computation over D as a continuous function from the domain VNat of "vertical" natural numbers (together with limit point ω) to D. Our ordering on computations then corresponds exactly to the pointwise ordering on such functions.

The least element of $T_1 D$ is the sequence $\bot^{\omega}$. The finite (isolated) elements of $T_1 D$ are just the eventually constant sequences all of whose elements are finite in D. It is easy to verify that T_1 maps domains to domains and is indeed a functor. The comonad laws hold: naturality of ϵ corresponds to continuity of f; naturality of δ states that the operation of applying a function componentwise to a sequence "commutes" with taking prefixes; every computation is the least upper bound of its prefixes; and every prefix of a prefix of t is also a prefix of t.

For illustration, let Two be the domain $\{\bot, \top\}$. The domain Two $\times$ Two is shown in Figure 3, and Figure 4 shows the six continuous functions from Two $\times$ Two to Two, ordered pointwise. We give these functions mnemonic names: $\bot$ and $\top$ are constant functions; l is strict in its left argument; r is strict in its right argument; b is strict in both arguments; poll returns $\top$ if either of its two arguments is $\top$, so that poll is not strict in either argument. Each function is depicted by a Hasse diagram corresponding to Figure 3, in which the nodes are shaded to indicate their image under the function being described: o corresponds to $\bot$, • to $\top$.

Figure 5 shows part of $T_1(\text{Two} \times \text{Two})$. Figure 6 shows some of the T_1-algorithms from Two $\times$ Two to Two, ordered pointwise. The notation for describing algorithms is based on Figure 5, again with o representing $\bot$ and • representing $\top$. In each case the intended algorithm is the least continuous

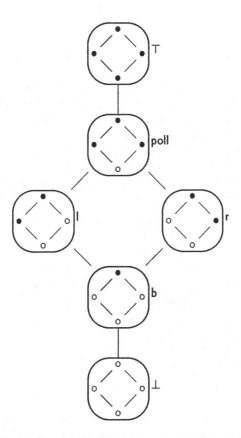

Figure 4. Continuous functions from **Two** × **Two** to **Two**, ordered pointwise.

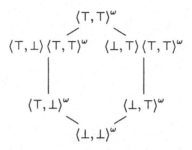

Figure 5. Some paths in $T_1(\mathtt{Two} \times \mathtt{Two})$.

function on paths consistent with this description. The nomenclature is intended to indicate (as yet only informally) the function computed by each algorithm and what computation strategy the algorithm uses. For instance, the algorithms **pb**, **lb**, **rb** and **db** all compute the function **b**; **pb** computes both arguments in parallel immediately, **lb** computes left-first and then right, **rb** computes right-first and then left, and **db** computes both arguments in either order. Since the diagram includes only one algorithm for **poll**, for ⊥, and for ⊤, in these cases we use the same name for the algorithm as for the function.

Since $T_1(\mathtt{Two} \times \mathtt{Two})$ includes paths with repeated steps, we can also make distinctions between algorithms which differ not in the order in which they evaluate their arguments, but in the amount of time they are prepared to wait for each successive increment to be achieved. For instance, for the function **b** there are algorithms pb_n, lb_n, rb_n and db_n for each $n \geq 0$, characterized as the least functions on paths such that

$$\mathsf{pb}_n(\langle \bot, \bot \rangle^n \langle \top, \top \rangle^\omega) = \top$$
$$\mathsf{lb}_n(\langle \bot, \bot \rangle^n \langle \top, \bot \rangle \langle \top, \bot \rangle^n \langle \top, \top \rangle^\omega) = \top$$
$$\mathsf{rb}_n(\langle \bot, \bot \rangle^n \langle \bot, \top \rangle \langle \bot, \top \rangle^n \langle \top.\top \rangle^\omega) = \top$$
$$\mathsf{db}_n(\langle \bot, \bot \rangle^n \langle \top, \bot \rangle \langle \top, \bot \rangle^n \langle \top, \top \rangle^\omega) = \top$$
$$\mathsf{db}_n(\langle \bot, \bot \rangle^n \langle \bot, \top \rangle \langle \bot, \top \rangle^n \langle \top, \top \rangle^\omega) = \top.$$

Informally, pb_n is the algorithm which needs to evaluate both arguments and returns ⊤ provided each evaluation succeeds (with result ⊤) in at most n time steps. Similarly, lb_n evaluates both arguments and returns ⊤ provided evaluation of the left argument succeeds in at most n time steps and evaluation of the right argument succeeds in at most $2n + 1$ time steps.

The following relationships hold, for all $n \geq 0$:

$$\mathsf{pb}_n \sqsubseteq \mathsf{lb}_n \sqsubseteq \mathsf{pb}_{2n+1}$$
$$\mathsf{pb}_n \sqsubseteq \mathsf{rb}_n \sqsubseteq \mathsf{pb}_{2n+1}$$
$$\mathsf{db}_n = \mathsf{lb}_n \sqcup \mathsf{rb}_n.$$

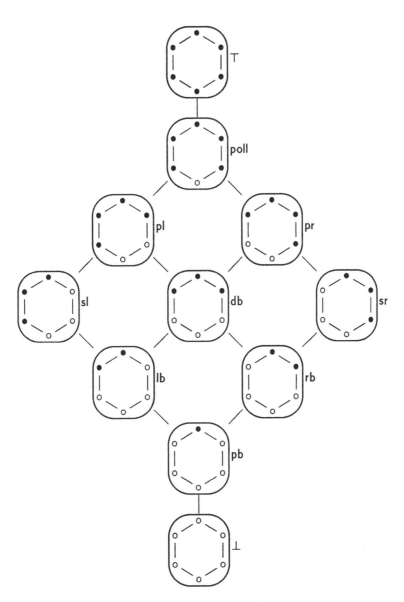

Figure 6. Some T_1-algorithms from Two × Two to Two, ordered pointwise.

Moreover, $\mathsf{pb}_n \sqsubseteq \mathsf{pb}_{n+1}$, $\mathsf{lb}_n \sqsubseteq \mathsf{lb}_{n+1}$, $\mathsf{r}_n \sqsubseteq \mathsf{r}_{n+1}$, and $\mathsf{db}_n \sqsubseteq \mathsf{db}_{n+1}$ for all $n \geq 0$. Each of these sequences of algorithms has the same least upper bound, characterized as the algorithm b_* that maps every path with lub $\langle \mathsf{T}, \mathsf{T} \rangle$ to T. Of course, in Figure 6, pb is just pb_0, and so on.

3.2   Strictly increasing paths

In the comonad T_1 a computation has a built-in measure of the number of time steps it takes between successive proper increments. We obtain a more abstract notion of computation by retaining only the increments themselves, so that we may still make distinctions on the basis of the order of evaluation of arguments. To do this we model a computation as a "strictly increasing path". We define the strictly increasing path comonad T_2 as follows.

- Let T_2D be the set of strictly increasing finite and infinite sequences over D. For convenience, we represent a finite sequence as an eventually constant infinite sequence. That is, the elements of T_2D are either of form $\langle d_n \rangle_{n=0}^{\infty}$, with $d_n \sqsubset_D d_{n+1}$ for all $n \geq 0$; or of form $d_0 \ldots d_{N-1}d_N^{\omega}$, where $N \geq 0$ and $d_n \sqsubset_D d_{n+1}$ for $0 \leq n < N$. Let $\sqsubseteq_{T_2D}$ be the least partial order on T_2D such that

$$d_0 \ldots d_{N-1}d_N^{\omega} \sqsubseteq_{T_2D} d_0 \ldots d_{N-1}t \quad \text{if } t \in T_2D \ \& \ d_N \sqsubseteq_D t_0.$$

  This ordering is based on the prefix ordering on sequences, but adjusted to take appropriate account of the underlying order on data values. The order $\sqsubseteq_{T_2D}$ is actually the stable ordering [2] on T_2D, when we regard the elements of T_2D as (strictly increasing, possibly eventually constant) stable functions from VNat to D. Note that every continuous function from VNat to D is also stable.

- For a continuous function $f : D \to D'$ we define T_2f to be the function which applies f componentwise and suppresses any repetitions (except for constant suffixes). That is, T_2f is the least continuous function such that for all $d \in D$, for all $d_0, d_1 \in D$ such that $d_0 \sqsubset d_1$, and for all $t \in T_2D$,

$$\begin{aligned} T_2f(d^{\omega}) &= (fd)^{\omega} \\ T_2f(d_0d_1t) &= (fd_0)(T_2f(d_1t)) \qquad \text{if } fd_0 \neq fd_1 \\ &= T_2f(d_1t) \qquad\qquad\; \text{if } fd_0 = fd_1. \end{aligned}$$

- For all $t \in T_2D$, let $\epsilon_D t$ be the lub of t. That is, $\epsilon \langle d_n \rangle_{n=0}^{\infty} = \bigsqcup_{n=0}^{\infty} d_n$.

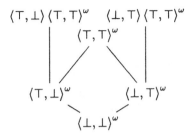

Figure 7. Part of $T_2(\text{Two} \times \text{Two})$.

- For all $t \in T_2 D$, let $\delta_D t$ be the sequence of prefixes of t. Again, if $t = \langle d_n \rangle_{n=0}^{\infty}$ then for each $n \geq 0$, $(\delta t)_n = d_0 \ldots d_n d_n^{\omega}$. Note that if t is strictly increasing, so is δt.

The least element of $T_2 D$ is the sequence $\perp^{\omega}$. The lub of a directed (or bounded) set of strictly increasing paths is again a strictly increasing path[2]. The finite elements of $T_2 D$ are the eventually constant sequences of form $\langle d_0 \ldots d_{N-1} \rangle d_N^{\omega}$, where $N \geq 0$ and d_N is a finite element of D. Every element of $T_2 D$ is the lub of its finite approximations. Thus, T_2 maps domains to domains. Functoriality is easily checked.

Although the order is not pointwise, it is still true that every $t \in T_2 D$ is the lub of its prefixes. The comonad laws hold for (T_2, ϵ, δ).

Figure 7 shows some of the paths in $T_2(\text{Two} \times \text{Two})$. Figure 8 shows some of the T_2-algorithms from $\text{Two} \times \text{Two}$ to Two, ordered pointwise, using a notation based on Figure 7. Again the nomenclature is chosen to indicate the function computed and the computation strategy. Each of the T_1-algorithms of Figure 6 has a corresponding T_2-algorithm, for which we use the same name; but because of the different ordering on paths, there are three additional T_2-algorithms. Note also that since $T_2 D$ does not include paths with repeated elements, only pb_0 and pb_1 of the family of pb_n algorithms defined above have corresponding T_2-algorithms.

3.3   Timed data

A simple notion of computation over domains is obtained by regarding a computation as a pair consisting of a data value and a natural number, intuitively representing the amount of time or the cost associated with the calculation

---

[2]However, this would not be the case if we ordered $T_2 D$ componentwise, since $T_2 D$ is not directed-complete under the componentwise ordering.

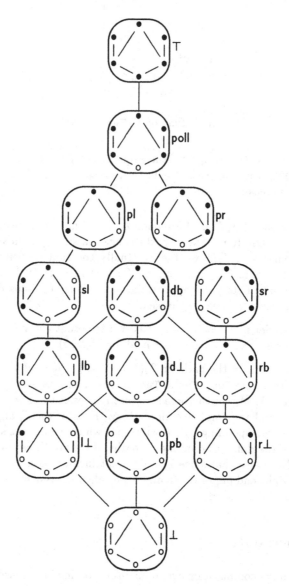

Figure 8. Some T_2-algorithms from Two × Two to Two, ordered pointwise.

of the value. With this intuitition it seems reasonable to regard one compu-
tation $\langle d, n \rangle$ as approximating another $\langle d', n' \rangle$ iff $d \sqsubseteq d'$ and $n' \leq n$; that is,
a better computation produces a more precise data value in less time. This
suggests the use of the following comonad:

- $T_3 D = D \times \text{VNat}^{\text{op}}$, ordered componentwise.

- For $f : D \to D'$, $(T_3 f) \langle d, n \rangle = \langle fd, n \rangle$.

- $\epsilon \langle d, n \rangle = d$.

- $\delta \langle d, n \rangle = \langle \langle d, n \rangle, n \rangle$.

Here VNat^{op} is the domain consisting of the natural numbers together with
ω, ordered by the reverse of the usual ordering, so that ω is the least element.
The least element of $T_3 D$ is $\langle \bot_D, \omega \rangle$.

We may define for each continuous function $f : D \to D'$ and each $n \in \text{VNat}$
an algorithm f_n from D to D':

$$
\begin{aligned}
f_n \langle d, k \rangle &= fd && \text{if } k \leq n \\
&= \bot && \text{otherwise.}
\end{aligned}
$$

Clearly, whenever $f \sqsubseteq g$ we also get $f_n \sqsubseteq g_n$. Moreover, because of our
ordering on $T_3 D$, we get $f_n \sqsubseteq f_{n+1}$ for each $n \geq 0$; and f_ω is simply $\lambda \langle d, n \rangle . fd$.
The lub of the f_n is the function $f_* = \lambda \langle d, n \rangle . (n = \omega \to \bot, fd)$, which is
below f_ω. Using this nomenclature, we show some of the T_3-algorithms from
$\text{Two} \times \text{Two}$ to Two in Figure 9.

It is also possible to define a comonad based on the functor $TD = D \times \text{VNat}$,
using the usual ordering on the integer component.

4 Relating algorithms and functions

4.1 Computational comonads

We will say that a comonad is *computational* if for each object A, every
data value in A can be regarded as a "degenerate" computation in TA, and
degenerate computations possess certain simple properties. This permits us to
extract from an algorithm a function from values to values, by looking at the
algorithm's effect when applied to degenerate computations. Two algorithms
are called extensionally equivalent iff they determine the same function.

More precisely, we define a computational comonad to be a comonad (T, ϵ, δ)
together with a natural transformation $\gamma : I_C \overset{\cdot}{\to} T$ satisfying some axioms

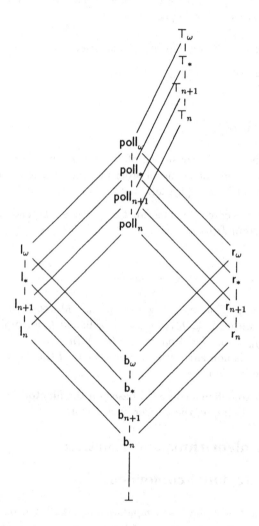

Figure 9. Some T_3-algorithms from Two × Two to Two, ordered pointwise.

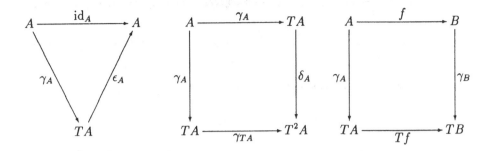

Figure 10. Properties of a computational comonad: these diagrams commute, for all A, B, and all morphisms $f : A \to^{\mathcal{C}} B$.

which capture formally what we mean by degeneracy. We then show that this permits us to define an "extensional equivalence" relation on each hom-set in $\mathcal{C}_T$. Extensional equivalence is preserved by composition, so that we actually have a congruence on $\mathcal{C}_T$. The underlying category $\mathcal{C}$ may then be recovered from $\mathcal{C}_T$ by taking a quotient.

Definition 4.1 A *computational comonad* over a category $\mathcal{C}$ is a quadruple $(T, \epsilon, \delta, \gamma)$ where (T, ϵ, δ) is a comonad over $\mathcal{C}$ and $\gamma : I_{\mathcal{C}} \xrightarrow{\cdot} T$ is a natural transformation such that, for every object A,

- $\epsilon_A \circ \gamma_A = \mathsf{id}_A$

- $\delta_A \circ \gamma_A = \gamma_{TA} \circ \gamma_A$.

Naturality guarantees that, for every morphism $f : A \to^{\mathcal{C}} B$,

- $Tf \circ \gamma_A = \gamma_B \circ f$.

Figure 10 shows these properties in diagrammatic form. •

As an immediate corollary of these properties, ϵ_A is epi and γ_A is mono, for every object A.

Proposition 4.2 *If $(T, \epsilon, \delta, \gamma)$ is a computational comonad, then there is a pair of functors* (alg, fun) *between $\mathcal{C}$ and $\mathcal{C}_T$ with the following definitions and properties:*

- **alg** : $\mathcal{C} \to \mathcal{C}_T$ *is the identity on objects, and* $\mathsf{alg}(f) = f \circ \epsilon_A$, *for every* $f : A \to^{\mathcal{C}} B$.

- **fun** : $\mathcal{C}_T \to \mathcal{C}$ *is the identity on objects, and* $\mathsf{fun}(a) = a \circ \gamma_A$, *for all* $a : A \to^{\mathcal{C}_T} B$.

- **fun** *induces an equivalence relation* $=^e$ *on* $\mathcal{C}_T$, *given by*

$$a_1 =^e a_2 \iff \mathsf{fun}(a_1) = \mathsf{fun}(a_2).$$

This relation is a congruence; that is, for all $a_1, a_2 : A \to^{\mathcal{C}_T} B$ *and* $a_1', a_2' : B \to^{\mathcal{C}_T} C$,

$$a_1 =^e a_2 \ \& \ a_1' =^e a_2' \quad \Rightarrow \quad a_1' \, \bar{\circ} \, a_1 =^e a_2' \, \bar{\circ} \, a_2.$$

- *The quotient category of* $\mathcal{C}_T$ *by* $=^e$ *is isomorphic to* $\mathcal{C}$.

- **fun** ∘ **alg** $= I_{\mathcal{C}}$. *That is, for all* $f : A \to^{\mathcal{C}} B$, $\mathsf{fun}(\mathsf{alg} \ f) = f$.

- **alg** ∘ **fun** $=^e I_{\mathcal{C}_T}$, *in that for all* $a : A \to^{\mathcal{C}_T} B$, $\mathsf{alg}(\mathsf{fun} \ a) =^e a$.

Proof: Functoriality of **alg** and **fun** are straightforward. For instance:

$$
\begin{aligned}
\mathsf{fun}(a' \, \bar{\circ} \, a) &= (a' \, \bar{\circ} \, a) \circ \gamma \\
&= (a' \circ Ta \circ \delta) \circ \gamma \\
&= a' \circ Ta \circ \gamma \circ \gamma \quad \text{since } \delta \circ \gamma = \gamma \circ \gamma \\
&= a' \circ \gamma \circ a \circ \gamma \quad \text{by naturality of } \gamma \\
&= \mathsf{fun}(a') \circ \mathsf{fun}(a).
\end{aligned}
$$

A similar calculation shows that $=^e$ is a congruence.

The quotient of $\mathcal{C}_T$ by $=^e$ has the same objects as $\mathcal{C}_T$ (and therefore the same objects as $\mathcal{C}$), and the morphisms in the quotient category from A to B are the $=^e$-equivalence classes of morphisms from A to B in $\mathcal{C}_T$. Let us write $[a]$ for the equivalence class of a. Clearly, the map $f \mapsto [\mathsf{alg}(f)]$ is an isomorphism of hom-sets, showing that $\mathcal{C}_T/_{=^e}$ is isomorphic to $\mathcal{C}$.

The facts that **fun** ∘ **alg** $= I_{\mathcal{C}}$ and **alg** ∘ **fun** $=^e I_{\mathcal{C}_T}$ are elementary consequences of the definitions.   □

We say that $\mathsf{fun}(a)$ is the extensional morphism computed by a. Since $\mathsf{fun}(\mathsf{alg} f) = f$, every extensional morphism f is computed by some (not necessarily unique) intensional morphism.

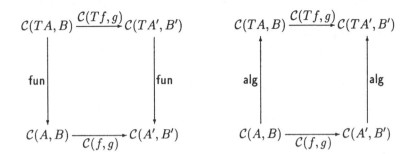

Figure 11. Naturality of fun and alg for a computational comonad: these diagrams commute, for all $f : A' \to^{\mathcal{C}} A$ and $g : B \to^{\mathcal{C}} B'$.

These results show that every computational comonad can be used to produce an intensional category that yields back the underlying extensional category when we take the extensional quotient. Next we show that fun and alg are natural transformations. Let Set be the category of sets and functions.

Definition 4.3 The two-variable hom-functor $\mathcal{C}(T(-), -)$ from $\mathcal{C}^{\mathrm{op}} \times \mathcal{C}$ to Set takes a pair (A, B) of objects to $\mathcal{C}(TA, B)$ and takes a pair of morphisms (f, g) with $f : A' \to^{\mathcal{C}} A$ and $g : B \to^{\mathcal{C}} B'$ to $\mathcal{C}(Tf, g) : \mathcal{C}(TA, B) \to \mathcal{C}(TA', B')$, where for all $a : TA \to^{\mathcal{C}} B$,

$$\mathcal{C}(Tf, g)(a) = g \circ a \circ Tf.$$

Similarly, the two-variable hom-functor $\mathcal{C}(-, -)$ takes (A, B) to $\mathcal{C}(A, B)$ and (f, g) to $\mathcal{C}(f, g)$, where for all $h : A \to^{\mathcal{C}} B$,

$$\mathcal{C}(f, g)(h) = g \circ h \circ f.$$

●

Proposition 4.4 Let $(T, \epsilon, \delta, \gamma)$ be a computational comonad over a category $\mathcal{C}$. Then fun and alg, as defined in Proposition 4.2, are natural transformations:

fun $: \mathcal{C}(T(-), -) \dashrightarrow \mathcal{C}(-, -)$
alg $: \mathcal{C}(-, -) \dashrightarrow \mathcal{C}(T(-), -)$.

That is, for all $f : A' \to^C A$ and $g : B \to^C B'$, the following identities hold:

$$\text{fun} \circ C(Tf, g) = C(f, g) \circ \text{fun}$$
$$C(Tf, g) \circ \text{alg} = \text{alg} \circ C(f, g).$$

Figure 11 expresses these properties in diagram form.

Proof: Straightforward, using naturality of γ and ϵ.   □

We now show that the three example comonads introduced earlier can be extended to become examples of computational comonads.

4.2   Examples.

1. For the increasing paths comonad T_1, let $\gamma_D : D \to T_1 D$ be defined by $\gamma_D d = d^\omega$, for all $d \in D$. Clearly γ_D is continuous, and γ is a natural transformation. Moreover, the computational comonad laws hold, since the lub of d^ω is d, and all prefixes of d^ω are equal to d^ω.

   The functor fun maps each algorithm from Two × Two to Two into a function from Two × Two to Two. In particular, $\text{fun}(\text{pb}) = \text{fun}(\text{lb}) = \text{fun}(\text{rb}) = \text{fun}(\text{db}) = \text{b}$. Similarly, $\text{fun}(\text{pb}_n) = \text{b}$ for all $n \geq 0$, and $\text{fun}(\text{b}_*) = \text{b}$. In fact, $\text{b}_* = \text{alg}(\text{b})$.

   Figure 12 illustrates the result of taking the extensional quotient of Figure 6. Boxes enclose equivalence classes of algorithms, arcs between boxes represent the quotient ordering, and within each box we retain the pointwise order to ease comparison with Figure 6. As expected, the quotient figure is isomorphic to Figure 4 when we identify each equivalence class with the function computed.

2. For the strictly increasing paths comonad, we again let $\gamma_D : D \to T_2 D$ be $\gamma_D d = d^\omega$. Again this is a continuous function, and γ is a natural transformation. Again the computational comonad laws hold. Figure 13 shows the quotient of Figure 8 by extensional equivalence. Note that $\text{fun}(\text{d}\bot) = \text{fun}(\text{l}\bot) = \text{fun}(\text{r}\bot) = \text{fun}(\bot) = \bot$.

   Again the quotient diagram is isomorphic to Figure 4.

3. For the timed data comonad, we obtain a suitable γ by deciding what cost to associate with a degenerate computation. For each $k \in \text{VNat}$ we may take $\gamma_k d = \langle d, k \rangle$ and obtain a natural transformation satisfying the requirements of a computational comonad. Define fun_k to be the functor whose action on algorithms is given by $\text{fun}_k(a) = a \circ \gamma_k$, and let

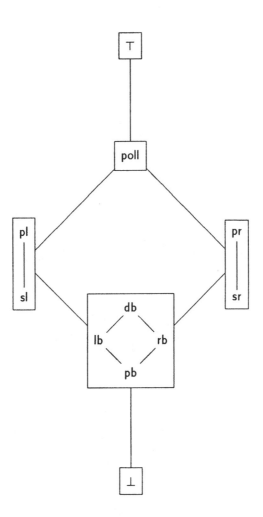

Figure 12. Quotient of Figure 6 by extensional equivalence.

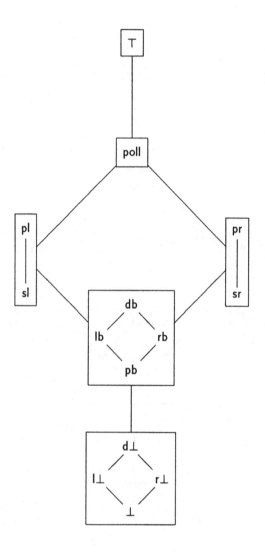

Figure 13. Quotient of Figure 8 by extensional equivalence.

$=^e{}_k$ be the equivalence relation induced by γ_k. For example, we have, for each $k \geq 0$:

$$
\begin{array}{lll}
\mathsf{fun}_k(\mathsf{b}_n) & = & \mathsf{b} \qquad\qquad \text{if } k \leq n \\
& = & \bot \qquad\qquad \text{if } k > n.
\end{array}
$$

Clearly, $\mathsf{b}_n =^e{}_k \mathsf{b}_{n+1}$ iff $k \neq n + 1$.

Again the Kleisli category quotients onto the underlying category under the congruence induced by γ_k. Figure 14 shows the quotient of Figure 9 under the equivalence induced by γ_{n+1}.

5 Products and Exponentiation

5.1 Products

Now suppose that the underlying category $\mathcal{C}$ has products, and for each pair of objects A_1 and A_2 there is a distinguished product, which we denote $A_1 \times A_2$, with π_i ($i = 1, 2$) being the projections. It is easy to show that distinguished product objects in $\mathcal{C}$ are also product objects in $\mathcal{C}_T$, with projections in $\mathcal{C}_T$ given by:

$$
\begin{array}{lll}
\hat{\pi}_i & : & A_1 \times A_2 \to^{\mathcal{C}_T} A_i \\
\hat{\pi}_i & = & \epsilon_{A_i} \circ T \pi_i \\
& = & \pi_i \circ \epsilon_{A_1 \times A_2}.
\end{array}
$$

Pairing of morphisms in $\mathcal{C}_T$ is the pairing of morphisms in $\mathcal{C}$, and the combination $\langle T \pi_1, T \pi_2 \rangle$ plays a special rôle in light of the following properties.

Proposition 5.1 *If $\mathcal{C}$ has products then $\sigma : T(- \times -) \dot\to T(-) \times T(-)$ defined by:*

$$
\begin{array}{l}
\sigma_{A,B} : T(A \times B) \to TA \times TB \\
\sigma_{A,B} = \langle T \pi_1, T \pi_2 \rangle
\end{array}
$$

is a natural transformation such that, for all objects A and B,

$$
\begin{array}{lll}
(\epsilon_A \times \epsilon_B) \circ \sigma_{A,B} & = & \epsilon_{A \times B} \\
(\delta_A \times \delta_B) \circ \sigma_{A,B} & = & \sigma_{TA,TB} \circ T\sigma_{A,B} \circ \delta_{A \times B} \\
\sigma_{A,B} \circ \gamma_{A \times B} & = & \gamma_A \times \gamma_B.
\end{array}
$$

Proof: Naturality of σ follows from the universal property of products and the functoriality of T. The remaining properties are easy consequences of naturality of ϵ, δ, and γ. Note the identity $\sigma \circ T \langle f, g \rangle = \langle Tf, Tg \rangle$. In particular, $\sigma \circ T\sigma = \langle T^2 \pi_1, T^2 \pi_2 \rangle$. $\qquad\square$

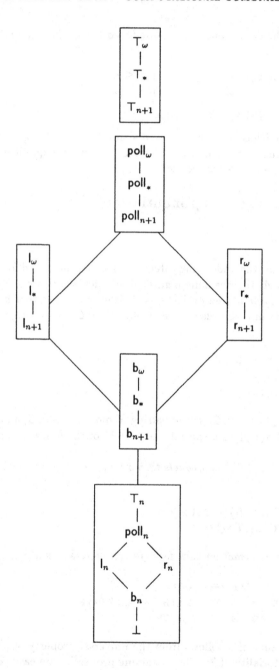

Figure 14. Quotient of Figure 9 by the equivalence induced by γ_{n+1}.

5.2   Exponentiation

Now suppose that the underlying category $\mathcal{C}$ is cartesian closed. That is, we assume that $\mathcal{C}$ has a distinguished terminal object and distinguished binary products, and that for every pair of objects B and C there is a distinguished exponentiation object $B \to C$ satisfying the usual requirements: for all B and C there is a morphism $\mathsf{app}_{B,C} : (B \to C) \times B \to^{\mathcal{C}} C$ such that, for all A and all morphisms $f : A \times B \to^{\mathcal{C}} C$ there is a unique morphism $\mathsf{curry}(f) : A \to^{\mathcal{C}} (B \to C)$ satisfying $\mathsf{app}_{B,C} \circ (\mathsf{curry}(f) \times \mathsf{id}_B) = f$.

Equivalently, a category is cartesian closed if it has finite products and there is a pair of natural isomorphisms

$$\mathsf{curry} : \mathcal{C}(- \times B, C) \dot{\to} \mathcal{C}(-, B \to C)$$
$$\mathsf{uncurry} : \mathcal{C}(-, B \to C) \dot{\to} \mathcal{C}(- \times B, C).$$

Here $\mathcal{C}(- \times B, C)$ and $\mathcal{C}(-, B \to C)$ are contravariant hom-functors from $\mathcal{C}^{\mathrm{op}}$ to the category Set, with the standard definitions [1]. This is the same as requiring that $\mathsf{curry}(\mathsf{uncurry}\, h) = h$ and $\mathsf{uncurry}(\mathsf{curry}\, g) = g$, together with the following naturality conditions: for all $f : A \to^{\mathcal{C}} A'$, $g : A' \times B \to^{\mathcal{C}} C$, and $h : A' \to^{\mathcal{C}} (B \to C)$,

$$\mathsf{curry}(g \circ (f \times \mathsf{id})) = (\mathsf{curry}\, g) \circ f$$
$$\mathsf{uncurry}(h) \circ (f \times \mathsf{id}) = \mathsf{uncurry}(h \circ f).$$

It follows easily from these conditions that one can choose $\mathsf{app} = \mathsf{uncurry}(\mathsf{id})$ as a suitable application morphism.

We want to give some general conditions under which analogous properties can be obtained for the Kleisli category $\mathcal{C}_T$. Assuming that $\mathcal{C}$ is cartesian closed, the obvious candidate in $\mathcal{C}_T$ for the exponential object of B and C is $TB \to C$. Moreover, we know that there is a natural isomorphism between $\mathcal{C}_T(A, TB \to C)$ and $\mathcal{C}(TA \times TB, C)$. Since $\mathcal{C}_T(A \times B, C)$ is $\mathcal{C}(T(A \times B), C)$, it is clear that we must make some assumptions about the relationship between $T(A \times B)$ and $TA \times TB$.

If $T(A \times B)$ and $TA \times TB$ are naturally isomorphic it is easy to show that $\mathcal{C}_T$ is cartesian closed whenever $\mathcal{C}$ is. This is apparently a "Folk Theorem". The comonad T_1 has this property, and we gave in [4] a proof using this property that the Kleisli category of T_1 is cartesian closed.

However, there are reasonable examples in which T does not preserve products, including T_2 and T_3 as described earlier. Instead, we will make a weaker assumption: that the comonad can be equipped with natural ways to move back and forth between $T(A \times B)$ and $TA \times TB$ that interact sensibly with the comonad operations ϵ, δ, and γ. This can be conveniently summarized by means of two natural transformations split and merge satisfying certain combinational laws.

Definition 5.2 Let $(T, \epsilon, \delta, \gamma)$ be a computational comonad. A *computational pairing* is a pair of natural transformations

$$\text{split} : T(- \times -) \xrightarrow{\cdot} T(-) \times T(-)$$
$$\text{merge} : T(-) \times T(-) \xrightarrow{\cdot} T(- \times -)$$

such that, for all objects A and B, the following properties hold:

$$(\epsilon_A \times \epsilon_B) \circ \text{split}_{A,B} = \epsilon_{A \times B}$$
$$\epsilon_{A \times B} \circ \text{merge}_{A,B} = \epsilon_A \times \epsilon_B$$
$$\text{split}_{A,B} \circ \gamma_{A \times B} = \gamma_A \times \gamma_B$$
$$\text{merge}_{A,B} \circ (\gamma_A \times \gamma_B) = \gamma_{A \times B}$$
$$(\delta_A \times \delta_B) \circ \text{split}_{A,B} = \text{split}_{TA,TB} \circ T \, \text{split}_{A,B} \circ \delta_{A \times B}$$
$$\text{merge}_{TA,TB} \circ (\delta_A \times \delta_B) = T \, \text{split}_{A,B} \circ \delta_{A \times B} \circ \text{merge}_{A,B} \, .$$

Naturality of split and merge requires that for $f : A \to^c A'$ and $g : B \to^c B'$,

$$\text{split}_{A',B'} \circ T(f \times g) = (Tf \times Tg) \circ \text{split}_{A,B}$$
$$\text{merge}_{A',B'} \circ (Tf \times Tg) = T(f \times g) \circ \text{merge}_{A,B} \, .$$

We summarize these properties in diagram form in Figure 15.   •

The properties listed above formalize the sense in which we require the splitting and merging operations to interact sensibly with ϵ, δ, and γ. In particular, the following properties follow immediately.

Corollary 5.3 *Let* $(T, \epsilon, \delta, \gamma)$ *be a computational comonad and let* split *and* merge *form a computational pairing. Then for all A and B,*

$$
\begin{aligned}
(\epsilon_A \times \epsilon_B) \circ \text{split}_{A,B} \circ \text{merge}_{A,B} &= (\epsilon_A \times \epsilon_B) \\
\text{split}_{A,B} \circ \text{merge}_{A,B} \circ (\gamma_A \times \gamma_B) &= (\gamma_A \times \gamma_B) \\
\epsilon_{A \times B} \circ \text{merge}_{A,B} \circ \text{split}_{A,B} &= \epsilon_{A \times B} \\
\text{merge}_{A,B} \circ \text{split}_{A,B} \circ \gamma_{A \times B} &= \gamma_{A \times B}.
\end{aligned}
$$

We have already seen that $\sigma = \langle T\pi_1, T\pi_2 \rangle$ qualifies as a suitable split operation (Proposition 5.1). Despite this fact, split (and merge) are not generally uniquely determined by the computational pairing laws and we wish to permit the use of comonads with "non-standard" choices of split. Moreover, naturality of split and merge does not by itself imply the computational pairing laws.

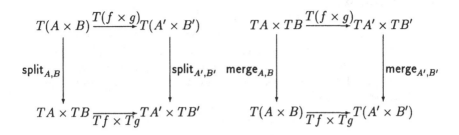

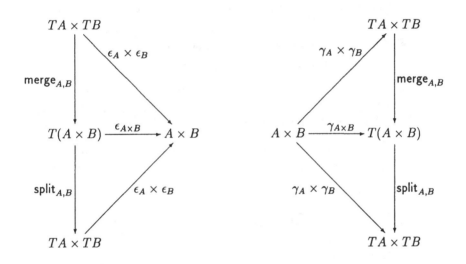

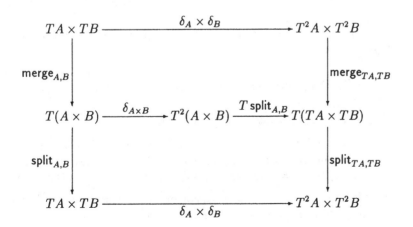

5.3   Examples

1. Return again to the increasing path comonad T_1. The natural transformation $\sigma = \langle T_1 \, \pi_1, T_1 \, \pi_2 \rangle$ is given by: $\sigma(u) = \langle \lambda n.\, \pi_1(u_n), \lambda n.\, \pi_2(u_n) \rangle$. This is actually an isomorphism, with inverse given by $\mathsf{merge}(\langle s, t \rangle) = \lambda n.\, \langle s_n, t_n \rangle$. Intuitively, each of these two operations works "in parallel" on the two components.

   Both σ and merge are natural transformations, and they satisfy the computational pairing laws, which state that:

   • Merging and splitting commute with componentwise application of functions.

   • Merging and splitting respect lubs of sequences.

   • Merging and splitting respect prefixes.

   • Merging two degenerate computations produces a degenerate computation, and splitting a degenerate computation produces a pair of degenerate computations.

2. There are other intuitively sensible ways to split and merge in the increasing paths comonad T_1. We can define a form of (left-first) interleaving by:

   $$\mathsf{lmerge}(\langle s, t \rangle) = \lambda n.\, \left\langle s_{\lceil n/2 \rceil}, t_{\lfloor n/2 \rfloor} \right\rangle.$$

   For example, this gives:

   $$\mathsf{lmerge}(s_0 s_1 s_2^\omega, t_0 t_1 t_2^\omega) = \langle s_0, t_0 \rangle \, \langle s_1, t_0 \rangle \, \langle s_1, t_1 \rangle \, \langle s_2, t_1 \rangle \, \langle s_2, t_2 \rangle^\omega.$$

   To go with lmerge, we define a split that operates only on alternate steps of a computation:

   $$\mathsf{split}_2(u) = \langle \lambda n.\, \pi_1(u_{2n}), \lambda n.\, \pi_2(u_{2n}) \rangle.$$

   We then obtain the identity $\mathsf{split}_2 \circ \mathsf{lmerge} = \mathsf{id}$.

   It is easy to verify that lmerge and split_2 are natural transformations, and that the computational pairing properties hold, making use of the equalities $\lfloor \min(i, 2j)/2 \rfloor = \min(\lfloor i/2 \rfloor, j)$ and $\lceil \min(i, 2j)/2 \rceil = \min(\lceil i/2 \rceil, j)$.

   There is clearly also a right-first version of interleaving rmerge and this also interacts sensibly with split_2 as given above.

3. For the strictly increasing paths comonad, each of the split-merge combinations above adapts in the obvious way, modified so as to ensure that the result of splitting a strictly increasing sequence of pairs is a pair of strictly increasing sequences. Thus, for example,

$$\sigma(\langle \perp, \perp \rangle \langle \mathsf{T}, \perp \rangle \langle \mathsf{T}, \mathsf{T} \rangle^\omega) = \langle \perp \mathsf{T}^\omega, \perp \mathsf{T}^\omega \rangle$$
$$\mathsf{merge}(\perp \mathsf{T}^\omega, \perp \mathsf{T}^\omega) = \langle \perp, \perp \rangle \langle \mathsf{T}, \mathsf{T} \rangle^\omega.$$

In fact $T_2(A \times B)$ and $T_2 A \times T_2 B$ are not generally isomorphic, because a strictly increasing sequence of pairs does not necessarily increase strictly in *both* components at each stage. Nevertheless, σ and merge are still natural transformations satisfying the requirements listed above for a computational pairing, and we have the identity $\sigma \circ \mathsf{merge} = \mathsf{id}$.

The (appropriately adjusted) lmerge and split_2 operations also form a computational pairing, and $\mathsf{split}_2 \circ \mathsf{lmerge} = \mathsf{id}$; similar properties hold for rmerge and split_2.

5.4 Pairing, currying and uncurrying on algorithms

Using the split operation of a computational pairing provides a way to combine a pair of algorithms into an algorithm on pairs. If split is taken to be σ, this is the standard way to form the product of two morphisms in the Kleisli category. We can also use split to define intensional analogues to the contravariant hom-functors $\mathcal{C}(- \times B, C)$ and $\mathcal{C}(-, B \to C)$.

Definition 5.4 Let $\mathcal{C}$ be a category with finite products, let $(T, \epsilon, \delta, \gamma)$ be a computational comonad, and let split be a natural transformation from $T(- \times -)$ to $T - \times T-$. For $f : A \to^{\mathcal{C}_T} A'$ and $g : B \to^{\mathcal{C}_T} B'$ we define $(f \hat{\times} g) : (A \times A') \to^{\mathcal{C}_T} (B \times B')$ by $f \hat{\times} g = (f \times g) \circ \mathsf{split}$.   •

Proposition 5.5 *Let $\mathcal{C}$ be a category with finite products, let $(T, \epsilon, \delta, \gamma)$ be a computational comonad and let split and* merge *form a computational pairing. Then there is a functor $\hat{\times}$ from $\mathcal{C}_T \times \mathcal{C}_T$ to $\mathcal{C}_T$ that maps a pair of objects (A, B) to $A \times B$ and maps a pair of morphisms (a, b) to $(a \hat{\times} b) = (a \times b) \circ \mathsf{split}$.*

Proof: To show that $\hat{\times}$ maps identity morphisms to identity morphisms:

$$(\widehat{\mathsf{id}}_A \hat{\times} \widehat{\mathsf{id}}_B) = (\epsilon_A \times \epsilon_B) \circ \mathsf{split} = \epsilon_{A \times B} = \widehat{\mathsf{id}}_{A \times B}.$$

To show that $\hat{\times}$ preserves composition, let $a : A \to^{\mathcal{C}_T} A'$, $b : B \to^{\mathcal{C}_T} B'$, $a' : A' \to^{\mathcal{C}_T} A''$, and $b' : B' \to^{\mathcal{C}_T} B''$. Then:

$$
\begin{aligned}
(a' \mathbin{\hat{\times}} b') \mathbin{\bar{o}} (a \mathbin{\hat{\times}} b) &= ((a' \times b') \circ \mathsf{split}) \mathbin{\bar{o}} ((a \times b) \circ \mathsf{split}) \\
&= (a' \times b') \circ \mathsf{split} \circ T(a \times b) \circ T\,\mathsf{split} \circ \delta \\
&= (a' \times b') \circ (Ta \times Tb) \circ \mathsf{split} \circ T\,\mathsf{split} \circ \delta \\
&= (a' \times b') \circ (Ta \times Tb) \circ (\delta \times \delta) \circ \mathsf{split} \\
&= ((a' \circ Ta \circ \delta) \times (b' \circ Tb \circ \delta)) \circ \mathsf{split} \\
&= ((a' \mathbin{\bar{o}} a) \times (b' \mathbin{\bar{o}} b)) \circ \mathsf{split} \\
&= (a' \mathbin{\bar{o}} a) \mathbin{\hat{\times}} (b' \mathbin{\bar{o}} b).
\end{aligned}
$$

□

Definition 5.6 Let $\mathcal{C}$ be a cartesian closed category, let $(T, \epsilon, \delta, \gamma)$ be a computational comonad, and let split and merge form a computational pairing. The contravariant functor $\mathcal{C}_T(- \mathbin{\hat{\times}} B, C)$ from $\mathcal{C}_T{}^{\mathrm{op}}$ to Set is defined as follows.

- The functor maps an object A to $\mathcal{C}_T(A \times B, C)$.

- The functor maps a morphism $f : A \to^{\mathcal{C}_T} A'$ to the function $\lambda g.g \mathbin{\bar{o}} (f \mathbin{\hat{\times}} \widehat{\mathsf{id}}_B)$ from $\mathcal{C}_T(A' \times B, C)$ to $\mathcal{C}_T(A \times B, C)$.

Similarly, the contravariant functor $\mathcal{C}_T(-, TB \to C)$ from $\mathcal{C}_T{}^{\mathrm{op}}$ to Set, is defined by:

- On objects the functor maps A to $\mathcal{C}_T(A, TB \to C)$.

- On morphisms the functor maps $f : A \to^{\mathcal{C}_T} A'$ to the function $\lambda h.h \mathbin{\bar{o}} f$ from $\mathcal{C}_T(A', TB \to C)$ to $\mathcal{C}_T(A, TB \to C)$.

•

Proposition 5.7 *Let $(T, \epsilon, \delta, \gamma)$ be a computational comonad and let* split *and* merge *be a computational pairing. Given $a : T(A \times B) \to^{\mathcal{C}} C$ and $b : TA \to^{\mathcal{C}} (TB \to C)$, define*

$$
\begin{array}{ll}
\widehat{\mathrm{curry}}(a) : TA \to^{\mathcal{C}} (TB \to C) & \widehat{\mathrm{uncurry}}(b) : T(A \times B) \to^{\mathcal{C}} C \\
\widehat{\mathrm{curry}}(a) = \mathrm{curry}(a \circ \mathsf{merge}) & \widehat{\mathrm{uncurry}}(b) = \mathrm{uncurry}(b) \circ \mathsf{split}.
\end{array}
$$

Then:

- $\widehat{\text{curry}}$ *and* $\widehat{\text{uncurry}}$ *are natural transformations:*

$$\widehat{\text{curry}} : \mathcal{C}_T(- \hat{\times} B, C) \to \mathcal{C}_T(-, TB \to C)$$
$$\widehat{\text{uncurry}} : \mathcal{C}_T(-, TB \to C) \to \mathcal{C}_T(- \hat{\times} B, C).$$

- *For all* $a : A \times B \to^{\mathcal{C}_T} C$, $\widehat{\text{uncurry}}(\widehat{\text{curry}}(a)) =^e a$.

- *For all* $f : A \times B \to^{\mathcal{C}} C$, $\widehat{\text{uncurry}}(\widehat{\text{curry}}(\text{alg } f)) = \text{alg } f$.

Proof:

- Naturality of $\widehat{\text{curry}}$ requires that $\widehat{\text{curry}}(g \bar{\circ} (f \hat{\times} \widehat{\text{id}}_B)) = (\widehat{\text{curry}} g) \bar{\circ} f$, for all $f : A \to^{\mathcal{C}_T} A'$ and $g : A' \times B \to^{\mathcal{C}_T} C$. This follows from naturality of currying in $\mathcal{C}$ and the properties of computational pairing:

$$
\begin{aligned}
\widehat{\text{curry}}(g \bar{\circ} (f \hat{\times} \widehat{\text{id}})) &= \widehat{\text{curry}}(g \circ T(f \times \epsilon) \circ T \text{ split} \circ \delta) \\
&= \widehat{\text{curry}}(g \circ T(f \times \epsilon) \circ T \text{ split} \circ \delta \circ \text{merge}) \\
&= \widehat{\text{curry}}(g \circ T(f \times \epsilon) \circ \text{merge} \circ (\delta \times \delta)) \\
&= \widehat{\text{curry}}(g \circ \text{merge} \circ (Tf \times T\epsilon) \circ (\delta \times \delta)) \\
&= \widehat{\text{curry}}(g \circ \text{merge} \circ ((Tf \circ \delta) \times (T\epsilon \circ \delta))) \\
&= \widehat{\text{curry}}(g \circ \text{merge} \circ ((Tf \circ \delta) \times \text{id})) \\
&= \widehat{\text{curry}}(g \circ \text{merge}) \circ (Tf \circ \delta) \\
&= \widehat{\text{curry}}(g) \circ Tf \circ \delta \\
&= \widehat{\text{curry}}(g) \bar{\circ} f.
\end{aligned}
$$

- Similarly, to show naturality of $\widehat{\text{uncurry}}$ we need $\widehat{\text{uncurry}}(h) \bar{\circ} (f \hat{\times} \widehat{\text{id}}) = \widehat{\text{uncurry}}(h \bar{\circ} f)$, for all $f : A \to^{\mathcal{C}_T} A'$ and $h : A' \to^{\mathcal{C}_T} (TB \to C)$. Again the proof is straightforward:

$$
\begin{aligned}
\widehat{\text{uncurry}}(h) \bar{\circ} (f \hat{\times} \widehat{\text{id}}) &= \text{uncurry}(h) \circ \text{split} \circ T(f \times \epsilon) \circ T \text{ split} \circ \delta \\
&= \text{uncurry}(h) \circ (Tf \times T\epsilon) \circ \text{split} \circ T \text{ split} \circ \delta \\
&= \text{uncurry}(h) \circ (Tf \times T\epsilon) \circ (\delta \times \delta) \circ \text{split} \\
&= \text{uncurry}(h) \circ ((Tf \circ \delta) \times (T\epsilon \circ \delta)) \circ \text{split} \\
&= \text{uncurry}(h) \circ ((Tf \circ \delta) \times \text{id}) \circ \text{split} \\
&= \text{uncurry}(h \circ Tf \circ \delta) \circ \text{split} \\
&= \text{uncurry}(h \bar{\circ} f) \circ \text{split} \\
&= \widehat{\text{uncurry}}(h \bar{\circ} f).
\end{aligned}
$$

- Let $a : A \times B \to^{\mathcal{C}_T} C$. Then

$$
\begin{aligned}
\widehat{\text{uncurry}}(\widehat{\text{curry}}(a)) \circ \gamma &= \text{uncurry}(\text{curry}(a \circ \text{merge})) \circ \text{split} \circ \gamma \\
&= (a \circ \text{merge}) \circ \text{split} \circ \gamma \\
&= a \circ (\text{merge} \circ \text{split} \circ \gamma) \\
&= a \circ \gamma,
\end{aligned}
$$

showing that $\widehat{\text{uncurry}}(\widehat{\text{curry}} \, a) =^e a$.

- Let $f : A \times B \to^{\mathcal{C}} C$. Then

$$
\begin{aligned}
\widehat{\text{uncurry}}(\widehat{\text{curry}}(\text{alg } f)) &= \text{uncurry}(\text{curry}(\text{alg}(f) \circ \text{merge})) \circ \text{split} \\
&= \text{alg}(f) \circ \text{merge} \circ \text{split} \\
&= f \circ \epsilon \circ \text{merge} \circ \text{split} \quad \text{by Corollary 5.3} \\
&= f \circ \epsilon \\
&= \text{alg } f.
\end{aligned}
$$

□

We have shown that an intensional pairing produces a weak form of exponentiation structure: we obtain notions of currying and uncurrying on algorithms that are natural transformations but satisfy a weaker condition than isomorphism. We may rephrase these properties in terms of the existence of a notion of "application" in the intensional category as follows.

Proposition 5.8 *Let $\mathcal{C}$ be a cartesian closed category, $(T, \epsilon, \delta, \gamma)$ be a computational comonad and let* split *and* merge *be a computational pairing. For all B and C there is an "application morphism"*

$$
\widehat{\text{app}}_{B,C} \; : \; [TB \to C] \times B \to^{\mathcal{C}_T} C
$$

such that, for all $a : A \times B \to^{\mathcal{C}_T} C$

$$
\widehat{\text{app}}_{B,C} \; \bar{\circ} (\widehat{\text{curry}}(a) \; \hat{\times} \; \hat{\text{id}}_B) =^e a.
$$

Proof: Define $\widehat{\text{app}}_{B,C} = \widehat{\text{uncurry}}(\hat{\text{id}}_{TB \to C}) = \widehat{\text{uncurry}}(\epsilon_{TB \to C})$. As a corollary of the naturality of $\widehat{\text{uncurry}}$ (Proposition 5.7), we get:

$$
\begin{aligned}
\widehat{\text{app}} \; \bar{\circ} (b \; \hat{\times} \; \hat{\text{id}}) &= \widehat{\text{uncurry}}(\hat{\text{id}}) \; \bar{\circ} \; (b \; \hat{\times} \; \hat{\text{id}}) \\
&= \widehat{\text{uncurry}}(\hat{\text{id}} \; \bar{\circ} \; b) \\
&= \widehat{\text{uncurry}}(b).
\end{aligned}
$$

Thus, in particular, $\widehat{\text{app}} \; \bar{\circ} (\widehat{\text{curry}}(a) \; \hat{\times} \; \hat{\text{id}}) = \widehat{\text{uncurry}}(\widehat{\text{curry}}(a)) =^e a$.

Note that although $\widehat{\text{curry}}(a)$ is not necessarily the only morphism h such that $\widehat{\text{app}} \; \bar{\circ} (h \; \hat{\times} \; \text{id}) =^e a$, all such morphisms satisfy the condition that $\widehat{\text{uncurry}}(h) =^e a$. □

Thus, we have a weak form of cartesian closedness: instead of the usual diagram for exponentiation we replace $=$ by $=^e$ and we relax the uniqueness condition. This is summarized in Figure 16.

Next we consider what happens if we make further assumptions on the relationship between split and merge.

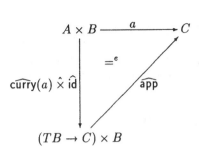

Figure 16. When T has computational pairing this diagram in $\mathcal{C}_T$ commutes up to extensional equivalence, for all $a : A \times B \to^{\mathcal{C}_T} C$.

Proposition 5.9 *Let $\mathcal{C}$ be a cartesian closed category and $(T, \epsilon, \delta, \gamma)$ be a computational comonad with a computational pairing* split *and* merge.

- *If* merge $\circ$ split $=$ id *then* $\widehat{\text{uncurry}} \circ \widehat{\text{curry}} =$ id.

- *If* split $\circ$ merge $=$ id *then* $\widehat{\text{curry}} \circ \widehat{\text{uncurry}} =$ id.

As a corollary we get the following version of the "Folk Theorem":

Corollary 5.10 *If $\mathcal{C}$ is cartesian closed and $(T, \epsilon, \delta, \gamma)$ is a computational comonad over $\mathcal{C}$, with a computational pairing such that* merge $\circ$ split $=$ id *and* split $\circ$ merge $=$ id, *then the category $\mathcal{C}_T$ is cartesian closed.*

Note the important fact that our definitions are parameterized by the choice of split and merge. Once these are chosen, $\widehat{\text{app}}$, $\widehat{\text{curry}}$ and $\widehat{\text{uncurry}}$ are determined uniquely. The Kleisli category itself is independent of split and merge; what happens, however, is that each choice of these two natural transformations induces a (weak form of) exponentiation structure on this category. The Kleisli category may possess many different notions of merging and splitting, and therefore many different ways to curry, uncurry and apply algorithms. This means that one may use the Kleisli category to give an interpretation to a functional programming language containing several syntactically and semantically distinct forms of application. This would be desirable, for instance, if the language included both a strict and a non-strict form of application.

5.5   Examples

1. The Kleisli category based on the increasing path comonad T_1 is cartesian closed, with exponentiation structure built from the standard split-merge combination, which form an isomorphism.

Using the computational pairing lmerge and split_2, we obtain intensional forms of currying, uncurrying, and application which we will call $\widehat{\mathsf{curry}}_l$, $\widehat{\mathsf{uncurry}}_2$ and $\widehat{\mathsf{app}}_l$. This provides a weak form of exponentiation: $\widehat{\mathsf{curry}}_l$ and $\widehat{\mathsf{uncurry}}_2$ are natural transformations, and for all $a : A \times B \rightarrow^{c_{T_1}} C$ we get

$$\widehat{\mathsf{app}}_l \, \bar{\mathsf{o}}(\widehat{\mathsf{curry}}_l(a) \,\hat{\times}\, \hat{\mathsf{id}}) =^e a.$$

Since $\mathsf{split}_2 \circ \mathsf{lmerge} = \mathsf{id}$, it follows that $\widehat{\mathsf{curry}}_l(\widehat{\mathsf{uncurry}}_2 h) = h$. However, $\widehat{\mathsf{uncurry}}_2(\widehat{\mathsf{curry}}_l g) =^e g$, and equality may fail here. For example,

$$\widehat{\mathsf{uncurry}}_2(\widehat{\mathsf{curry}}_l \mathsf{pb}) = \widehat{\mathsf{uncurry}}_2(\widehat{\mathsf{curry}}_l \mathsf{lb}) = \mathsf{pb},$$

and $\widehat{\mathsf{uncurry}}_2(\widehat{\mathsf{curry}}_l \, \mathsf{rb})$ is the least algorithm mapping the path $\langle \bot, \top \rangle \langle \bot, \top \rangle \langle \top, \top \rangle^\omega$ to $\top$. This algorithm of course also computes the function b.

Similar properties hold for the computation pairing rmerge and split_2, with the derived operations $\widehat{\mathsf{curry}}_r$, $\widehat{\mathsf{uncurry}}_2$ and $\widehat{\mathsf{app}}_r$.

2. The strictly increasing paths comonad T_2, with computational pairing σ and merge, again has operations $\widehat{\mathsf{curry}}$, $\widehat{\mathsf{uncurry}}$ and $\widehat{\mathsf{app}}$ that provide weak forms of exponentiation. As we remarked earlier, σ and merge are not isomorphisms. Instead, $\sigma \circ \mathsf{merge} = \mathsf{id}$ and for all $u \in T_2(A \times B)$ the computation $\mathsf{merge}(\sigma(u))$ is pointwise above u. Hence, $\widehat{\mathsf{curry}}(\widehat{\mathsf{uncurry}} \, h) = h$ and $\widehat{\mathsf{uncurry}}(\widehat{\mathsf{curry}} \, g) =^e g$. As an example, we have:

$$\widehat{\mathsf{uncurry}}(\widehat{\mathsf{curry}} \, \mathsf{lb}) = \widehat{\mathsf{uncurry}}(\widehat{\mathsf{curry}} \, \mathsf{rb}) = \widehat{\mathsf{uncurry}}(\widehat{\mathsf{curry}} \, \mathsf{pb}) = \mathsf{pb}.$$

The lmerge and split_2 computational pairing also gives rise to a weak form of exponentiation, as does the rmerge and split_2 computational pairing.

6   Ordered categories

So far, although our principal examples were based on a cartesian closed category of domains, we have not fully exploited the order structure. This permitted us to state and prove results that hold in a more general category-theoretic setting. Next we suppose that the underlying category is an ordered category: each hom-set is equipped with a complete partial order, and composition is continuous. A functor T of ordered categories is required to respect the ordering, in that for all $f, g : A \rightarrow^c B$ if $f \leq g$ then $Tf \leq Tg$. Moreover, T must also be continuous (in its action on morphisms). All of our examples so far satisfy these conditions.

Suppose $(T, \epsilon, \delta, \gamma)$ is a computational comonad over an ordered category $\mathcal{C}$. Then clearly $\mathcal{C}_T$ is again an ordered category. All of the results of the previous sections go through in the ordered setting. In particular, the functors fun and alg introduced earlier respect the ordering; and the proof of Proposition 4.2 can be adapted to show that the extensional quotient of the ordering on $\mathcal{C}_T(A, B)$ is just the order on $\mathcal{C}(A, B)$.

We can also obtain some slightly stronger results by taking advantage of the ordering. We omit most of the proofs, which may be easily obtained from the results above, using monotonicity and continuity of composition.

Proposition 6.1 *Let $\mathcal{C}$ be an ordered ccc and let $(T, \epsilon, \delta, \gamma)$ be a computational comonad over $\mathcal{C}$ with a computational pairing.*

- *If* split $\circ$ merge $\geq$ id *then* $\widehat{\text{curry}} \circ \widehat{\text{uncurry}} \geq$ id.

- *If* split $\circ$ merge $\leq$ id *then* $\widehat{\text{curry}} \circ \widehat{\text{uncurry}} \leq$ id.

- *If* merge $\circ$ split $\geq$ id *then* $\widehat{\text{uncurry}} \circ \widehat{\text{curry}} \geq$ id.

- *If* merge $\circ$ split $\leq$ id *then* $\widehat{\text{uncurry}} \circ \widehat{\text{curry}} \leq$ id.

Next we introduce a simple generalization of the notion of cartesian closed ordered category, obtained by relaxing the requirement that currying and uncurrying form an isomorphism. Instead we allow currying and uncurrying to form an adjunction in each homset, so that we have an example of a *local adjunction* (see for example [9]) with additional properties. The relevance of "lax" notions of adjunction such as these in computational settings (albeit with different motivations) has been pointed out in different contexts by other authors, for instance in [14].

Definition 6.2 An ordered category $\mathcal{C}$ is *cartesian up-closed* if and only if it has finite products and for all pairs of objects B and C there is an object $B \rightarrow C$ and a pair of lax natural transformations curry, uncurry between $\mathcal{C}(- \times B, C)$ and $\mathcal{C}(-, B \rightarrow C)$ satisfying:

$$
\begin{aligned}
\text{curry(uncurry } h) &\leq h \\
\text{uncurry(curry } g) &\geq g \\
\text{curry}(g \circ (f \times \text{id})) &\leq \text{curry}(g) \circ f \\
\text{uncurry}(h) \circ (f \times \text{id}) &\geq \text{uncurry}(h \circ f).
\end{aligned}
$$

Similarly, we say that C is *cartesian down-closed* iff it has finite products and there is a pair of lax natural transformations curry, uncurry satisfying:

$$
\begin{aligned}
\mathsf{curry}(\mathsf{uncurry}\,h) &\geq h \\
\mathsf{uncurry}(\mathsf{curry}\,g) &\leq g \\
\mathsf{curry}(g \circ (f \times \mathsf{id})) &\geq \mathsf{curry}(g) \circ f \\
\mathsf{uncurry}(h) \circ (f \times \mathsf{id}) &\leq \mathsf{uncurry}(h \circ f).
\end{aligned}
$$

•

Definition 6.3 Let C be an ordered category with finite products, and let B and C be objects of C.

- An *up-exponential* for B and C is an object $B \to C$ of C together with a morphism $\mathsf{app}_{B,C} : (B \to C) \times B \to^{\mathcal{C}} C$ such that for every $f : A \times B \to^{\mathcal{C}} C$ there is a least morphism $\mathsf{curry}(f) : A \to^{\mathcal{C}} (B \to C)$ satisfying:

  $$\mathsf{app} \circ (\mathsf{curry}(f) \times \mathsf{id}) \geq f.$$

- A *down-exponential* for B and C is an object $B \to C$ of C together with a morphism $\mathsf{app}_{B,C} : (B \to C) \times B \to^{\mathcal{C}} C$ such that for every $f : A \times B \to^{\mathcal{C}} C$ there is a greatest morphism $\mathsf{curry}(f) : A \to^{\mathcal{C}} (B \to C)$ satisfying:

  $$\mathsf{app} \circ (\mathsf{curry}(f) \times \mathsf{id}) \leq f.$$

•

The following result may be shown by adapting the usual proof that the two alternative definitions of cartesian closed categories are equivalent.

Proposition 6.4 *An ordered category C is cartesian up-closed iff it has finite products and up-exponentials.*

An ordered category is cartesian down-closed iff it has finite products and down-exponentials.

Note that in general up-exponentials and down-exponentials need not be unique (even up to isomorphism), since their definition relies on the choice of computational pairing. However, if the same object $B \to C$ and morphism $\mathsf{app}_{B,C}$ qualifies simultaneously as an up- and a down-exponential then it forms the usual notion of exponentiation and the category is cartesian closed in the usual sense.

Proposition 6.5 *Let C be a cartesian up-closed category and $(T, \epsilon, \delta, \gamma)$ be a computational comonad with a computational pairing such that*

$$\text{split} \circ \text{merge} \leq \text{id}$$
$$\text{merge} \circ \text{split} \geq \text{id}$$
$$(\delta \times \delta) \circ \text{split} \leq \text{split} \circ T \text{ split} \circ \delta$$
$$\text{merge} \circ (\delta \times \delta) \geq T \text{ split} \circ \delta \circ \text{merge}.$$

Then C_T is cartesian up-closed.

Proof:

- Let $a : A \times B \to^{C_T} C$ and $b : A \to^{C_T} (TB \to C)$. Then:

$$
\begin{aligned}
\widehat{\text{uncurry}}(\widehat{\text{curry}}\, a) \;&=\; \text{uncurry}(\text{curry}(a \circ \text{merge})) \circ \text{split} \\
&\geq\; a \circ \text{merge} \circ \text{split} \geq a \\
\widehat{\text{curry}}(\widehat{\text{uncurry}}\, b) \;&=\; \text{curry}(\text{uncurry}(b) \circ \text{split} \circ \text{merge}) \\
&\leq\; \text{curry}(\text{uncurry}\, b) \leq b.
\end{aligned}
$$

- To show that $\widehat{\text{curry}}$ is a lax natural transformation, let $f : A \to^{C_T} A'$ and $g : A' \times B \to^{C_T} C$. Then:

$$
\begin{aligned}
\widehat{\text{curry}}(g \,\bar{\circ}\, (f \,\hat{\times}\, \widehat{\text{id}})) \;&=\; \widehat{\text{curry}}(g \circ T(f \times \epsilon) \circ T \text{ split} \circ \delta) \\
&=\; \text{curry}(g \circ T(f \times \epsilon) \circ T \text{ split} \circ \delta \circ \text{merge}) \\
&\leq\; \text{curry}(g \circ T(f \times \epsilon) \circ \text{merge} \circ (\delta \times \delta)) \\
&=\; \text{curry}(g \circ \text{merge} \circ (Tf \times T\epsilon) \circ (\delta \times \delta)) \\
&=\; \text{curry}(g \circ \text{merge} \circ ((Tf \circ \delta) \times (T\epsilon \circ \delta))) \\
&=\; \text{curry}(g \circ \text{merge} \circ ((Tf \circ \delta) \times \text{id})) \\
&\leq\; \text{curry}(g \circ \text{merge}) \circ (Tf \circ \delta) \\
&=\; \widehat{\text{curry}}(g) \circ Tf \circ \delta \\
&=\; \widehat{\text{curry}}(g) \,\bar{\circ}\, f.
\end{aligned}
$$

- To show that $\widehat{\text{uncurry}}$ is a lax natural transformation, suppose that $f : A \to^{C_T} A'$ and $h : A' \to^{C_T} (TB \to C)$. Then

$$
\begin{aligned}
\widehat{\text{uncurry}}(h) \,\bar{\circ}\, (f \,\hat{\times}\, \widehat{\text{id}}) \;&=\; \text{uncurry}(h) \circ \text{split} \circ T(f \times \epsilon) \circ T \text{ split} \circ \delta \\
&\geq\; \text{uncurry}(h) \circ (Tf \times T\epsilon) \circ \text{split} \circ T \text{ split} \circ \delta \\
&=\; \text{uncurry}(h) \circ (Tf \times T\epsilon) \circ (\delta \times \delta) \circ \text{split} \\
&=\; \text{uncurry}(h) \circ ((Tf \circ \delta) \times (T\epsilon \circ \delta)) \circ \text{split} \\
&=\; \text{uncurry}(h) \circ ((Tf \circ \delta) \times \text{id}) \circ \text{split} \\
&\geq\; \text{uncurry}(h \circ Tf \circ \delta) \circ \text{split} \\
&=\; \text{uncurry}(h \,\bar{\circ}\, f) \circ \text{split} \\
&=\; \widehat{\text{uncurry}}(h \,\bar{\circ}\, f).
\end{aligned}
$$

$$\square$$

A similar result holds for a cartesian down-closed category with a computational pairing satisfying reversed inequalities.

7   Examples

We now return to the third comonad introduced earlier, after which we will introduce briefly some related notions of computation on different categories of domains and functions.

7.1   Timed data

In the timed data comonad T_3, the standard split operation is:

$$\text{split} \left\langle \left\langle a, b \right\rangle, n \right\rangle = \left\langle \left\langle a, n \right\rangle, \left\langle b, n \right\rangle \right\rangle .$$

Given our interpretation of $\langle d, n \rangle$ as a computation yielding d at cost n, an obvious choice for a merge operation is:

$$\text{merge}(\langle a, m \rangle, \langle b, n \rangle) = \langle \langle a, b \rangle, \max(m, n) \rangle .$$

Both of these operations are natural transformations, and we obtain the following properties:

$$\begin{aligned}
&\text{split} \circ \text{merge} \sqsubseteq \text{id} \\
&\text{merge} \circ \text{split} = \text{id} \\
&(\delta \times \delta) \circ \text{split} = \text{split} \circ T_3 \text{ split} \circ \delta \\
&\text{merge} \circ (\delta \times \delta) \sqsupseteq T_3 \text{ split} \circ \delta \circ \text{merge} .
\end{aligned}$$

The underlying category is cartesian closed, hence also cartesian up-closed. It follows from Proposition 6.5 that the Kleisli category of T_3 is cartesian up-closed.

7.2   Strict algorithms

The category of domains and strict continuous functions is not cartesian closed, although the category does have products. For each pair of domains D and D', the set of strict continuous functions $D \to_s D'$, ordered pointwise, is again a domain. The usual uncurrying operation on functions preserves strictness, but the usual currying does not. Instead, we may define a variant form of currying by:

$$\text{curry}_s : (A \times B \to_s C) \to (A \to_s (B \to_s C))$$
$$\text{curry}_s(f) = \lambda x.\lambda y.(x = \bot \vee y = \bot \to \bot, f(x,y)).$$

When f is strict, $\text{curry}_s(f)$ is the best strict function approximating $\text{curry}(f)$ pointwise. For instance, let lor, ror and sor be the left-strict, right-strict, and doubly-strict or-functions. Then

$$\text{uncurry}(\text{curry}_s \, \text{lor}) = \text{uncurry}(\text{curry}_s \, \text{ror}) = \text{uncurry}(\text{curry}_s \, \text{sor}) = \text{sor} \, .$$

It is easy to check that curry_s is a natural transformation (and so is uncurry).

For all $f : A \times B \to_s C$ and all $g : A \to_s (B \to_s C)$, we have:

$$\text{curry}_s(\text{uncurry} \, g) = g$$
$$\text{uncurry}(\text{curry}_s \, f) \sqsubseteq f.$$

Hence, the category of domains and strict continuous functions is cartesian down-closed.

Let $T_1 D$ be the set of increasing paths over D (not just the strict continuous maps from VNat to D), ordered pointwise. The maps ϵ, δ and γ are all strict, as are all of the split and merge operations above. We may therefore use the Kleisli construction to build a model of strict parallel algorithms. To illustrate this model, note that all of the algorithms of Figure 6 also belong in this category, with the exception of $\top$, which is non-strict.

Since the underlying category is cartesian down-closed, each of the computational pairings discussed earlier for T_1 gives rise to a down-exponentiation structure, so that the category of domains and strict algorithms is again cartesian down-closed.

We may also adapt the T_2 and T_3 comonads to this category.

7.3   Computation on effectively given domains

The category of effectively given domains and computable functions is cartesian closed. A domain is effectively given iff its finite elements are recursively enumerable (hence, countable), it is decidable whether two finite elements are consistent, and (an index for) the lub of two consistent finite elements is decidable (as a function of their indices). An element of D is computable iff the set of (indices of) its finite approximations is recursively enumerable.

The functor T_1 can be adapted to this category, by defining T_1D for an effectively given domain D to be the computable increasing paths over D (equivalently, the computable continuous functions from VNat to D, ordered pointwise). All of the auxiliary operations (ϵ, δ, γ, and so on) are computable. Hence we obtain a category of effectively given domains and computable algorithms, and this category quotients onto the underlying category of effectively given domains and computable functions. This algorithms category is again cartesian closed.

The functor T_3 maps effectively given domains to effectively given domains, and again the auxiliary operations are computable. We therefore obtain a category of effectively given domains and T_3-algorithms that quotients onto the underlying category and is cartesian up-closed.

The functor T_2 preserves algebraicity but not ω-algebraicity, since T_2D may have uncountably many finite elements. The T_2 comonad therefore does not adapt to the category of effectively given domains and computable functions.

7.4 Computation on pre-domains

We use the term *pre-domain* for a "bottomless" domain: a directed-complete, bounded complete, algebraic partial order with no requirement that there be a least element. The category of pre-domains and continuous functions is cartesian closed.

Let T_4D be the set of non-empty finite or infinite increasing sequences over D, ordered by the usual prefix ordering. Clearly this forms a pre-domain, and the finite elements are just the finite sequences. T_4D is generally a pre-domain rather than a domain, even if D has a least element, because the prefix ordering does not relate sequences with different first elements. We make T into a functor by specifying that (as usual) T_4f applies f componentwise.

Again we let ϵ be the lub operation and let δt be the sequence of (non-empty) prefixes of t. Then (T_4, ϵ, δ) forms a comonad.

We may regard a computation of length 1 as degenerate, and this corresponds to defining the function γ from D to T_4D by $\gamma d = \langle d \rangle$. Although this function γ is not continuous, so that we cannot claim that T_4 is a computational comonad, we still obtain a congruence relation on algorithms by defining

$$a =^{e} a' \iff \forall d \in D.a\langle d \rangle = a'\langle d \rangle.$$

Note that for all $f, g : A \to B$ we have

$$(f \circ \epsilon) =^{e} (g \circ \epsilon) \Rightarrow f = g.$$

It is then easy to modify the proof of Proposition 4.2 to show that the Kleisli category of this comonad quotients onto the underlying category under $=^e$.

We may define splitting and merging operations as follows. The standard way to split is:

$$\mathsf{split}(\langle x_0, y_0 \rangle \ldots \langle x_k, y_k \rangle) = \langle x_0 \ldots x_k, y_0 \ldots y_k \rangle$$
$$\mathsf{split}(\langle x_n, y_n \rangle_{n=0}^{\infty}) = \langle \langle x_n \rangle_{n=0}^{\infty}, \langle y_n \rangle_{n=0}^{\infty} \rangle .$$

Let merge be the least continuous function satisfying:

$$\mathsf{merge}(x_0 \ldots x_k, y_0 \ldots y_m) = \langle x_0, y_0 \rangle \ldots \langle x_n, y_n \rangle \qquad (n = \min(m, k))$$
$$\mathsf{merge}(x_0 \ldots x_k, \langle y_i \rangle_{i=0}^{\infty}) = \langle x_0, y_0 \rangle \ldots \langle x_k, y_k \rangle$$
$$\mathsf{merge}(\langle x_i \rangle_{i=0}^{\infty}, y_0 \ldots y_m) = \langle x_0, y_0 \rangle \ldots \langle x_m, y_m \rangle$$
$$\mathsf{merge}(\langle x_n \rangle_{n=0}^{\infty}, \langle y_n \rangle_{n=0}^{\infty}) = \langle x_n, y_n \rangle_{n=0}^{\infty} .$$

Clearly, $\mathsf{merge} \circ \mathsf{split} = \mathsf{id}$ and $\mathsf{split} \circ \mathsf{merge} \sqsubseteq \mathsf{id}$. These two operations are obviously natural and satisfy the computational pairing properties, except that $\epsilon \circ \mathsf{merge} \sqsubseteq \epsilon \times \epsilon$. The Kleisli category is cartesian down-closed.

8   Conclusions

We have described a category-theoretic approach to intensional semantics, based on the idea that a notion of computation or intensional behavior may be modelled by means of a computational comonad, and that the Kleisli category thus obtained can be viewed as an intensional model. The morphisms in this category map computations to values, and from such a morphism one may recover a map from values to values. One may define an equivalence relation that identifies all algorithms that compute the same function, and this equivalence relation can be used to collapse the Kleisli category onto the underlying category.

We have identified a set of conditions under which the Kleisli category possesses exponentiations or weaker types of exponentiation, based on the existence of natural ways to pair computations. We described a series of examples to illustrate the applicability of our definitions and results. In doing so, we have placed our recent work [4] in a wider context.

Our work arose out of an attempt, begun in [3], to generalize an earlier intensional model of Berry and Curien [6]. They defined a category of deterministic concrete data structures and sequential algorithms, showed that this category is cartesian closed, and that it collapses onto the category of deterministic concrete data structures and sequential functions under an obvious notion of extensional equivalence. The sequential functions category is not cartesian closed, and their construction of sequential algorithms was not based on a comonad. The operational semantics implicit in their work was coroutine-like, demand-driven, and lazy, but with the restriction that computation should proceed *sequentially*, with at most one argument being evaluated at a time. In our generalization of their model, we relax the sequentiality restriction so as to permit parallel computation, and we adopt an operational semantics based on parallel coroutine-like lazy evaluation. The *query model* of parallel algorithms between deterministic concrete data structures, described in [3], contains algorithms for non-sequential functions such as parallel-or. However, the model's construction was rather complex and we were unable to formulate a suitable notion of composition for algorithms. Instead, in [4] we presented a much more streamlined form of algorithm between Scott domains and for the first time we cast our construction in terms of a comonad. Of the comonads introduced in this paper, T_1 corresponds to the comonad used in [4]; T_2 is closer in spirit to the query model of [3], but we are able here to go considerably further. In [5] we show that the Berry-Curien sequential algorithms may be embedded in the T_1 model in a manner that respects operational behavior: the image of every sequential algorithm is a parallel algorithm that operates sequentially. Our model can thus be used to provide a semantics for the Berry-Curien language CDS0 of sequential algorithms. An extension of this to a parallel algorithmic language should be fairly straightforward.

Moggi has developed an abstract view of programming languages in which a notion of computation is modelled as a monad [12, 13]. Examples of notions of computation as monads include: computation with side-effects, computation with exceptions, partial computations, and non-deterministic computations. In this view, the meaning of a program is taken as a function from values to computations, and an intuitive operational semantics is that a program from A to B takes an input of type A and returns an output computation. This point of view is consistent with an *input-driven* lazy operational semantics. Our "opposite" point of view based on comonads (which are, after all, monads on the opposite category) is consistent with a *demand-driven* lazy operational semantics, as discussed above. Moggi states in [12] that his view of programs corresponds to call-by-value parameter passing, and he says that there is an alternative view of "programs as functions from computations to computations" corresponding to call-by-name. Our work shows that there is also a third alternative: programs as functions from computations to values.

Common to these approaches is the realization that values should be distinguished from computations. Both approaches use an endofunctor T to embody this distinction, TA representing a datatype of computations over A. Moggi shows that many such endofunctors arising in denotational semantics can be usefully equipped with the extra structure needed to form a monad, and we show that certain endofunctors embodying intensional semantics can be usefully equipped with the structure of a comonad. The motivations and the operational intuitions behind the monad approach and the comonad approach are different, and the two approaches should be regarded as orthogonal or complementary. We plan to explore to what extent (and to what effect) the two approaches can be combined. For instance, given a comonad T and a monad P over the same category $\mathcal{C}$ one might obtain (assuming that T and P satisfy certain properties) a category of (T, P)-algorithms, in which a morphism from A to B is a morphism in $\mathcal{C}$ from TA to PB.

We plan to investigate notions of computation in further domain-theoretic settings. As noted above, we are already working on categories of algorithms on (generalized) concrete data structures [5]. It would be interesting to see if the Berry-Curien sequential algorithms category could be embedded in the Kleisli category of a suitable comonad over a sequential functions category, but no cartesian closed category based on sequential functions is yet known. We intend to investigate notions of computation on the category of dI-domains and stable functions [2], and on the category of qualitative domains and linear functions [7].

Acknowledgements

The diagrams in this paper were drawn using macros designed by John Reynolds.

References

[1] M. Barr and C. Wells. *Category Theory for Computing Science*. Prentice-Hall International, 1990.

[2] G. Berry. Stable models of typed λ-calculi. In *Proc. 5th Coll. on Automata, Languages and Programming*, number 62 in Lecture Notes in Computer Science, pages 72–89. Springer-Verlag, July 1978.

[3] S. Brookes and S. Geva. Towards a theory of parallel algorithms on concrete data structures. In *Semantics for Concurrency, Leicester 1990*, pages 116–136. Springer Verlag, 1990.

[4] S. Brookes and S. Geva. A cartesian closed category of parallel algorithms between Scott domains. Technical Report CMU-CS-91-159, Carnegie Mellon University, School of Computer Science, 1991. Submitted for publication.

[5] S. Brookes and S. Geva. Continuous functions and parallel algorithms on concrete data structures. Technical Report CMU-CS-91-160, Carnegie Mellon University, School of Computer Science, 1991. To appear in the Proceedings of MFPS'91.

[6] P.-L. Curien. *Categorical Combinators, Sequential Algorithms and Functional Programming*. Research Notes in Theoretical Computer Science. Pitman, 1986.

[7] J.-Y. Girard. *Proofs and Types*. Cambridge University Press, 1989.

[8] C. Gunter and D. Scott. Semantic domains. In *Handbook of Theoretical Computer Science, Volume B: Formal Models and Semantics*. MIT Press/Elsevier, 1990.

[9] C. B. Jay. Local adjunctions. *Journal of Pure and Applied Logic*, 53:227–238, 1988.

[10] G. Kahn and D. B. MacQueen. Coroutines and networks of parallel processes. In *Information Processing 1977*, pages 993–998. North Holland, 1977.

[11] S. Mac Lane. *Categories for the Working Mathematician*. Springer-Verlag, 1971.

[12] E. Moggi. Computational lambda-calculus and monads. In *Fourth Annual Symposium on Logic in Computer Science*. IEEE Computer Society Press, June 1989.

[13] E. Moggi. Notions of computation and monads. *Information and Computation*, 1991.

[14] R. A. G. Seely. Modeling computations: A 2-categorical framework. In *Symposium on Logic in Computer Science*. IEEE Computer Society Press, June 1987.

Weakly distributive categories

J.R.B. Cockett[*]   R.A.G. Seely[†]

Abstract

There are many situations in logic, theoretical computer science, and category theory where two binary operations—one thought of as a (tensor) "product", the other a "sum"—play a key role, such as in distributive categories and in ∗-autonomous categories. (One can regard these as essentially the AND/OR of traditional logic and the TIMES/PAR of (multiplicative) linear logic, respectively.) In the latter example, however, the distributivity one often finds is conspicuously absent: in this paper we study a "linearisation" of distributivity that *is* present in this context. We show that this weak distributivity is precisely what is needed to model Gentzen's cut rule (in the absence of other structural rules), and show how it can be strengthened in two natural ways, one to generate full distributivity, and the other to generate ∗-autonomous categories.

0. Introduction

There are many situations in logic, theoretical computer science, and category theory where two binary operations, "tensor products" (though one may be a "sum"), play a key role. The multiplicative fragment of linear logic is a particularly interesting example as it is a Gentzen style sequent calculus in which the structural rules of contraction, thinning, and (sometimes) exchange are dropped. The fact that these rules are omitted considerably simplifies the derivation of the cut elimination theorem. Furthermore, the proof theory of this fragment is interesting and known [Se89] to correspond to ∗-autonomous categories as introduced by Barr in [Ba79].

In the study of categories with two tensor products one usually assumes a distributivity condition, particularly in the case when one of these is either the product or sum. The multiplicative fragment of linear logic (*viz.* ∗-autonomous categories) is a significant exception to this situation; here the two tensors "times" ($\otimes$) and "par" ($\invamp$, in this paper denoted $\oplus$—note that this conflicts with Girard's notation) do not distribute one over the other.

[*]Department of Computer Science, The University of Calgary, 2500 University Drive N.W., Calgary, Alberta, Canada T2N 1N4. This author was partially supported by the Sydney Category Seminar, the Australian Research Council, and NSERC, Canada.

[†]Department of Mathematics, McGill University, 805 Sherbrooke St. W., Montréal, Québec, Canada H3A 2K6. This author was partially supported by the Sydney Category Seminar, Le Fonds FCAR, Québec, and NSERC, Canada.

However, *-autonomous categories are known to satisfy a weak notion of distributivity. This weak distributivity is given by maps of the form:

$$A \otimes (B \oplus C) \longrightarrow (A \otimes B) \oplus C$$

$$A \otimes (B \oplus C) \longrightarrow B \oplus (A \otimes C)$$

(and two other versions should the tensors lack symmetry.)

These maps, interpreted as entailments, are also valid in what might be considered the minimal logic of two such tensors, namely the classical Gentzen sequent calculus with the left and right introduction rules for conjunction and disjunction and with cut as the only structure rule. This Gentzen style proof theory has a categorical presentation already in the literature, *viz.* the polycategories of Lambek and Szabo [Sz75]. It should therefore be possible to link *-autonomous categories and polycategories. However, this begs a wider question of precisely what properties a category must satisfy to be linked in this manner to the logical superstructure provided by a polycategory.

It turns out that these weak distributivity maps, when present coherently, are precisely the necessary structure required to construct a polycategory superstructure, and whence a Gentzen style calculus, over a category with two tensors. The weak distributivity maps allow the expression of the Gentzen cut rule in terms of ordinary (categorical) composition.

We call categories with two tensors linked by coherent weak distribution *weakly distributive categories*. They can be built up to be the proof theory of the full multiplicative fragment of classical linear logic[1] by coherently adding maps

$$\top \longrightarrow A \oplus A^{\perp}$$

$$A \otimes A^{\perp} \longrightarrow \perp$$

(and symmetric duals as necessary), or to the proof theory of the $\wedge, \vee$ fragment of intuitionistic propositional logic by coherently adding contraction, thinning, and exchange. The former corresponds to *-autonomous categories and the latter to distributive categories. In fact, weakly distributive categories lie at the base of a rich logical hierarchy, unifying several hitherto separate developments in the logics of theoretical computer science.

One point must be made about the connection with linear logic. A novel feature of our presentation is that we have considered the two tensor structure separately from the structure given by linear negation $(-)^{\perp}$. We show how to obtain the logic of *-autonomous categories from that of weakly distributive categories, giving, in effect, another presentation of *-autonomous categories. It sometimes happens

---

[1]The system *FILL* (full intuitionistic linear logic) of dePaiva [dP89] amounts to having just the second of these (families of) maps. From the autonomous category viewpoint, these are the more natural maps, as they correspond to evaluations. The symmetry of the *-autonomous viewpoint then suggests the first (family of) maps.

that it is easier to verify *-autonomy this way; for example, verifying that a lattice with appropriate structure is *-autonomous becomes almost trivial if one checks the weak distributivity first (see [Ba91].)

In this short version of our paper, we shall not have space to say as much as we would like about models of this structure. However, one should note that *-autonomous categories, distributive categories, braided monoidal categories, among others, are all weakly distributive. In addition, the opposite of a weakly distributive category is weakly distributive (with the tensors changing roles), so, for example, co-distributive categories are weakly distributive. One has frequently been struck by the strangeness of the distributivity in such co-distributive categories as the category of commutative rings, or the category of distributive lattices, and so on— they may now be seen as weakly distributive in the standard manner.

We have also been very brief about coherence questions here; further, we plan to show how to associate a term calculus to these categories. (This term calculus is based on a calculus developed in [Co89] for symmetric monoidal categories.) We plan to elaborate on all these matters elsewhere. (Added in proof: Coherence has been treated in the recent work of Blute and Seely [BS91], as well as a partial answer to the question of the conservativity of the extension to *-autonomous categories. In that paper, coherence is completely settled, but conservativity is only shown for the fragment without units.)

Acknowledgements: The authors benefited from many helpful discussions with (to mention only a few) Mike Barr, Yves Lafont, and Ross Street. We also must acknowledge the hospitality (and tolerance) of each other's family, who put up with our long deliberations during our visits to each other's homes. The diagrams in this paper were produced with the help of the *diagram* macros of F. Borceux.

1.   Polycategories

We shall begin with a review of Szabo's notion of a polycategory:

Definition 1.1 *A polycategory* **C** *consists of a set* $Ob(\mathbf{C})$ *of objects and a set* $M\varphi(\mathbf{C})$ *of morphisms, (also called arrows, polymorphisms, ...) just like a category, except that the source and target of a morphism are finite sequences of objects*

$$source: M\varphi(\mathbf{C}) \longrightarrow Ob(\mathbf{C})^*$$

$$target: M\varphi(\mathbf{C}) \longrightarrow Ob(\mathbf{C})^*$$

where $X^* =$ *the free monoid generated by* X.

There are identity morphisms $i_A \colon A \to A$ *between singleton sequences only and a notion of composition given by the* **cut** *rule:*

$$\frac{\Gamma_1, A, \Gamma_2 \xrightarrow{g} \Gamma_3 \quad \Delta_1 \xrightarrow{f} \Delta_2, A, \Delta_3}{\Gamma_1, \Delta_1, \Gamma_2 \xrightarrow{g \; {}_i\circ_j \; f} \Delta_2, \Gamma_3, \Delta_3}$$

where the length of Γ_1 is i and the length of Δ_2 is j. When the subscripts are clear from the context they shall be dropped.

We have the following equations:

1.
$$\Gamma_1 \xrightarrow{f} \Gamma_2, A, \Gamma_3 \quad = \quad \frac{A \xrightarrow{i_A} A \quad \Gamma_1 \xrightarrow{f} \Gamma_2, A, \Gamma_3}{\Gamma_1 \xrightarrow{i_A \; 0 \circ_j \; f} \Gamma_2, A, \Gamma_3}$$

2.
$$\Gamma_1, A, \Gamma_2 \xrightarrow{f} \Gamma_3 \quad = \quad \frac{\Gamma_1, A, \Gamma_2 \xrightarrow{f} \Gamma_3 \quad A \xrightarrow{i_A} A}{\Gamma_1, A, \Gamma_2 \xrightarrow{f \; i \circ_0 \; i_A} \Gamma_3}$$

3.
$$\frac{\Phi_1, B, \Phi_2 \xrightarrow{h} \Phi_3 \quad \dfrac{\Delta_1, A, \Delta_2 \xrightarrow{g} \Delta_3, B, \Delta_4 \quad \Gamma_1 \xrightarrow{f} \Gamma_2, A, \Gamma_3}{\Delta_1, \Gamma_1, \Delta_2 \xrightarrow{g \; i \circ_j \; f} \Gamma_2, \Delta_3, B, \Delta_4, \Gamma_3}}{\Phi_1, \Delta_1, \Gamma_1, \Delta_2, \Phi_2 \xrightarrow{h \; l \circ_{j+m} \; (g \; i \circ_j \; f)} \Gamma_2, \Delta_3, \Phi_3, \Delta_4, \Gamma_3}$$

$$= \quad \frac{\dfrac{\Phi_1, B, \Phi_2 \xrightarrow{h} \Phi_3 \quad \Delta_1, A, \Delta_2 \xrightarrow{g} \Delta_3, B, \Delta_4}{\Phi_1, \Delta_1, A, \Delta_2, \Phi_2 \xrightarrow{h \; l \circ_m \; g} \Delta_3, \Phi_3, \Delta_4} \quad \Gamma_1 \xrightarrow{f} \Gamma_2, A, \Gamma_3}{\Phi_1, \Delta_1, \Gamma_1, \Delta_2, \Phi_2 \xrightarrow{(h \; l \circ_m \; g) \; l+i \circ_j \; f} \Gamma_2, \Delta_3, \Phi_3, \Delta_4, \Gamma_3}$$

4.
$$\frac{\dfrac{\Phi_1, A, \Phi_2, B, \Phi_3 \xrightarrow{h} \Phi_4 \quad \Gamma_1 \xrightarrow{f} \Gamma_2, A, \Gamma_3}{\Phi_1, \Gamma_1, \Phi_2, B, \Phi_3 \xrightarrow{h \; i \circ_j \; f} \Gamma_2, \Phi_4, \Gamma_3} \quad \Delta_1 \xrightarrow{g} \Delta_2, B, \Delta_3}{\Phi_1, \Gamma_1, \Phi_2, \Delta_1, \Phi_3 \xrightarrow{(h \; i \circ_j \; f) \; i+k+l \circ_m \; g} \Delta_2 | \Gamma_2, \Phi_4, \Delta_3 | \Gamma_3}$$

$$= \quad \frac{\dfrac{\Phi_1, A, \Phi_2, B, \Phi_3 \xrightarrow{h} \Phi_4 \quad \Delta_1 \xrightarrow{g} \Delta_2, B, \Delta_3}{\Phi_1, A, \Phi_2, \Delta_1, \Phi_3 \xrightarrow{h \; i+1+l \circ_m \; g} C} \quad \Gamma_1 \xrightarrow{f} \Gamma_2, A, \Gamma_3}{\Phi_1, \Gamma_1, \Phi_2, \Delta_1, \Phi_3 \xrightarrow{(h \; i+1+l \circ_m \; g) \; i \circ_j \; f} \Delta_2 | \Gamma_2, \Phi_4, \Delta_3 | \Gamma_3}$$

provided at least one of Γ_2, Δ_2 is empty, and at least one of Γ_3, Δ_3 is empty, so $\Delta_i | \Gamma_i$ represents the trivial concatenation of a sequence and an empty sequence.

5.
$$\frac{\Phi_1, B, \Phi_2 \xrightarrow{h} \Phi_3 \quad \dfrac{\Delta_1, A, \Delta_2 \xrightarrow{g} \Delta_3 \quad \Gamma_1 \xrightarrow{f} \Gamma_2, A, \Gamma_3, B, \Gamma_4}{\Delta_1, \Gamma_1, \Delta_2 \xrightarrow{g \; i \circ_j \; f} \Gamma_2, \Delta_3, \Gamma_3, B, \Gamma_4}}{\Phi_1 | \Delta_1, \Gamma_1, \Phi_2 | \Delta_2 \xrightarrow{h \; m \circ_{j+k+l} \; (g \; i \circ_j \; f)} \Gamma_2, \Delta_3, \Gamma_3, \Phi_3, \Gamma_4}$$

$$= \quad \frac{\Delta_1, A, \Delta_2 \xrightarrow{g} \Delta_3 \quad \dfrac{\Phi_1, B, \Phi_2 \xrightarrow{h} \Phi_3 \quad \Gamma_1 \xrightarrow{f} \Gamma_2, A, \Gamma_3, B, \Gamma_4}{\Phi_1, \Gamma_1, \Phi_2 \xrightarrow{h \; m \circ_{j+1+l} \; f} \Gamma_2, A, \Gamma_3, \Phi_3, \Gamma_4}}{\Phi_1 | \Delta_1, \Gamma_1, \Phi_2 | \Delta_2 \xrightarrow{g \; i \circ_j \; (h \; m \circ_{j+1+l} \; f)} \Gamma_2, \Delta_3, \Gamma_3, \Phi_3, \Gamma_4}$$

where again at least one of Φ_1, Δ_1 is empty, and at least one of Φ_2, Δ_2 is empty.

Remark 1.2 (Planar Polycategories)

There is a certain amount of permutation built into the cut rule and this results in the restrictions we have had to place on some of the equations. Lambek [La90] has given several weaker variants in which the restrictions are built directly into the cut rule so that this permutation is avoided: a similar system was presented by G.L. Mascari at the Durham Symposium. In the weakest system, **BL1**, cut is restricted to instances where either $\Gamma_1 = \Gamma_2 = \phi$ or $\Delta_2 = \Delta_3 = \phi$; this corresponds to having no weak distributivities, in the sense of the next section. A stronger system, **BL2**, in addition also allows cuts where either $\Gamma_1 = \Delta_3 = \phi$ or $\Gamma_2 = \Delta_2 = \phi$, corresponding in our setting to having only the "non permuting" distributivities δ_L^L and δ_L^R. We shall call the notion of polycategory based on **BL2**, allowing only cuts where either Γ_1 or $\Delta_2 = \phi$, and either Γ_2 or $\Delta_3 = \phi$, a "planar polycategory" (though it is not necessarily a polycategory!), as it corresponds to planar non commutative linear logic. (Note that in a planar polycategory, the restrictions on the equations are unnecessary.)

By allowing unrestricted cut in our setup, we are in effect introducing two "permuting" weak distributivities, δ_R^L and δ_L^R, as defined in the next section. So, in a sense, we are not dealing with a strictly non commutative logic. For example, if the weak distribution rules are inverted, the permuting ones will give a braiding structure in general. Thus we are generalizing braided monoidal categories (rather than symmetric or general non symmetric monoidal categories). We shall return to this point elsewhere. □

Next, we define a polycategory with two tensors: this amounts to having two binary operations $\otimes, \oplus$ on objects, extended to morphisms according to the following inference rules:

$$(\otimes\ \text{L})\ \frac{\Gamma_1, A, B, \Gamma_2 \xrightarrow{f} \Gamma_3}{\Gamma_1, A \otimes B, \Gamma_2 \xrightarrow{f^{\otimes i}} \Gamma_3} \qquad (\otimes\ \text{R})\ \frac{\Gamma_1 \xrightarrow{f} \Gamma_2, A, \Gamma_3 \qquad \Delta_1 \xrightarrow{g} \Delta_2, B, \Delta_3}{\Gamma_1, \Delta_1 \xrightarrow{f\ _i\otimes_j\ g} \Gamma_2|\Delta_2, A \otimes B, \Gamma_3|\Delta_3}$$

provided (in $(\otimes\ \text{R})$) at least one of Γ_2, Δ_2 is empty and at least one of Γ_3, Δ_3 is empty. In $(\otimes\ \text{L})$, $i = $ length of Γ_1; in $(\otimes\ \text{R})$, $i = $ length of Γ_2, $j = $ length of Δ_2, (so $ij = 0$).

$$(\oplus\ \text{L})\ \frac{\Gamma_1, A, \Gamma_2 \xrightarrow{f} \Gamma_3 \qquad \Delta_1, B, \Delta_2 \xrightarrow{g} \Delta_3}{\Gamma_1|\Delta_1, A \oplus B, \Gamma_2|\Delta_2 \xrightarrow{f\ _i\oplus_j\ g} \Gamma_3, \Delta_3} \qquad (\oplus\ \text{R})\ \frac{\Gamma_1 \xrightarrow{f} \Gamma_2, A, B, \Gamma_3}{\Gamma_1 \xrightarrow{f^{\oplus i}} \Gamma_2, A \oplus B, \Gamma_3}$$

provided (in ($\oplus$ L)) at least one of Γ_1, Δ_1 is empty and at least one of Γ_2, Δ_2 is empty. In ($\oplus$ L), $i =$ length of Γ_1, and $j =$ length of Δ_1, (so $ij = 0$); in ($\oplus$ R), $i =$ length of Γ_2.

(Note that we have indexed the labels as we did with cut; when clear from the context, we shall drop these subscripts.)

There are many further equivalences of derivations as in Definition 1.1. These can be considerably simplified if we give the following equivalent formulation of the tensor rules:

Definition 1.3 *A two-tensor-polycategory is a polycategory with two binary operations $\otimes, \oplus$ on objects, with morphisms*

$$m_{AB} : A, B \longrightarrow A \otimes B$$

$$w_{AB} : A \oplus B \longrightarrow A, B$$

and the rules of inference ($\otimes$ L) and ($\oplus$ R) above. These rules are to represent bijections stable under cut, so the following equations must hold:

- $m^\otimes = i$

- $g \circ f^\otimes = (g \circ f)^\otimes$ *for* $g : \Delta_1, C, \Delta_2 \longrightarrow \Delta_3$ *and* $f : \Gamma_1 \longrightarrow \Gamma_2, C, \Gamma_3$, *and where Γ_1 contains the sequence A, B.*

- $f^\otimes \circ g = (f \circ g)^\otimes$ *for* $g : \Delta_1 \longrightarrow \Delta_2, C, \Delta_3$ *and* $f : \Gamma_1, C, \Gamma_2 \longrightarrow \Gamma_3$, *and where one of Γ_1, Γ_2 contains the sequence A, B.*

- $f = f^{\otimes i} \circ m$ *for* $f : \Gamma_1, A, B, \Gamma_2 \longrightarrow \Gamma_3$

- $w^\oplus = i$

- $g \circ f^\oplus = (g \circ f)^\oplus$ *for* $g : \Delta_1, C, \Delta_2 \longrightarrow \Delta_3$ *and* $f : \Gamma_1 \longrightarrow \Gamma_2, C, \Gamma_3$, *and where one of Γ_2, Γ_3 contains the sequence A, B.*

- $f^\oplus \circ g = (f \circ g)^\oplus$ *for* $g : \Delta_1 \longrightarrow \Delta_2, C, \Delta_3$ *and* $f : \Gamma_1, C, \Gamma_2 \longrightarrow \Gamma_3$, *and where Γ_3 contains the sequence A, B.*

- $f = w \circ f^{\oplus i}$ *for* $f : \Gamma_1 \longrightarrow \Gamma_2, A, B, \Gamma_3$

We shall leave it as an exercise to show that this is equivalent to the other presentation. However, we must stress that cut elimination does *not* hold for the second presentation of two-tensor-polycategories; the amount of cut built into the rules ($\otimes$ R) and ($\oplus$ L) is necessary to prove cut elimination.

It is straightforward to define the category of polycategories (just keep in mind that we interpret sequents $\Gamma \longrightarrow \Delta$ as maps $\otimes \Gamma \longrightarrow \oplus \Delta$, and functors should preserve the tensors.) So a functor $F : \mathbf{C} \longrightarrow \mathbf{D}$ is a map $Ob(\mathbf{C}) \longrightarrow Ob(\mathbf{D})$ and a map $M\varphi(\mathbf{C}) \longrightarrow M\varphi(\mathbf{D})$ so that this and the induced map $Ob(\mathbf{C})^* \longrightarrow$

$Ob(\mathbf{D})^*$ commute with *source* and with *target*. A functor between two-tensor-polycategories must preserve the two tensors.

A natural transformation $\alpha\colon F \longrightarrow G$ assigns a $\mathbf{D}$ morphism $\alpha_A\colon F(A) \longrightarrow G(A)$ to each singleton sequence A from $\mathbf{C}$, satisfying the usual naturality condition.

We shall denote the 2–category of polycategories by **PolyCat**, and the 2–category of two-tensor-polycategories by $\mathbf{PolyCat}_{\otimes\oplus}$. We then note that the latter is a conservative extension of the former:

Proposition 1.4 *There is a 2–adjunction* $F \dashv U$

$$\mathbf{PolyCat} \underset{U}{\overset{F}{\rightleftarrows}} \mathbf{PolyCat}_{\otimes\oplus}$$

whose unit $\mathbf{C} \longrightarrow UF(\mathbf{C})$ *is full for each polycategory* $\mathbf{C}$.

Proof. Given a polycategory $\mathbf{C}$, $F(\mathbf{C})$ is the free two-tensor-polycategory generated by $\mathbf{C}$. That is, close the set $Ob(\mathbf{C})$ under the tensors $\otimes, \oplus$ to obtain the objects of $F(\mathbf{C})$, and take the sequents of $\mathbf{C}$ as non logical axioms, closing under the inference rules to obtain the morphisms of $F(\mathbf{C})$ (actually, you must factor out by the appropriate equivalences first). For a two-tensor-polycategory, U just forgets the two tensor structure.

For a two-tensor-polycategory $\mathbf{D}$, the counit $FU(\mathbf{D}) \longrightarrow \mathbf{D}$ collapses the new tensor structure onto the old. For a polycategory $\mathbf{C}$, the unit $\mathbf{C} \longrightarrow UF(\mathbf{C})$ is the usual inclusion into the free structure. To see that this map is full, we use the cut elimination theorem for two-tensor-polycategories. $F(\mathbf{C})$ has only the sequents of $\mathbf{C}$ as its non logical axioms, so by cut elimination any derivation in $F(\mathbf{C})$ is equivalent to one with cuts restricted to sequents from $\mathbf{C}$. If $\Gamma \longrightarrow \Delta$ is a tensor-free sequent of $F(\mathbf{C})$, (for example, is in the image of the unit,) then any derivation of $\Gamma \longrightarrow \Delta$ is equivalent to a derivation in $\mathbf{C}$, since with the cuts restricted to the tensor-free part of $F(\mathbf{C})$, none of the left or right introduction rules could be used in the derivation (they introduce tensors that could never be eliminated). □

Remark: We believe more: *viz.* that the unit is faithful as well, but as of this writing some details remain to be checked. (For if two derivations of a tensor-free sequent are equivalent in $F(\mathbf{C})$, then this equivalence must be true in $\mathbf{C}$ as well.)

We have not considered the question of units $\top, \bot$ for the tensors $\otimes, \oplus$—these can be added if wanted, together with the obvious rules and equations (as done in [Se89]). Since the point of these units is that they represent "empty" places in the sequents, they are rather redundant in the polycategory context; however, they are useful when we consider weak distributive categories, and we shall feel free to consider $\mathbf{PolyCat}_{\otimes\oplus}$ enriched with these units when this makes matters technically simpler. For an alternate treatment in terms of proof nets, including the units, see [BS91]. Using nets rather than sequents, that paper also manages to solve the coherence problem for weakly distributive categories with units.

2. Weakly distributive categories

2.1. Definition

A **weakly distributive category** C is a category with two tensors and four weak distribution natural transformations. The two tensors will be denoted by $\otimes$ and $\oplus$ and we shall call $\otimes$ the **tensor** and $\oplus$ the **cotensor**. Each tensor comes equipped with a unit object, an associativity natural isomorphism, and a left and right unit natural isomorphism:

$$(\otimes, \top, a_\otimes, u_\otimes^L, u_\otimes^R)$$
$$a_\otimes \;:\; (A \otimes B) \otimes C \longrightarrow A \otimes (B \otimes C)$$
$$u_\otimes^R \;:\; A \otimes \top \longrightarrow A$$
$$u_\otimes^L \;:\; \top \otimes A \longrightarrow A$$
$$(\oplus, \perp, a_\oplus, u_\oplus^L, u_\oplus^R)$$
$$a_\oplus \;:\; (A \oplus B) \oplus C \longrightarrow A \oplus (B \oplus C)$$
$$u_\oplus^R \;:\; A \oplus \perp \longrightarrow A$$
$$u_\oplus^L \;:\; \perp \oplus A \longrightarrow A$$

The four weak distribution transformations shall be denoted by:

$$\delta_L^L : A \otimes (B \oplus C) \;\longrightarrow\; (A \otimes B) \oplus C$$
$$\delta_R^L : A \otimes (B \oplus C) \;\longrightarrow\; B \oplus (A \otimes C)$$
$$\delta_L^R : (B \oplus C) \otimes A \;\longrightarrow\; (B \otimes A) \oplus C$$
$$\delta_R^R : (B \oplus C) \otimes A \;\longrightarrow\; B \oplus (C \otimes A).$$

This data must satisfy certain coherence conditions which we shall discuss shortly. Before doing so we remark that there are three independent symmetries which arise from this data:

[op′] reverse the arrows and swap both the $\otimes$ and $\oplus$ and $\top$ and $\perp$.

[$\otimes$′] reverse the tensor $\otimes$ (so $A \otimes' B = B \otimes A$).

[$\oplus$′] reverse the cotensor $\oplus$.

The notion of a weakly distributive category is preserved by all these symmetries and we shall use this fact to give an economical statement of the required commuting conditions, which are generated by the following diagrams.

Tensors: The two tensor products must each satisfy the usual conditions of a tensor product. (This gives four equations.)

Unit and distribution:

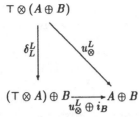

$$\top \otimes (A \oplus B)$$

with δ_L^L on the left and $u_\otimes^L$ on the right, leading to

$$(\top \otimes A) \oplus B \xrightarrow[u_\otimes^L \oplus i_B]{} A \oplus B$$

(Under the symmetries, this gives eight more equations.)

Associativity and distribution:

$$
\begin{array}{ccc}
(A \otimes B) \otimes (C \oplus D) & \xrightarrow{\ a_\otimes\ } & A \otimes (B \otimes (C \oplus D)) \\
\downarrow{\scriptstyle \delta_L^L} & = & \downarrow{\scriptstyle i_A \otimes \delta_L^L} \\
 & & A \otimes ((B \otimes C) \oplus D) \\
 & & \downarrow{\scriptstyle \delta_L^L} \\
((A \otimes B) \otimes C) \oplus D & \xrightarrow{\ a_\otimes \oplus i_D\ } & (A \otimes (B \otimes C)) \oplus D
\end{array}
$$

(Under the symmetries, this gives eight more equations.)

Coassociativity and distribution:

$$
\begin{array}{ccc}
A \otimes ((B \oplus C) \oplus D) & \xrightarrow{\ i_A \otimes a_\oplus\ } & A \otimes (B \oplus (C \oplus D)) \\
\downarrow{\scriptstyle \delta_L^L} & = & \downarrow{\scriptstyle \delta_R^L} \\
(A \otimes (B \oplus C)) \oplus D & & B \oplus (A \otimes (C \oplus D)) \\
\downarrow{\scriptstyle \delta_R^L \oplus i_D} & & \downarrow{\scriptstyle i_B \oplus \delta_L^L} \\
(B \oplus (A \otimes C)) \oplus D & \xrightarrow{\ a_\oplus\ } & B \oplus ((A \otimes C) \oplus D)
\end{array}
$$

(Under the symmetries, this gives four more equations.)

Distribution and distribution:

$$(A \oplus B) \otimes (C \oplus D)$$

with δ_L^L on the left and δ_R^R on the right,

$$
\begin{array}{ccc}
((A \oplus B) \otimes C) \oplus D & & A \oplus (B \otimes (C \oplus D)) \\
\downarrow{\scriptstyle \delta_R^R \oplus i_D} & = & \downarrow{\scriptstyle i_A \oplus \delta_L^L} \\
(A \oplus (B \otimes C)) \oplus D & \xrightarrow{\ a_\oplus\ } & A \oplus ((B \otimes C) \oplus D)
\end{array}
$$

(Under the symmetries, this gives four more equations, for a total of twenty eight.)

2.2. Weakly distributive categories and polycategories

Now we can make the connection between weakly distributive categories and two-tensor-polycategories; essentially these are the same thing. (With Proposition 1.4 this justifies our claiming that weakly distributive categories constitute the essential content of polycategories.) We shall denote the category of weakly distributive categories and functors preserving the tensor and cotensor by **WkDistCat**. (We suppose here that we are using the version of two-tensor-polycategories with units, to correspond to the units in the weakly distributive categories.)

Theorem 2.1 *There is an equivalence of 2-categories*

$$\mathbf{PolyCat}_{\otimes\oplus} \underset{P}{\overset{W}{\rightleftarrows}} \mathbf{WkDistCat}$$

Proof. Given a weakly distributive category $\mathbf{W}$, $P(\mathbf{W})$ is the polycategory with the same set of objects as $\mathbf{W}$, and with morphisms given by: $\Gamma \longrightarrow \Delta$ is a morphism if and only if $\otimes \Gamma \longrightarrow \oplus \Delta$ is a morphism of $\mathbf{W}$. To check that the cut rule, and the left and right introduction rules, are valid, we use the weak distributivities; for example, we shall illustrate the following instance of cut. Given maps (in $\mathbf{W}$) $C_1 \otimes A \otimes C_2 \overset{g}{\to} C_3$ and $D_1 \overset{f}{\to} D_2 \oplus A \oplus D_3$, we can construct $C_1 \otimes D_1 \otimes C_2 \overset{g\,\mathbf{1}\circ_1\,f}{\longrightarrow} D_2 \oplus C_3 \oplus D_3$ as follows (ignoring some instances of associativity for simplicity):

$$
\begin{aligned}
C_1 \otimes D_1 \otimes C_2 &\xrightarrow{i\otimes f\otimes i} & C_1 \otimes (D_2 \oplus A \oplus D_3) \otimes C_2 \\
&\xrightarrow{\delta_R^L\otimes i} & (D_2 \oplus (C_1 \otimes (A \oplus D_3))) \otimes C_2 \\
&\xrightarrow{\delta_R^R} & D_2 \oplus ((C_1 \otimes (A \oplus D_3)) \otimes C_2) \\
&\xrightarrow{i\oplus(\delta_L^L\otimes i)} & D_2 \oplus (((C_1 \otimes A) \oplus D_3) \otimes C_2) \\
&\xrightarrow{i\oplus\delta_L^R} & D_2 \oplus ((C_1 \otimes A \otimes C_2) \oplus D_3) \\
&\xrightarrow{i\oplus g\oplus i} & D_2 \oplus C_3 \oplus D_3
\end{aligned}
$$

(Other ways of introducing the distributivities to move the C's next to A are equivalent, by the coherence conditions on the interaction of distributivity with itself and with associativity.)

We must then check that all the equivalences of two-tensor-polycategories follow from the coherence diagrams of weakly distributive categories. This is a frightful but routine exercise. (But note that the extra structure due to the two tensors is easy since $(\otimes$ L) and $(\oplus$ R) are identities. So we really only need check the five equivalences of Definition 1.1.)

Next, given a two-tensor-polycategory $\mathbf{P}$, the weakly distributive category $W(\mathbf{P})$ is just the category part of $\mathbf{P}$, *viz.* those morphisms whose source and target are singletons. The distributivities are essentially given by the (cut) rule and the axioms ($\otimes$ R) and ($\oplus$ L). For instance, δ_R^L is given as (note the "exchange"):

$$\frac{A, C \longrightarrow A \otimes C \quad B \oplus C \longrightarrow B, C}{A, B \oplus C \longrightarrow B, A \otimes C}$$

And the coherence conditions follow from the equivalences for polycategories.

It is clear from the constructions above that $WP(\mathbf{W})$ is isomorphic to $\mathbf{W}$; indeed they are the same category. And essentially for the same reason, $\mathbf{P}$ is isomorphic to $PW(\mathbf{P})$. (Essentially, this just depends on the bijection

$$\frac{\Gamma \longrightarrow \Delta}{\otimes \Gamma \longrightarrow \oplus \Delta}$$

which means that the category part of a two-tensor-polycategory carries all the information of the polycategory.)   □

3.   Distributive categories

How are weakly distributive categories related to distributive categories? It turns out that they are very close indeed—if the tensors are the cartesian product and coproduct (nicely), then the two notions coincide. So in a natural sense, weak distributivity is the natural notion for general tensors.

A weakly distributive category is **symmetric** (resp. $\otimes$–symmetric, $\oplus$–symmetric) in case the tensors are symmetric (resp. the tensor is symmetric with $s_\otimes$, the cotensor with $s_\oplus$) and

$$
\begin{array}{ccccccc}
A \otimes (B \oplus C) & \xrightarrow{\ i \otimes s_\oplus\ } & A \otimes (C \oplus B) & \xrightarrow{\ s_\otimes\ } & (C \oplus B) \otimes A & \xrightarrow{\ s_\oplus \otimes i\ } & (B \oplus C) \otimes A \\
\downarrow{\scriptstyle \delta_L^L} & & \downarrow{\scriptstyle \delta_R^L} & & \downarrow{\scriptstyle \delta_R^R} & & \downarrow{\scriptstyle \delta_L^R} \\
(A \otimes B) \oplus C & \xrightarrow{\ s_\oplus\ } & C \oplus (A \otimes B) & \xrightarrow{\ i \oplus s_\otimes\ } & C \oplus (B \otimes A) & \xrightarrow{\ s_\oplus\ } & (B \otimes A) \oplus C
\end{array}
$$

commuting in all squares (resp. those squares which exist).

A weakly distributive category is **cartesian** (resp. $\otimes$--cartesian or $\oplus$–cartesian) if the category is symmetric (resp. $\otimes$–symmetric, $\oplus$–symmetric) with the tensor a product (with $\top$ the final object) and the cotensor a coproduct (with $\bot$ the initial object).

A source of motivation for the study of weak distributivity is the fact that distributive categories are (cartesian) examples. This means that the category of **Sets** (or any topos) is a cartesian weakly distributive category.

We now verify that distributive categories are a source of examples. An **elementary distributive category** [Co90] has finite products and coproducts such that the comparison map from the coproduct

$$\langle i \times b_0 | i \times b_1 \rangle : A \times B + A \times C \longrightarrow A \times (B + C)$$

is an isomorphism. We shall denote the inverse of $\langle i \times b_0 | i \times b_1 \rangle$ by δ.

Proposition 3.1 *Elementary distributive categories are cartesian weakly distributive categories.*

Proof. Let

$$\delta_L^L : A \times (B + C) \xrightarrow{\delta} A \times B + A \times C \xrightarrow{i + p_1} A \times B + C$$

then, as $+$ and $\times$ are symmetric the other weak distributions can be obtained from this. Due to the symmetry of product and coproduct it suffices to prove that the four basic diagrams hold together with their op$'$ duals. This gives eight diagrams to check. However, the use of symmetry reduces this to six, *viz.* that the two canonical ways of expressing each of the following arrows are equal:

$$\top \times (A + B) \longrightarrow A + B$$
$$(\bot + A) \times B \longrightarrow A \times B$$
$$(A \times B) \times (C + D) \longrightarrow A \times (B \times C) + D$$
$$(A \times B) \times (C + D) \longrightarrow A \times C + B \times D$$
$$A \times ((C + D) + E) \longrightarrow A \times C + (D + E)$$
$$(A + B) \times (C + D) \longrightarrow A + (B \times C) + D$$

For the first of these consider:

$$
\begin{array}{ccc}
\top \times (A + B) & \xrightarrow{\ p_1\ } & A + B \\
{\scriptstyle \delta}\big\downarrow & & \big\uparrow{\scriptstyle p_1 + i} \\
\top \times A + \top \times B & \xrightarrow[\ i + p_1\ ]{} & \top \times A + B
\end{array}
$$

As $\delta = \langle i \times b_0 | i \times b_1 \rangle^{-1}$ to obtain commutativity it suffices to show:

$$\top \times A \xrightarrow{b_0} \top \times A + \top \times B \xrightarrow{\langle i \times b_0 | i \times b_1 \rangle} \top \times (A + B) \xrightarrow{p_1} A + B = \top \times A \xrightarrow{p_1} A \xrightarrow{b_0} A + B$$

$$\top \times B \xrightarrow{b_1} \top \times A + \top \times B \xrightarrow{\langle i \times b_0 | i \times b_1 \rangle} \top \times (A + B) \xrightarrow{p_1} A + B = \top \times B \xrightarrow{p_1} B \xrightarrow{b_1} A + B$$

which are clear.

For the op$'$ dual of this we have:

$$(\bot + A) \times B \xrightarrow{\;\delta\;} \bot \times B + A \times B \xrightarrow{\;p_0 + i\;} \bot + A \times B$$

$$b_1^{-1} \times i \downarrow \qquad\qquad\qquad\qquad\qquad\qquad\qquad \downarrow b_1^{-1}$$

$$A \times B \;=\!=\!=\!=\!=\!=\!=\!=\!=\!=\!=\!=\; A \times B$$

which commutes as $b_1 \times i.\delta.p_0 + i = b_1.p_0 + i = b_1$.

The remaining equations are checked using full distribution applied in two different ways. One can check that the diagram commutes for the components of the coproducts, using the inverses of these distributions, and finally project to obtain the weak distributions.   □

It is of some interest to wonder what conditions must be added to a cartesian weakly distributive category to force it to be (elementary) distributive. Demanding that it is cartesian is not sufficient: this can be seen in two ways.

First, an abelian category is a cartesian weakly distributive category as it is a symmetric tensor category on the biproduct. This follows as any braided monoidal category is a weakly distributive category by letting the non permuting weak distributions be the associativity of the tensor and the permuting weak distributions be given by the braiding. Thus, certainly any symmetric monoidal category is a weakly distributive category. Finally, an abelian category is not distributive.

Second, the dual of a distributive category (a codistributive category) is clearly cartesian weakly distributive as the latter is a self-dual notion. However, a codistributive category is not distributive. Indeed, a codistributive category which is simultaneously a distributive category must be a preorder (as the final object is costrict).

In order to obtain a distributive category there must, therefore, be some relationship required between the distribution, projection, and embedding maps. Our first attempt to pin this down is as follows:

Lemma 3.2 *A cartesian weakly distributive category is distributive if and only if the following diagrams*

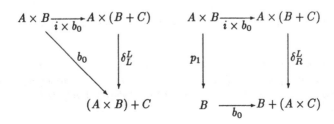

commute.

Proof. It is easy to check that a distributive category satisfies the two diagrams. For the converse, we must construct the inverse δ of $\langle i \times b_0 | i \times b_1 \rangle$.

We set $\delta = A \times (B + C) \xrightarrow{\Delta \times i} (A \times A) \times (B + C) \xrightarrow{a_\times} A \times (A \times (B + C)) \xrightarrow{i \times \delta_R^L}$ $A \times (B + A \times C) \xrightarrow{\delta_L^L} A \times B + A \times C$. To show that this is the inverse of $\langle i \times b_0 | i \times b_1 \rangle$ we precompose with $i \times b_0$ (by symmetry the same thing will happen on precomposing with $i \times b_1$) and show the result is b_0:

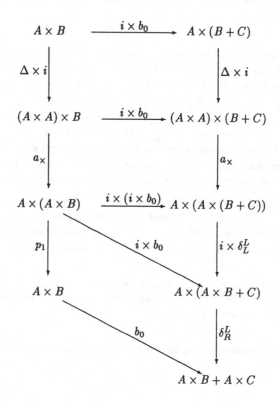

where the triangle and parallelogram are the two conditions added.   □

An initial object is **strict** in case every map to it is an isomorphism. Notice that, as an abelian category has a zero, it cannot have a strict initial object without being trivial. The initial object of an elementary distributive category, however, is necessarily strict (see [Co90]). This is a difference we now exploit:

Theorem 3.3 *A cartesian weakly distributive category is an elementary distributive category if and only if it has a strict initial object.*

Proof. It suffices to show that the two diagrams above commute in the presence of a strict initial object. To see this consider the two naturality diagrams

$$
\begin{array}{ccc}
A \times (B + \bot) & \xrightarrow{\;i \times (i + \bot)\;} & A \times (B + C) \\
\delta_L^L \downarrow & & \downarrow \delta_L^L \\
A \times B + \bot & \xrightarrow[\;i + \bot\;]{} & (A \times B) + C
\end{array}
\qquad
\begin{array}{ccc}
A \times (B + \bot) & \xrightarrow{\;i \times (i + \bot)\;} & A \times (B + C) \\
\delta_R^L \downarrow & & \downarrow \delta_R^L \\
B + (A \times \bot) & \xrightarrow[\;i + (i \times \bot)\;]{} & B + (A \times C)
\end{array}
$$

The first immediately yields the first condition of the lemma. The second due to strictness has the bottom left object isomorphic to B and the horizontal map is then the coproduct embedding. It suffices to prove that the vertical map is essentially a projection. For this consider

$$
\begin{array}{ccc}
A \times (B + \bot) & \xrightarrow{\;! \times i\;} & \top \times (B + \bot) \\
\delta_R^L \downarrow & & \downarrow \delta_R^L \\
B + A \times \bot & \xrightarrow[\;i + (! \times i)\;]{} & B + \top \times \bot
\end{array}
$$

The lower horizontal map is an isomorphism due to the strictness of the initial object. However, the map across the square is clearly equivalent to a projection. □

4. Adding negation

Definition 4.1 *We define a weakly distributive category with negation to be a weakly distributive category with an object function* $(\_)^\perp$, *together with the following parameterized families of maps ("contradiction" and "tertium non datur"):*

$$
A^\perp \otimes A \xrightarrow{\;\gamma_A^L\;} \bot \qquad\qquad A \otimes A^\perp \xrightarrow{\;\gamma_A^R\;} \bot
$$
$$
\top \xrightarrow{\;\tau_A^L\;} A^\perp \oplus A \qquad\qquad \top \xrightarrow{\;\tau_A^R\;} A \oplus A^\perp
$$

which satisfy the following coherence condition

$$
\begin{array}{ccc}
A \otimes \top & \xrightarrow{\;i \otimes \tau^R\;} & A \otimes (A \oplus A^\perp) \\[4pt]
& & \Big\downarrow \delta_R^L \\[4pt]
u_\otimes^R \Big\downarrow & = & A \oplus (A \otimes A^\perp) \\[4pt]
& & \Big\downarrow i \oplus \gamma^R \\[4pt]
A & \xleftarrow[\;u_\oplus^R\;]{} & A \oplus \bot
\end{array}
$$

and its eight symmetric forms. Note that the op′ dual should be modified to switch A and $A^\perp$.

Notice that we have not required that $(\_)^\perp$ be a contravariant functor, but merely that it is defined on objects. Of course, $(\_)^\perp$ is a contravariant functor but this is a consequence of the axioms as we shall see. First we note:

Lemma 4.2 *In a weakly distributive category with negation we have the following adjunctions*

$$
\begin{array}{cc}
A \otimes -\dashv A^\perp \oplus - & A^\perp \otimes -\dashv A \oplus - \\
- \otimes B \dashv - \oplus B^\perp & - \otimes B^\perp \dashv - \oplus B
\end{array}
$$

corresponding to the following bijections

$$
\frac{A \otimes B \longrightarrow C}{B \longrightarrow A^\perp \oplus C}
\qquad
\frac{A^\perp \otimes B \longrightarrow C}{B \longrightarrow A \oplus C}
$$

$$
\frac{A \otimes B \longrightarrow C}{A \longrightarrow C \oplus B^\perp}
\qquad
\frac{A \otimes B^\perp \longrightarrow C}{A \longrightarrow C \oplus B}
$$

Proof. We shall treat just the adjunction $- \otimes B \dashv - \oplus B^\perp$ as an illustration. Given a map $A \otimes B \longrightarrow C$, we derive the corresponding map as $A \longrightarrow A \otimes \top \longrightarrow A \otimes (B \oplus B^\perp) \longrightarrow (A \otimes B) \oplus B^\perp \longrightarrow C \oplus B^\perp$. Conversely, given $A \longrightarrow C \oplus B^\perp$, we have $A \otimes B \longrightarrow (C \oplus B^\perp) \otimes B \longrightarrow C \oplus (B^\perp \otimes B) \longrightarrow C \oplus \bot \longrightarrow C$.

In particular, the unit $\eta_A : A \longrightarrow (A \otimes B) \oplus B^\perp$ is given by

$$
\eta_A : A \xrightarrow{\;u_\otimes^{R-1}\;} A \otimes \top \xrightarrow{\;i \otimes \tau^R\;} A \otimes (B \oplus B^\perp) \xrightarrow{\;\delta_L^L\;} (A \otimes B) \oplus B^\perp
$$

and the counit $\epsilon_A : (A \oplus B^\perp) \otimes B \longrightarrow A$ is given (as the symmetries might suggest) by

$$
\epsilon_A : (A \oplus B^\perp) \otimes B \xrightarrow{\;\delta_R^R\;} A \oplus (B^\perp \otimes B) \xrightarrow{\;i \oplus \gamma^L\;} A \oplus \bot \xrightarrow{\;u_\oplus^R\;} A
$$

We leave checking the triangle identities as an exercise (or see the fuller version of this paper).   □

We may now use the adjunctions to define the effect of $(-)^\perp$ on maps:

$$\frac{\dfrac{\dfrac{\dfrac{A \longrightarrow B}{\top \otimes A \longrightarrow B}}{\top \longrightarrow B \oplus A^\perp}}{B^\perp \otimes \top \longrightarrow A^\perp}}{B^\perp \longrightarrow A^\perp}$$

it is then a matter of verifying that this is functorial, by explicitly giving the "formula" for $B^\perp \longrightarrow A^\perp$ in terms of $A \longrightarrow B$, and verifying that the appropriate diagrams commute (again this is in the fuller version of this paper).

Furthermore, notice that $(\_)^\perp$ is full and faithful as there is a bijection $Hom(A, B) \simeq Hom(B^\perp, A^\perp)$.

Theorem 4.3 *The notions of symmetric weakly distributive categories with negation and $*$-autonomous categories coincide.*

Proof. One direction is more or less automatic now in view of Barr's characterization of $*$-autonomous categories in [Ba79]. That is to say, symmetric weakly distributive categories with negation are $*$-autonomous. Of course, to make the translation to Barr's framework, we must make the following (standard) definition: $A \multimap B = A^\perp \oplus B$.

The involutive nature of $(-)^\perp$ follows from the lemma straightforwardly: *viz.* the iso $A = A^{\perp\perp}$ is induced by the adjunctions:

$$\frac{\dfrac{\dfrac{\dfrac{A \longrightarrow B}{\top \otimes A \longrightarrow B}}{\top \longrightarrow B \oplus A^\perp}}{\top \otimes A^{\perp\perp} \longrightarrow B}}{A^{\perp\perp} \longrightarrow B}$$

Then we can conclude that $A \multimap B = (A \otimes B^\perp)^\perp$ also.

In either case, it is now easy to verify the essential bijection:

$$\frac{\dfrac{\dfrac{\dfrac{\dfrac{\dfrac{A \longrightarrow (B \multimap C^\perp)}{A \longrightarrow B^\perp \oplus C^\perp}}{A \otimes C \longrightarrow B^\perp}}{C \otimes A \longrightarrow B^\perp}}{C \longrightarrow B^\perp \oplus A^\perp}}{C \longrightarrow (B \multimap A^\perp)}$$

Note the use of symmetry here. Of course, if the tensors are not symmetric, then we would have two internal hom's, the other being $B \mathbin{\circ\!\!-} A = B \oplus A^\perp$, and the bijection above would end with $C \longrightarrow (A^\perp \mathbin{\circ\!\!-} B)$.

Next the other half of the proof: here we give just a brief sketch. It is a straight-forward verification to check that *-autonomous categories are weakly distributive, though the diagrams can be pretty horrid. We shall just indicate how the weak distribution δ_L^L is obtained, leaving the rest to the faith of the reader.

Defining $A \oplus B = A^\perp \multimap B$, we need $\delta_L^L: A \otimes (B^\perp \multimap C) \longrightarrow (A \otimes B)^\perp \multimap C$. While it is possible to give a formula for this morphism, it is perhaps more instructive to give its derivation:

First note that under the functor $(-)^\perp$, the internal hom bijection becomes

$$\frac{C^\perp \longrightarrow (A \otimes B)^\perp}{B \otimes C^\perp \longrightarrow A^\perp}$$

From this it is easy to derive maps $A \otimes (A \otimes B)^\perp \longrightarrow B^\perp \longrightarrow (B^\perp \multimap C) \multimap C$. Then we can use the bijection

$$\frac{A \otimes X \longrightarrow Y \multimap C}{A \otimes Y \longrightarrow X \multimap C}$$

to derive the map $A \otimes (B^\perp \multimap C) \longrightarrow (A \otimes B)^\perp \multimap C$ as needed.   □

Remark 4.4 (Planar non commutativity)

The above suggests that (non symmetric) weakly distributive categories with nega-tion provide a natural notion of non symmetric *-autonomous categories, and hence of non commutative linear logic (rather, the multiplicative fragment thereof).

However, there is another such notion, building on the planar polycategories of Remark 1.2. In this variant, two different negations are used, $^\perp A$ and $A^\perp$. In his presentation at this Symposium, G.L. Mascari described such a system, with the inference rules

$$\frac{\Gamma \vdash \Delta, B}{\Gamma, B^\perp \vdash \Delta} \qquad \frac{A, \Gamma \vdash \Delta}{\Gamma \vdash A^\perp, \Delta}$$

$$\frac{\Gamma \vdash B, \Delta}{^\perp B, \Gamma \vdash \Delta} \qquad \frac{\Gamma, A \vdash \Delta}{\Gamma \vdash \Delta, {}^\perp A}$$

A full account of this syntax appears in [Ab90].

We can modify our presentation to account for this variant. We replace γ_A^L with $\gamma_A^L: {}^\perp A \otimes A \longrightarrow \perp$ and τ_A^R with $\tau_A^R: \top \longrightarrow A \oplus {}^\perp A$. (And modify the coherence conditions as well, dropping those that no longer make sense.) Then we can derive the adjunctions

$$\begin{array}{cc} - \otimes B^\perp \dashv - \oplus B & A \otimes - \dashv A^\perp \oplus - \\ {}^\perp B \otimes - \dashv B \oplus - & - \otimes A \dashv - \oplus {}^\perp A \end{array}$$

corresponding to these rules, and the (natural) isomorphisms $(^{\perp}A)^{\perp} \simeq A$, $^{\perp}(A^{\perp}) \simeq A$. In this context, we would have that $A \multimap B \simeq (A \otimes {}^{\perp}B)^{\perp}$, $B \circ\!\!- A = B \oplus {}^{\perp}A \simeq {}^{\perp}(B^{\perp} \otimes A)$.

Our original presentation arose in an attempt to describe commutative linear logic: it displays some of the features of the planar non commutative form as well as the commutative form. At this time we feel it is very premature to pronounce definitively on the "best" degree of non commutativity in linear logic, and so we offer only these comments: First, our main observation is that the core of the multiplicative fragment of linear logic may be found in the two tensors, connected by weak distributivity. (We do not believe that the central role played by the weak distributivities—permutative or not—has been sufficiently observed before[2].) Second, to include negation and internal hom, one need only add negation in the most simple minded manner (the internal hom structure follows naturally). Third, the various versions of this fragment may be classified by the degree of the weak distributivity assumed and the nature of the negation added.   □

5.   Some posetal examples

To conclude, we shall briefly consider some simple examples of weakly distributive categories which are preorders. The beauty of the posetal weakly distributive categories is that one need not check the coherence conditions as all diagrams commute. It suffices to have the weak distributions present. Notice first that, when such a category is cartesian, the initial object is necessarily strict giving:

Lemma 5.1 *All cartesian weakly distributive categories which are preorders are equivalent to distributive lattices.*

Thus, the interesting posetal examples occur when one or both tensors are non cartesian. There are plenty of examples of these. Here are two sources:

- (Droste) Let L be a lattice ordered monoid (that is a set having a commutative, associative, and idempotent operation $x \wedge y$, and an associative operation $x \cdot y$ with unit 1 such that $z \cdot (x \wedge y) = (z \cdot x) \wedge (z \cdot y)$ and $(x \wedge y) \cdot z = (x \cdot z) \wedge (y \cdot z)$) in which every element is less than 1 (so this is the unit of $\wedge$ too) then L is a posetal weakly distributive category. This because:

$$x \cdot (y \wedge z) = (x \cdot y) \wedge (x \cdot z) \le (x \cdot y) \wedge (1 \cdot z) = (x \cdot y) \wedge z$$

and similarly for the other weak distributions.

An example of such an L is the negative numbers. In general one may take the negative portion of any lattice ordered group (free groups can be lattice ordered so that the multiplication need not be commutative).

---

[2]An exception is recent ongoing work of V. dePaiva and J.M.E. Hyland, unpublished but partially presented at Category Theory 1991, Montréal (June 1991), which has among other things pointed out some of the aspects of the distributivities we have in mind here.

- A shift monoid is a commutative monoid $(M, 0, +)$ with a designated invertible element a. This allows one to define a second "shifted" multiplication $x \cdot y = x + y - a$ with unit a for which we have the following identity:

$$x \cdot (y + z) = (x \cdot y) + z$$

which clearly is a weak distribution. In this manner a shift monoid becomes a discrete weakly distributive category. Furthermore, it is not hard to show that every discrete weakly distributive category must be a shift monoid.

This example is also of interest as it suggests that when one inverts the weak distributions (which produces braidings on the tensors), the tensors, which need not be equivalent, are related by a $\oplus$ invertible object. This is, in fact, what happens in general.

It is also of interest to specialize our presentation of $*$-autonomous categories to the case of preorders. Again, only the existence of the maps themselves must be ensured, which gives:

Proposition 5.2 *A preorder is a $*$-autonomous category if and only if it has two symmetric tensors $\otimes$ and $\oplus$ and an object map $(-)^{\perp}$ such that*

(i) $x \otimes (y \oplus z) \leq (x \otimes y) \oplus z$,

(ii) $x \otimes x^{\perp} \leq \perp$,

(iii) $\top \leq x \oplus x^{\perp}$.

Suppose that M is a shift monoid equipped with a map $(-)^{\perp}$ such that $x + x^{\perp} = a$ ("tertium non datur") then we have

$$x \cdot x^{\perp} = x + x^{\perp} - a = a - a = 0$$

which is "contradiction". So M is a discrete $*$-autonomous category. Note that moreover M is a group, with $-x = x^{\perp} - a$; in fact shift groups (shift monoids with M a group) are the same as discrete $*$-autonomous categories in this way: $\top = a$, $x^{\perp} = a - x$, and conversely, a discrete $*$-autonomous category is a group (with respect to $\oplus$, with inverse given by $-x = x^{\perp} \otimes \perp$), and so a shift group (with $\top$ as designated invertible element). (A curiosity about this example: the initial shift group (also the initial shift monoid) is $\mathcal{Z}$, the integers, under addition with $\top = 1$. This structure also arises when checking the validity of proof nets.)

We can construct similar examples with ordered shift monoids, (for example, $\mathcal{Z}$ as above with the standard order), to get examples of $*$-autonomous posets. Note that a $*$-autonomous ordered shift monoid must be a group, since $x \cdot x^{\perp} \leq 0$ and $a \leq x + x^{\perp}$ imply that $x + x^{\perp} = a$, and so we are in the context above. Note also that by a suitable choice of a we can arrange for the poset to satisfy the mix rule, $x \otimes y \leq x \oplus y$, or its opposite $x \otimes y \geq x \oplus y$, or to be compact $x \otimes y = x \oplus y$.

References

[Ab90] Abramsky S. *Computational interpretations of linear logic.* Preprint, Imperial College, 1990.

[Ab90] Abrusci, V.M. *Phase semantics and sequent calculus for pure noncommutative classical linear logic.* Università degli Studi di Bari, Rapporto scientifico N. 11, 1990.

[Ba79] Barr M. **-Autonomous Categories.* Lecture Notes in Mathematics 752, Springer-Verlag, Berlin, Heidelberg, New York, 1979.

[Ba91] Barr M. *Fuzzy models of linear logic.* Preprint, McGill University, 1991.

[BS91] Blute, R., Seely R.A.G. *Natural deduction and coherence for weakly distributive categories.* Preprint, McGill University, 1991.

[Co89] Cockett J.R.B. *Distributive logic.* Preprint, University of Tennessee, 1989.

[Co90] Cockett J.R.B. *Introduction to distributive categories.* Preprint, Macquarie University, 1990.

[dP89] dePaiva V.C.V. *A Dialectica-like model of linear logic.* in D.H.Pitt et al., eds., *Category Theory and Computer Science,* Lecture Notes in Computer Science 389, Springer-Verlag, Berlin, Heidelberg, New York, 1989.

[Gi87] Girard J.-Y. *Linear logic.* Theoretical Computer Science 50 (1987), 1 - 102.

[JS86] Joyal A., Street R. *Braided monoidal categories.* Macquarie Mathematics Report 860081 (1986).

[La90] Lambek J. *From categorial grammar to bilinear logic,* to appear in K. Došen and P. Schroeder-Heister, eds., *Substructural logics.* Proceedings of Tübingen Conference, 1990.

[Se89] Seely R.A.G. *Linear logic, *-autonomous categories and cofree coalgebras,* in J. Gray and A. Scedrov, eds., *Categories in Computer Science and Logic,* (Proc. A.M.S. Summer Research Conference, June 1987), Contemporary Mathematics 92, (Am. Math. Soc. 1989).

[Sz75] Szabo M.E. *Polycategories.* Comm. Alg. 3 (1975), 663 - 689.

Sequentiality and full abstraction

P.-L. Curien[1]

Abstract

This paper collects observations about the two issues of sequentiality and full abstraction for programming languages. The format of the paper is that of an *extended lecture*. Some old and new results are hinted at, and references are given, without any claim to be exhaustive. We assume that the reader knows something about λ-calculus and about domain theory.

1 Introduction

Sequentiality and full abstraction have been often considered as related topics. More precisely, the quest of full abstraction led to the idea that sequentiality is a key issue in the semantics of programming languages.

In vague terms, full abstraction is the property that a mathematical semantics captures exactly the operational semantics of a specific language under study. Following the tradition of the first studies on full abstraction [Milner,Plotkin], the languages considered here are PCF, an extension of λ-calculus with arithmetic operators and recursion, and variants thereof. The focus on λ-calculus is amply justified by its rôle, either as a kernel (functional) programming language, or as a suitable metalanguage for the encoding of denotational semantics of a great variety of (sequential) programming languages.

It was Gérard Berry's belief that only after a detailed sudy of the syntax could one conceive the semantic definitions appropriate for reaching full abstraction. I always considered this as an illuminating idea, and this will be the starting point of this paper.

In section 2, we shall state Berry's Sequentiality Theorem: this will require us first to recall Wadsworth-Welch-Lévy's Continuity Theorem, and then to introduce a general notion of sequential function, due to Kahn-Plotkin.

In section 3, we shall abandon sequentiality for a while to define full

---

[1]CNRS-LIENS, 45 rue d'Ulm 75230 Paris Cedex 05, curien@dmi.ens.fr

abstraction and quote some results.

In section 4, we shall summarize the basics of sequential algorithms on concrete data structures, which form a non-extensional model of PCF. The extensive references for this topic are [Curien 86] and the joint papers [BC1] and [BC2].

In section 5, we shall briefly mention a variant of this model which was suggested by recent insights of M. Felleisen and C. Cartwright. By incorporating a notion of error in the framework of sequential algorithms on concrete data structures, a larger category with the same objects is built, which turns out to be an order-extensional model of PCF. It is actually a fully abstract model of an extension of PCF with *error* and *catch* constructs. More details are to be found in the forthcoming [Curien 91,CCF], and in the original [CF].

The paper updates in some respects, but does not substitute for the ten year old survey paper [BCL]. The paper does not suppose a previous reading of [BCL]. Other general references are [Meyer] (short) and [Stoughton1] (a monograph).

2 Syntactic sequentiality

It is often quoted that continuity in domain theory stems from the important result of Myhill-Shepherdson which states that the computable functionals of type $(\iota \to \iota) \to (\iota \to \iota)$ are continuous (w.r.t. the natural "less defined than" partial order; ι stands for the type of natural numbers) [Scott 76]. Hence continuity, in the order-theoretic (or T_0-topological) sense arises naturally from computability theory.

The Syntactic Continuity Theorem of λ-calculus appears to me as another strong foundation for continuous semantics, and is apparently less well known. The rather involved proof by Lévy uses techniques of labelling, and a preliminary result which is important per se: inside-out reductions are complete. We shall content ourselves here with the statement of the necessary definitions and of the theorem itself, and will refer the interested reader to [Lévy 75] or [Barendregt, section 14.3].

We briefly recall the definition of Böhm trees [Barendregt, Chapter 10]. First one defines the set $\omega\lambda$ of partial λ-terms, or $\omega\lambda$-terms, by the following syntax:

$$M ::= \Omega \mid x \mid MM \mid \lambda x.M .$$

The idea is that Ω stands for a part of an expression which has not been computed (typically a β-redex). Next we define the classical ω-match ordering by the following clauses:

$\Omega \leq M$,

$(M \leq N \Rightarrow MP \leq NP)$, $(M \leq N \Rightarrow PM \leq PN)$, $(M \leq N \Rightarrow \lambda x.M \leq \lambda x.N)$.

This order allows one to extend the branches of M ending with an Ω. This corresponds to pursuing the computation of the corresponding part of the program being computed. We can make this formal by introducing the following notion of *immediate value* $\omega(M)$ of an $(\omega)\lambda$-term M:

- $\omega(\lambda x.yM_1...M_n) \triangleq \lambda x.y \, \omega(M_1) ... \omega(M_n)$ (and the same with Ω in place of y),

- $\omega(M) \triangleq \Omega$ if M is not a head normal form, i.e. is not of the form $\lambda x.yM_1...M_n$.

(We write $\lambda x.N$ as a shorthand for $\lambda x_1...\lambda x_n.N$, when the length n of the vector x does not matter.) Notice that $\omega(M)$ is a normal form, as is immediately checked by induction on M. Also, if M is a normal form, then $\omega(M) \equiv M$. If M is not a normal form, we can define its (possibly infinite) value by:

$$BT(M) \triangleq \bigcup\{\omega(N)| \, M \to_\beta^* N\} \, .$$

BT(M) is well-defined, because:

i) The set of $\omega\lambda$-terms admits the set of (possibly infinite) $\omega\lambda$-trees as directed complete completion.

ii) The set $\{\omega(N)| \, M \to_\beta^* N\}$ is directed. This follows from the following facts:

- $(M \to N \Rightarrow \omega(M) \leq \omega(N))$,
- β is confluent: $(M \to_\beta^* N, M \to_\beta^* P) \Rightarrow (N \to_\beta^* Q, P \to_\beta^* Q$ for some Q) .

BT(M) is called the *Böhm tree* of M. If follows from the remarks that if M has a normal form, then BT(M) is the normal form of M. Notice that all terms have a Böhm tree. Among non normalizable terms, some have finite Böhm trees (like $(\lambda x.xx)(\lambda x.xx)$), some others have infinite Böhm trees (like the fixed point combinator $\lambda f.(\lambda x.f(xx)) \, (\lambda x.f(xx)))$.We denote by **BT** the subset of the set of (possibly infinite) $\omega\lambda$-trees which have the shape (formal definitions will come later) shown on Figure 1. It should be clear that **BT** contains the image of BT.

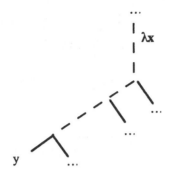

Figure 1

Before we move on towards the statement of the Continuity Theorem, let us pause to point out an important variant of the notion of approximation. The treatment of [Lévy 76] is based on the following *different* definition of immediate approximation (which we stress by using ω' instead of ω):

- $\omega'(\lambda x.M) \triangleq \lambda x.\omega'(M)$,
- $\omega'(yM_1...M_n) \triangleq y\,\omega'(M_1)\,...\,\omega'(M_n)$ (and the same with Ω in place of y)
- $\omega'((\lambda x.M)NM_1...M_n) \triangleq \Omega$.

The main difference between ω and ω' is that if $\omega(M)=\omega'(M)=\Omega$, then $\omega(\lambda x.M)=\Omega$ whereas $\omega'(\lambda x.M)=\lambda x.\Omega$. This difference does not affect much the proof of continuity, but considering ω or ω' leads to *distinct* theorems. We state it with ω, but the definition of ω' contained a thoughtful idea, which has been rediscovered later by Abramsky (see the heading Lazy λ-calculus in the next section). This closes our digression.

We need one more notion to formulate the Continuity Theorem. *Contexts* are terms with holes:

M ::= [] | x | MM | λx.M

Notice that this is the same syntax as the syntax of $\omega\lambda$-terms, up to renaming of ω into []. But here there is no emphasis about order. A context, usually written C[], is a program with holes, which can be filled in the straightforward way: the term C[M] is defined by:

C[M] = M if C[]=[]
C[M] = x if C[]=x
C[M] = C_1[M]C_2[M] if C[]=C_1[]C_2[]

$C[M] = \lambda x.C_1[M]$ if $C[] = \lambda x.C_1[]$

(regardless of captures, i.e. $C[x] = \lambda x.x$ where $C[] = \lambda x.[]$).

With any context $C[]$ we associate a function $\underline{C}$ from **BT** to **BT** defined by:

$\underline{C}(A) = \cup\{BT(C[N])| N \leq A\}$

where A, N range over Böhm trees, $\omega\lambda$-terms, respectively. $\underline{C}(A)$ is well-defined, by monotonicity of BT and of the context operation $M \mapsto C[M]$, and by directedness of $\{N| N \leq A\}$.

Syntactic continuity theorem (Wadsworth-Welch-Lévy)

Let $\underline{C}:BT \to BT$ be as above. Then the following diagram commutes:

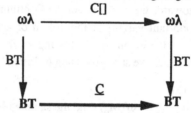

or, in a formula:

$\forall\ C[], M\ \ BT(C[M]) = \cup\{BT(C[N])| N \leq BT(M)\}$.

We turn to sequentiality. The most common (approximation of the) notion of sequentiality arises with strict functions. Strict functions between complete partial orders (where complete means: "directed complete with a minimum element (denoted $\bot$)") are the functions f preserving $\bot$, i.e. s.t. $f(\bot) = \bot$. Strictness says: there is no way to get an information on the result without knowing at least something of the input.

A second example is provided by the conditional function <u>cond</u>. We have:

    <u>cond</u>$(\bot, x, y) = \bot$
    <u>cond</u>$(true, \bot, y) = \bot$
    <u>cond</u>$(false, x, \bot) = \bot$

which we can read as follows:

    - we know that <u>cond</u> is strict in its first argument (no non-bottom result can be produced if the first argument is $\bot$),

    - we also have further strictness information at non bottom inputs: specifically, at $(true, \bot, \bot)$, <u>cond</u> is strict in its second argument, while at $(false, \bot, \bot)$, <u>cond</u> is strict in its third argument.

In contrast the function <u>por</u> which is s.t. <u>por</u>($\perp$,true)= <u>por</u>(true,$\perp$) = true is not strict in either argument, and hence is *not* sequential: there is no way of executing this function on a sequential machine, unless this machine is set up to simulate (say by interleaving) a parallel machine. This link between non sequentiality and the need for dovetailing (Turing machines) has been made precise in [Asperti].

A more sophisticated example of a function of three arguments which is strict in none was proposed by Berry. Take a function <u>Gustave</u> s.t.

<u>Gustave</u>($\perp$,true,false) = <u>Gustave</u>(true,false,$\perp$) = <u>Gustave</u>(false,$\perp$,true) = true.

This is beginning to shape the structure of the definition of sequentiality. Sequentiality is relative to an input state, and points to a needed "argument". The main difficulty is to find a suitable framework for a general notion of "argument" position. To make evident the need for such a general account, let us state the Sequentiality Theorem.

Syntactic Sequentiality Theorem (Berry)

The function BT: $\omega\lambda \rightarrow$ **BT** is sequential. For any context C[], the mapping $\underline{C}$: **BT** $\rightarrow$ **BT** is sequential.

In order to make sense of this statement, we must say what an "argument" position in a tree is. The answer is general, for all (finite and infinite) $\omega\Sigma$-terms, for any (first-order) signature Σ. A well known representation of (possibly infinite) trees is by means of partial mappings from $\mathbb{N}^*$ to Σ. We prefer to view these mappings through their graph. We arrive at the following. Given a signature Σ, a Σ-tree t (or Σ-term, especially if it is finite) is a set of pairs (u,f), where $u \in \mathbb{N}^*$ (the set of words of natural numbers) and $f \in \Sigma$, satisfying the two following conditions:

- (consistence): if (u,f),(u,g)$\in$t, then f=g,
- (safety): if (u,f)$\in$t and u=vi, then (v,g)$\in$t for some g with arity >i.

The u is nothing but the familiar notion of *occurrence*: (u,f)$\in$t reads: the symbol f occurs in t at occurrence u. The most standard notation for this in rewriting theory is: t/u=f. To illustrate the definition, let $\Sigma=\{g^2,f^3,a^0,b^0\}$ be a signature, where superscripts show the arity of the respective symbols. The ω-term f(a,g(Ω,b),Ω) is represented by the set $\{(\varepsilon,f),(0,a),(1,g),(11,b)\}$, where ε is the empty word. The reason for consistence is to ensure that the set describes the graph of a functional relation. The reason for safety is to respect arities: it avoids illegal terms such as g(a,b,b). Following tradition in algebraic semantics, whe shall write $T_\Omega^\infty(\Sigma)$ for the set of all finite and infinite $\omega\Sigma$-terms.

In this more general setting, an "argument" is an occurrence u. We anticipate the further generalization to concrete data structures (next section) and shall call an occurrence u a *cell*, a symbol f a *value* , and a pair (u,f) an *event*.

We also anticipate some important notation:

- u∈F(t) means (u,f)∈t for some f (we say: u is *filled* in t),
- u∈E(t) means u=vi for some v, i, and (v,g)∈t for some g with arity >i (we say: u is *enabled* in t),
- A(t) ≜ E(t)\F(t) (we say: u is *accessible* from t) .

Here is the definition of sequential function. Read "$T_\Omega^\infty(\Sigma)$" instead of "concrete data structure", "$\omega\Sigma$-term" instead of "state", "u" instead of "c" and "f" instead of "v".

Definition of Sequential Function   (Kahn-Plotkin)

Let f be a function between concrete data structures $M=(C,V,E,\vdash)$ and $M'=(C',V',E',\vdash)$. Let x be a state of M and c' be accessible from f(x). We say that f is *sequential* at (x,c') if the following holds (see figure 2):

$$(\exists z \geq x \ c' \in F(f(z))) \implies [\exists c \in A(x) \ \forall y \geq x \ (c' \in F(f(y)) \implies c \in F(y))]$$

Such a c is called a *sequentiality index* at (x,c'). A function f is sequential for short whenever it is sequential at any pair (x,c') s.t. $c' \in A(f(x))$.

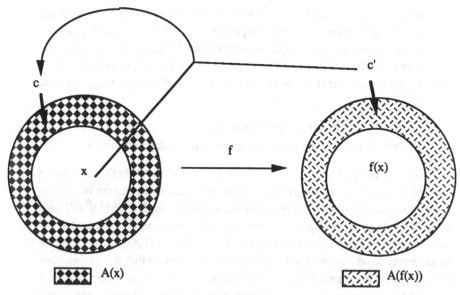

Figure 2

More appropriately, we should have drawn on Figure 2 several curved lines pointing to cells of A(x), starting from the junction of the x and c' lines, since we did not impose that there should be only one sequentiality index. For example, an addition needs both its arguments. But the picture anticipates the notion of sequential algorithm (section 4), which differs from that of sequential function by a choice of sequentiality indexes.

Let us illustrate the sequentiality theorem with an example. Take $M=(\lambda x.x\Omega)(\lambda y.z\underline{\Omega})$. We have $BT(M)=z\Omega$. The only index of BT at $(M,1)$ is 101, i.e. the underlined occurrence in M. This can be interpreted for $N = (\lambda x.x((\lambda t.P)Q))\ (\lambda y.z((\lambda u.R)S))$ as: $(\lambda u.R)S$ is a *needed* redex of N.

We refer to [Berry 77] for the original proof of the Sequentiality Theorem, which may also be found in [Barendregt 14.4]. A simpler proof of the first part of the statement may be found in [Curien]. We briefly show here how to derive the second part of the statement from the first part and from the Continuity Theorem, as an exercise in the kind of reasoning involved in proofs of sequentiality.

<u>Proof of</u> (BT sequential $\Rightarrow$ $\underline{C}$ sequential):

Suppose $\underline{C}[A]/u=\Omega$, $A<A'$, $\underline{C}[A']/u\neq\Omega$. Then $BT(C[M])/u \neq \Omega$ for some $M\leq A'$, by definition of $\underline{C}$. Set $N=M\cap A$. Then $BT(C[N])\leq\underline{C}[A]$ (by definition of $\underline{C}$), hence $BT(C[N])/u=\Omega$ .Let v be an index of BT at $(C[N],u)$. We can write $v=v'w$, where w is an occurrence of N. We claim that w is an index of $\underline{C}$ at (A,u). We have $A/w=\Omega$, since by sequentiality $C[M]/v\neq\Omega$ and $C[N]/v=\Omega$ .Let $B>A$ s.t. $\underline{C}[B]/u\neq\Omega$, and $P\leq B$ s.t. $BT(C[P])/u\neq\Omega$. Set $Q=P\cup N$ ($\leq B$). We have:

- $Q\geq P \Rightarrow BT(C[Q])/u\neq\Omega$.
- $C[Q]\geq C[N] \Rightarrow$ by sequentiality $(C[Q]/v\neq\Omega$, i.e. $Q/w\neq\Omega) \Rightarrow B/w\neq\Omega$. $\square$

3 Full abstraction for PCF and related languages

The language PCF consists in the following syntax:

(types)	$s::= \iota \mid s{\to}s$
(terms)	$M::= x \mid \underline{n} \mid f \mid Y(M) \mid \lambda x.M \mid MM$ (n natural number)
(fconstants)	$f::= \underline{add1} \mid \underline{sub1} \mid \underline{if0}$

The two inessential differences with the original PCF is that we take here Y as unary, and encode booleans as natural numbers. PCF is a *typed* language. Here are the typing rules of the constants:

$$\Gamma \vdash \underline{n} : \iota , \Gamma \vdash \underline{add1} : \iota {\rightarrow} \iota , \Gamma \vdash \underline{sub1} : \iota {\rightarrow} \iota , \Gamma \vdash \underline{if0} : \iota {\rightarrow} \iota {\rightarrow} \iota {\rightarrow} \iota$$

$$\frac{\Gamma \vdash M : t {\rightarrow} t}{\Gamma \vdash Y(M) : t}$$

We specify a normal order reduction strategy for closed terms of type ι, thereafter called *programs*, via Felleisen's formalism of *evaluation contexts* . Evaluation contexts are contexts constructed according to the following restricted syntax:

E::= [] | fE | EM

Then we specify *reduction axioms* written in the form M→M', by the following rules:

$$\underline{add1}\ \underline{n} \twoheadrightarrow \underline{n{+}1}$$
$$\underline{sub1}\ \underline{n{+}1} \twoheadrightarrow \underline{n}$$
$$\underline{if0}\ \underline{0} \twoheadrightarrow \lambda xy.x$$
$$\underline{if0}\ \underline{n{+}1} \twoheadrightarrow \lambda xy.y$$
$$Y(M) \twoheadrightarrow M(Y(M))$$
$$(\lambda x.M)N \twoheadrightarrow M[N/x]$$

Finally, the reduction steps of full programs are given by the following single rule:

$$\frac{M \twoheadrightarrow M'}{E[M] \mapsto E[M']}$$

Evaluation contexts point to the active place where the next reduction will occur. Determinism of computation is guaranteed by the following property.

Unique Decomposition Property

    Given a program M that is not a numeral, there exists only one pair of an evaluation context E[] (as in section 2 we emphasize contexts by writing them E[]) and a reducible term N s.t. M is E[N] (by reducible, we mean N→N' for some N'). □

    Models of PCF are defined according to the various equivalent definitions of simply typed λ-calculus: λ-models, cartesian closed categories with enough points, etc (see [Barendregt]). *Standard* models are those where ι is interpreted as the flat domain $\mathbb{N}_\perp = \{\perp, 0, 1, \dots\}$, where the basic constants receive the expected interpretation, and where moreover the interpretation of any type is a cpo and the interpretation of Y is the minimal fixpoint operator.

The following result states a first correspondence between syntax and semantics. It is proved by means of the so-called computability, or reduciblity, or realizability technique [Plotkin 77].

Computational Adequacy Theorem (Plotkin)
For any standard model of PCF with associated meaning function $[\![\ ]\!]$, the following equivalence holds, for any program P, and any natural number n:

$$[\![P]\!]=n \quad \Leftrightarrow \quad P\mapsto^*\underline{n} .$$

Computational adequacy is a statement about programs only. We want to be able to reason about procedures, i.e. expressions with free variables. For this, the following notion of *observational* preorder $\leq_{op}$ was introduced:

$$M \leq_{op} N \quad \Leftrightarrow_\Delta \quad [\![C[M]]\!] \leq [\![C[N]]\!], \text{ for any } \textit{program} \text{ context } C[]$$

(a program context is a context C[] s.t. C[M] and C[N] are programs). Notice that because of computational adequacy, the notion of operational preorder is purely syntactic. Now we can state the definition of full abstraction.

Definition of Full Abstraction
Let M be a standard model of PCF, with associated meaning function $[\![\ ]\!]$. M is called (inequationally) *fully abstract* if for any M,N the following equivalence holds:

$$M \leq_{op} N \quad \Leftrightarrow \quad [\![M]\!] \leq [\![N]\!] .$$

If "$\leq$" is replaced by "=", we get instead the notion of *equational* full abstraction.

Notice that, thanks to the Adequacy theorem, full abstraction can be considered as a purely semantic problem. One direction of the equivalence ($\Leftarrow$) comes almost for free: if $[\![M]\!]\leq[\![N]\!]$, then by the structural definition of the meaning function, we have $[\![C[M]]\!]\leq[\![C[N]]\!]$.

Another quite easy property is the following one.

The Definability Lemma
Let M be a standard model of PCF. Suppose that for any type s, any isolated element e of the associated cpo is definable, i.e. there exists a closed term of that type s.t. $[\![M]\!]=e$. Suppose moreover that the model is order-extensional (i.e. functions are ordered *pointwise* : $f\leq g \Leftrightarrow_\Delta \forall x\ f(x)\leq g(x)$) and consists of algebraic cpos. Then M is fully abstract.

The easy proof goes as follows. For any isolated elements $e_1,\dots,e_1$ of the appropriate types, we have $[\![M]\!]e_1\dots e_n = [\![C[M]]\!]$, where $C[]=[]P_1\dots P_n$ and $[\![P_i]\!]=e_i$ ($i=1\dots n$). Hence

$$M\leq_{op}N \Rightarrow_{\Delta} ([\![C[M]]\!]\leq[\![C[N]]\!]) , \text{ i.e. } [\![M]\!]e_1\dots e_n \leq [\![N]\!]e_1\dots e_n) \Rightarrow [\![M]\!]\leq[\![N]\!] .$$

Notice that by weakening order-extensionality to extensionality, we get the weaker equational full abstraction property.

The following theorem is a kind of converse to the Definability Lemma.

The Definability Theorem (Milner)

Let $\mathbb{M}$ be an extensional, fully abstract model of PCF. Then

- $\mathbb{M}$ is order-extensional,
- all the domains of the model are algebraic (in fact they are bifinite, or SFP in Plotkin's original terminology),
- all isolated elements are definable.

Instrumental in the proof of this theorem is the following lemma.

The (inequational) Context Lemma (Milner)

In PCF, $M\leq_{op}N$ iff $[\![C[M]]\!]\leq[\![C[N]]\!]$, for any *applicative* program context $C[]$, where applicative means $C[]=[]M_1\dots M_n$ for some $M_1,\dots,M_n$.

The lemma is used as follows in the proof (by contradiction) of the Definability Theorem. One builds with the help of a minimally chosen non-definable isolated element, which is called a *separator* , two terms whose semantics differ on the separator, but which are operationally equivalent. To prove their operational equivalence, checking applicative contexts is enough. This pattern was the original pattern used by Plotkin to show that the continuous model is not fully abstract (see below for a different argument).

The Existence Theorem

There exists an extensional, fully abstract model of PCF.

The two last theorems, combined, give a characterization: there exists a unique (up to isomorphism) extensional fully abstract model of PCF.

The Existence Theorem was proved by Milner for a combinatory logic version of PCF, and reproved by Berry for PCF. The construction is based on suitable completions of sets of equivalence classes of terms. The SFP constructions of

Plotkin are essential.

The fully abstract model has been later recast in more semantic terms by K. Mulmuley [Mulmuley], via the following main steps.

1) A *logical relation* $\leq_{op}$ is defined between elements of the continuous model and (closed) PCF terms. Logical relations are relations over base types which are extended to higher orders by the already quoted *realizability* paradigm which the unfamiliar reader will see at play in the definiton of $\leq_{op}$ which follows:

- at type ι: $d \leq_{op} M \Leftrightarrow_\Delta d \trianglelefteq [\![M]\!]$,
- at type $s \rightarrow t$: $d \leq_{op} M \Leftrightarrow_\Delta \forall e, N \; (e \leq_{op} N \Rightarrow de \leq_{op} MN)$.

2) a preorder relation is defined on the domains of the continuous model in the following way: set

$$d_1 \leq_{op} d_2 \Leftrightarrow_\Delta \forall M \; (d_1 \leq_{op} M \Rightarrow d_2 \leq_{op} M).$$

3) The model is quotiented by the equivalence associated with the preorder. Again SFP constructions are used to make sure that the quotient is a (family of) cpo(s).

For this construction to work, it is essential to start with a model of *lattices*, i.e. to assume that the domains have a maximum element T. This allowed Mulmuley to turn the quotient into a retraction. In this way a fully abstract model is constructed which is not standard, hence is not Milner's unique model. At the end of the construction, the T's can be removed, yielding the fully abstract model of Milner.

This reconstruction of Milner's model is surely mathematically interesting, but does not seem to give any more insight than the syntactic model into the structure of the fully abstract model.

Are the known models of PCF fully abstract? The Definability Theorem and the Sequentiality Theorem of last section allow us to conclude negatively for the following models:

-The continuous (standard) model of PCF is not fully abstract. This was proved, independently, by Plotkin [Plotkin 77] and Sazonov [Sazonov].

- The bistable model of Berry is not fully abstract.

Before we say a word about (bi)stability, we sketch one argument for these negative results. It is the same for both, thus we give it for the continuous model. If the standard continuous model (there is only one, since we have no choice for the base domain) is fully abstract, then por is definable. But the definability of por contradicts the Sequentiality Theorem.

What is the (bi)stable model? Berry has defined the notion of stable function (independently, essentially the same notion was already at the heart of Girard's theory of dilators [Girard 1], and later led the latter to stable models of system F [Girard 2]). Stable functions, for suitable domains, are defined equivalently as being

- the functions preserving finite compatible meets,
- the functions s.t. for any x, y s.t. y≤f(x), the set {z≤x| y≤f(z)} has a minimum.

Berry has often called stability "sequentiality after" ("après coup") . This is made precise by the following technical statement:

A function is stable iff its restriction to all principal ideals x↓ is sequential

(which we leave as an exercise). Intuitively, in a sequential program P, demand is output driven: the input x is not known completely; it is rather evaluated on request, and this will in turn activate another sequential program Q which outputs the input of P. If we had completed all of Q eagerly before computing P, then we would know all of x, and stability would be as good as sequentiality. But the computation of Q may lead to loops for the evaluation of irrelevant parts of x.

In contrast to stability, sequentiality indicates an underlying completely demand-driven computational flow. These observations form the basis of the operational semantics of the language CDS (see end of the next section).

Now what does bistable stand for? The order which works for stable functions is not the pointwise ordering, but the stable ordering, defined by

$$f \leq_{st} g \Leftrightarrow_\Delta \forall x,y \ (x \leq y \Rightarrow f(x) = f(y) \cap g(x)) \ .$$

A quite involved notion of structures carrying both a stable ordering and an extensional ordering has been developped by Berry to produce order-extensional models that he called bistable.

We complete the series of negative results by the following ones:

- The bisequential model is not fully abstract [Curien 86] (see next section).
- Strongly stable functions do not lead to a fully abstract model either.

Here the counterexamples have to be taken of order 2, since all first-order sequential functions are definable (exercise).

What is strong stability? It is a new notion, due to Bucciarelli and Ehrhard [BE2], which is

- defined in terms of much stronger meet preservation properties than stability,

- coincides at first order with sequentiality of functions.

The strongly stable model, like the stable model, is ordered by the stable ordering. Bucciarelli and Ehrhard use a different technique from Berry [BCL,Berry 79] to obtain an extensional model from their strongly stable model. They carry around pairs of domains of continuous and of strongly stable functions, the second being embedded in the first [BE3]. This new construction seems to have some advantages: its theory (the set of equalities between PCF terms that it validates) lies strictly between the continuous theory and the observational equivalence (a property which is not enjoyed by the bistable model, as was remarked in [Jim-Meyer]).

Full abstraction for order up to 3 has been shown by Sieber for a model obtained by retaining the points of the continuous model which are invariant through a class of logical relations called *sequential* [Sieber 91]. An extensional collapse is needed to turn this into a model.

While PCF resisted, and still resists a "natural" semantic construction of its fully abstract extensional model, authors have turned attention to *extensions* of PCF, or to *different notions of observation* for PCF (or both together). Full abstraction has also been studied for other λ-calculus related languages. Some authors have also investigated general language independent theorems. We briefly mention some results.

Extensions

Plotkin has shown that the continuous model of PCF is fully abstract for a simple parallel extension of PCF with a parallel conditional operator pif [Plotkin 77]. The author was able to adapt the technique to prove the same result for the more natural extension of PCF with a por operator [Curien 86]. Later, Stoughton has shown that the two operators are interdefinable [Sto3]. Further, adding moreover a "continuous existential" operator, Plotkin showed that all computable elements of the continuous model are definable. Another recent result worth mentioning is that the continuous model is the only extensional model of PCF+por [Sto4].

The idea that continuity provided full abstraction for a language with parallel facilities was pursued by Hennessy and Plotkin [Hennessy-Plotkin], who adapted powerdomain constructions to yield a fully abstract model of a simple (imperative) parallel language. But the continuous model fails to be fully abstract for random assignment [Apt-Plotkin] and for infinite streams [Abramsky1].

Sitaram and Felleisen consider an extension of PCF with control operators [Sitaram-Felleisen], in a call-by-value setting (see below). Further work of Cartwright and Felleisen is discussed in section 5.

Extensions of PCF with recursively defined types are considered in [Mulmuley] and [MC] at least.

Changes in the operational semantics

Call-by-value PCF

After the revisiting of denotational semantics which started with the work of Plotkin on partial continuous function to model call-by-value, much emphasis is now given to the use of various monads to tailor denotational semantics to various programming features for which the wild β-equality is too crude. It was thus natural to look at the call-by-value versions of PCF. [Sieber 90] seems to be the first published account of full abstraction in the call-by-value setting.

Lazy PCF

Bloom and Riecke have considered a more generous notion of observation for PCF: not only results of basic type can be observed, but also convergence of functions (in the weak sense discussed below under the heading of Lazy λ-calculus) [Bloom-Riecke].

The three branches: lazy PCF, call-by-value PCF, call-by-value PCF+control, give rise to similar results. The continuous model is different in each case:

- in the first case, all functions spaces are *lifted*,
- in the second case, only *strict* continuous functions are in the model,
- in the third case *exception values* are added.

Each of these models is adequate for the respective language, and fully abstract when pif is added (in the case of lazy PCF, a convergence testing operator up? is needed as well, see the heading Lazy λ-calculus below).

Sieber shows how the full-abstraction results for lazy PCF+pif+up? and call-by-value PCF+control+pif can be derived from the results on call-by-calue PCF, using syntactic translations [Sieber 90].

Full abstraction in other λ-calculus related languages

Classical untyped λ-calculus

By classical we mean here the theory of λ-calculus based on (unrestricted) β, head normal forms and Böhm trees, i.e. the λ-calculus presented in [Barendregt]. The D_∞-interpretation [Barendregt, Chapter 18] is fully abstract for the following notion of observation (regardless of which non trivial D one started with). The basic test for a term M, is "M has a head normal form (hnf)". Write $M \leq_{op} N$ when C[N] has a hnf whenever C[M] has a hnf, for every context C[]. This seems to be the first published full abstraction result (the result is due independently to Wadsworth and Hyland, and can be found as Theorem 19.2.9 in [Barendregt]). But, quoting from [Boudol], "the D_∞-interpretation is quite strange, since no finite non-trivial element of D_∞ is λ-definable".

Lazy λ-calculus

More sense seems to be made of observations on untyped λ-calculus based on *weak* head normal forms rather than head normal forms. Weak head normal forms are expressions of the form λx.M or $xM_1...M_n$ (hence a *closed* whnf is an abstraction, or an expression of the form $fM_1...M_n$ for some constant f). The observation in the untyped λ-calculus is now limited to convergence to abstraction (cf. the digression in section 2, and lazy PCF).

The version of the D_∞-construction adapted to this calculus consists in lifting the domains of functions. The equation solved is not $D=(D \to D)$, but $D=(D \to D)_\perp$. There is now a canonical "D_∞"-model: the model constructed from the trivial $D=\{\perp\}$. This model has been shown to be fully abstract for an extension of the lazy λ-calculus with a parallel convergence testing operator [Abramsky2,Abramsky-Ong]. The result was then revisited in [Boudol], who decomposed this operator into a non-deterministic choice operator and a convergence testing operator... until he discovered that call-by-value abstractions sitting in the language together with call-by-name abstractions allowed him to simulate the convergence test. Boudol also obtains a fully abstract model of the *call-by-value* lazy λ-calculus. Here *lazy* qualifies only the way *functions* are treated (whnf's versus hnf's).

One of the attractive points about Abramsky's model is that it offers a nice example of *Stone duality*. Under Stone duality, domains can be recast as the sets of *points* of algebraic structures (called *frames*). Such structures are typically sets of formulas (*Lindenbaum* algebras). Points are recovered as suitable filters of formulas. This is precisely one of the ways of defining Abramsky's model. The formulas are essentially the *conjunctive* types of Coppo ($\phi ::= \omega \mid \phi \to \phi \mid$

$\phi \cap \phi$).

Stone duality can explain, at least in the setting of the lazy λ-calculus, why the hard part of full abstraction seems to go the other way around from logic: in logic, usually *completeness* theorems may be hard: "if something is valid, then it is provable". The apparently corresponding side of full abstraction is the easy one: "if M≤N is valid, then M≤$_{op}$N". But under Stone duality, the validity of M≤N is on the proof system side, since the meaning of M consists of formulas, while M≤$_{op}$N corresponds to a realizability *interpretation* ⊨. More specifically:

- "M≤N valid" amounts to "⊢M:A implies ⊢N:A, for all type A",
- M≤$_{op}$N amounts to "M⊨A implies N⊨A, for all type A" .

Hence Stone duality teaches us that the operational order is on the semantic side, and the semantic order on the syntactic side. If this is not too confusing...

<u>ALGOL</u>

Various semantics of local storage have been investigated by A. Meyer and coauthors, and partial full abstraction results have been obtained [MeSi].

General language independent theorems

The translation method of Sieber, already discussed, falls under this heading.

Stoughton has given general necessary and sufficient conditions for a language to have fully abstract models (which he constructs syntactically, like Milner) [Sto1]. These conditions concern the satisfaction by the operational preorder of some least fixed point constraints.

Jim and Meyer [Jim-Meyer], improving on [Bloom], prove a general theorem which states that any extension of PCF with "PCF-like rules" must satisfy the inequational Context Lemma. In both papers, the satisfaction of the Context Lemma is exploited for the proof of negative results of an absolute nature. A fixed model, like the continuous lattice model (already mentioned about Mulmuley's work) or the stable model are not only not fully abstract for PCF, but not fully abstract for any "PCF-like"-extension of PCF (there is an additional restriction for the continuous lattice model: the extension must be confluent). The proof of the negative results follows the pattern of Plotkin sketched after the statement of the Context Lemma. A separator is used as a guide for the definition of two terms whose meanings differ on the separator, and which are operationnally equivalent, which by the Context lemma may be checked on applicative contexts only. In the stable case, the Context lemma is also used for showing that the separator is non-definable in any extension satisfying it. In the lattice continuous case, the additional restriction is used to ensure the non-definability of the separator. In contrast, the lattice continuous model turns out to be fully abstract for a simple *non-deterministic* extension of PCF [Bloom].

4 Sequential algorithms on concrete data structures

We shall recall the definition of (deterministic) concrete data structure, and show how the domains considered so far fit into the definition. We shall then define the exponential of two concrete data structures, which leads to the definition of the cartesian closed category of sequential algorithms. We refer the interested reader to [Curien 86] for details.

Definition of Concrete Data Structure (Kahn-Plotkin)

A concrete data structure (cds) is a 4-tuple $M=(C,V,E,\vdash)$ consisting of

- a set C of *cells* ,
- a set V of *values* ,
- a specified subset E of C×V of *events* ,
- an *enabling* relation $\vdash$ between finite sets of events on one hand and cells on the other hand.

A *state* of **M** is a subset x of E which is

- *consistent* : $(c,v),(c,w)\in x \Rightarrow v=w$,
- *safe* : $(c,v)\in x \Rightarrow X\vdash c$ for some $X\subseteq x$.

The cds used for sequential algortihms satisfy moreover the important following *stability* condition:

- if x is a state, then each cell filled in x has a unique enabling in x.

The following other condition is more a matter of convenience for carrying inductive proofs than a serious restriction. A cds is *well-founded* if the relation R defined by

R(c,d) iff c is filled in X and $X\vdash d$ for some X

is well-founded. We always assume that cds's are stable and well founded (such cds where called dcds in [Curien], and are called here cds for short).

We can now recast the various domains used in section 2 as concrete data structures. The flat domains of natural numbers or booleans (used for cond) are defined as follows (say for natural numbers):

$C=\{?\}$, $V=\mathbb{N}$, $E=C\times V$, $\vdash$ is specified by $\varnothing\vdash?$, which we abbreviate as $\vdash?$.

Here "?" serves as the name of the unique cell. We shall use **T** to denote the cds for the flat domain of booleans, whose states are $\varnothing$, true=$\{(?,\underline{t})\}$ and false=$\{(?,\underline{f})\}$.

A product of (flat) domains, say of the domains of natural numbers and booleans is represented by:

$C=\{?.1,?.2\}$, $V=\mathbb{N}\cup\{\underline{t},\underline{f}\}$, $E=\{(?.1,n)|n\in\mathbb{N}\}\cup\{(?.2,\underline{t})\}\cup\{(?.2,\underline{f})\}$, $\vdash?.1$, $\vdash?.2$

More generally, the *product* of two cds $M=(C,V,E,\vdash)$ and $M'=(C',V',E',\vdash)$ is the cds $(C'',V'',E'',\vdash)$ where C'' is the disjoint union of C and C' (whose elements are tagged), $V'' = V \cup V'$, $E'=\{(c.1,v)|\ (c,v)\in E\} \cup \{(c'.1,v')|\ (c',v')\in E'\}$, and similarly for the enablings.

The cds of $\omega\Sigma$-terms is defined by

$C=\mathbb{N}^*$, $F=\Sigma$, $E=C\times V$, $(v,g^n)\vdash vi$ if $i<n$ (where n is the arity of g).

We also confirm the terminology introduced in section 2:

(filled) $c\in F(x)$ means $(c,v)\in x$ for some v,
(enabled) $c\in E(x)$ means $X\vdash c$ for some $X\subseteq x$,
(accessible) $A(x) \triangleq E(x)\backslash F(x)$.

Exponential of concrete data structures (Berry-Curien)

Given two cds $M=(C,V,E,\vdash)$ and $M'=(C',V',E',\vdash)$, the cds $M\Rightarrow M' = (C'',V'',E'',\vdash)$ is defined as follows:

- C'' consists of pairs (x,c') (written xc') of a *finite* state of M and a cell of M',
- $V'' = C \cup V'$ (*disjoint* union),
- $E'' = \{(xc',c)|\ c\in A(x)\} \cup \{(xc',v')|(c',v')\in E'\}$,
- $(xc',c) \vdash yc'$ if $y= (x\cup\{(c,v)\})$ for some v s.t. $(c,v)\in E$,
- $(x_1c'_1,v'_1),...,(x_nc'_n,v'_n) \vdash xc'$ if $x= x_1\cup ... \cup x_n$ and $(c'_1,v'_1),...,(c'_n,v'_n)\vdash c'$.

This definition is quite abrupt. The reader may want to come back to it after reading the section further.

The *category of sequential algorithms and concrete data structures* has as objects the cds's; the arrows in Hom(M,M') are the states of $M\Rightarrow M'$. This category is cartesian closed, and gives rise to a model of PCF. The closed structure is easy to see. We invite the reader to check in detail that currying and uncurrying are just a matter of shifting parentheses: the generic cells of $(M\times M')\Rightarrow M''$ and $M\Rightarrow (M'\Rightarrow M'')$ are $(x,x')c''$ and $x(x'c'')$, respectively.

It is harder to check that we have a category . Composition cannot be conveniently described directly from the definition of the exponential alone. There is an alternative, more "abstract", or more "functional" definition of sequential algorithms, as pairs (f,i), where f is a sequential function, and i is a partial function, which when defined on pair (x,c') where f has a sequentiality index, picks out one of them. These pairs (f,i) are required to satisfy some axioms (see [Curien, Definition 2.5.4], and [Curien 91]).

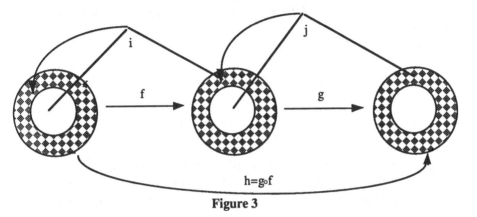

Figure 3

The compositon (h,k) of (f,i) and (g,j) is defined by (Figure 3 may help):

- h=g∘f ,
- k(x,c")=i(x,j(f(x),c")), whereby we mean "k(...) is defined and equal to i(...) iff i(...) is defined".

The abstract definition provides a more intuitive understanding than the definition of the exponential cds: a sequential algorithm is a pair of a sequential function and a computation strategy. For example, there are two strict "and" algorithms from T×T to T: the *left* strict "and" and the *right* strict "and". Here is the description of the left strict "and" as a state:

<u>left strict and</u> = {(∅?,?.1),({(?.1,ṭ)}?,?.2),({(?.1,ḟ)}?,?.2)
({(?.1,ṭ),(?.2,ṭ)}?,ṭ),({(?.1,ṭ),(?.2,ḟ)}?,ḟ,
({(?.1,ḟ),(?.2,ṭ)}?,ḟ),({(?.1,ḟ),(?.2,ḟ)}?,ḟ)}

The reader will probably prefer the following friendlier presentation of the right strict "and":

<u>right strict and</u>   =   valof ?.2 is
                              ṭ : valof ?.1 is
                                   ṭ : output ṭ

f : output f

f : valof ?.1 is

t : output f

f : output f

Such a tree representation is actually not peculiar to the sequential algorithms: every state of a *stable* and *filiform* cds has an underlying tree (or forest) structure: we illustrate the description by a forest of the state $\{(c1,v1),(c2,v2),(d1,w1),(d2,w2)\}$, in a cds where $\vdash c1$, $\vdash c2$, $(c1,v1)\vdash d1$ and $(c1,v1)\vdash d2$:

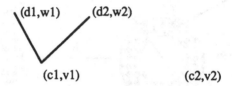

A *filiform* cds is a cds s.t. each enabling contains at most one event. Filiform cds's yield a cartesian closed full subcategory of the category of cds and sequential algorithms. Since flat domains are represented by filiform cds's, this subcategory is enough for our purposes.

The following figure should help the reader to go from the presentation of right strict and given above to a state similar to the presentation we gave of left strict and:

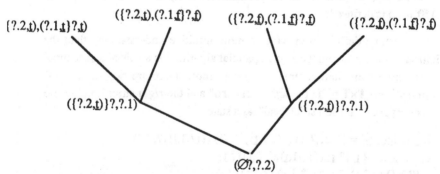

We shall not give any proof here. We content ourselves with a listing of the key properties, and refer the reader to [Curien 86] or [BC1].

    - the cds $M \Rightarrow M'$ is stable as soon as M' is (stability is essential to guarantee that sequential algorithms define functions),

    - the two definitions of sequential algorithm (as a state, or as a pair (f,i)) are equivalent,

- the function a ↦ (x ↦ a.x), where a.x ≜ {(c',v')| ∃y≤x (yc',v')∈a} maps surjectively sequential algorithms to sequential functions; it is also monotone, taking inclusion as the order on algorithms, and *stable* order as the order on functions (cf. previous section),

- the set of states of a cds **M** is a cpo (it is more than that: it is a concrete domain, and there is a representation theorem linking those domains with cds, see [KP] and [Curien 86]).

Now, what are sequential algorithms good for in the full abstraction problem? Sections 2 and 3 carried the message that "the fully abstract model of PCF should be sequential", since all the isolated elements should be definable, and the definable functions of order 1 are the sequential functions. We have a "sequential" model, but it is not extensional to start with: left strict and and right strict and define the *same* function from (the set of states of) T×T to T. The author was enthusiastic enough at the time to find a *bisequential* model, a sort of quotient of the model which is order-extensional (sections 4.2 and 4.3 of [Curien 86]). But at the end, counterexamples showed up which were not definable in PCF, hence the model was not fully abstract. These examples are well captured by the treatment proposed by Sieber, which we discussed in the previous section.

Thus, in the early eighties, the hope was lost to make sequential algorithms or a variant into a fully abstract model of PCF (and the new developments hinted at in the last section do not change that branch of the research tree). But another branch of exploration proved successful. We (G. Berry and the author) changed the language: we took advantage of the very syntactic nature of our model (the non-extensional one) to design a programming language called CDS, which is described in [BC2] and [Curien 86] (whose Chapter 3 is a slightly improved version of [BC2]), where the notion of observation is much closer to the denotational semantics: the operational semantics allows the user to observe in a call-by-need mode as much of the meaning of *any* program piece as he wants to examine. As an illustration, suppose that we are computing the identity λx.x of integer type. Its meaning is the identity algorithm. The top level loop of evaluation consists in the following steps.

- The user sends requests which are cell names of the exponential dcds. To start with, the initial request ∅? has to be sent. A request xc' may be construed as: "what does the algorithm at input x when trying to fill c' in the output?".

- The system returns the value of the cell in the meaning of the expression. For instance, the answer to the request ∅? is (valof) ?, and the answer to the request {(?,n)}? is (output) n. Answers valof c, output v' may be construed as: "the algorithm needs to know more about its input, namely the value of c", "the algorithm has enough information to ouput the value v' in c' ", respectively.

The most important feature of the operational semantics of CDS is that it allows to observe the *order of evaluation* of arguments (more generally the evaluation strategy). The language's expressiveness allows the user not only to observe these intensional aspects, but also to define higher-order procedures that actually can *distinguish* programs differing only intensionally. For example, [BC2] and [Curien 86, section 3.2.8] show how to construct a CDS program called GOUTEUR_DE_ET : $((T \times T) \Rightarrow T) \to T$ which is such that

> GOUTEUR_DE_ET . right strict and = $\{(?,\underline{t})\}$
> GOUTEUR_DE_ET . left strict and = $\{(?,\underline{f})\}$

For the notion of observation of CDS, a full abstraction result has been obtained [BC2,Curien 86]. But let us insist that there is a strong departure from the kind of observation present in PCF (and in many languages), which relies on a subset of *observable* types. In CDS, all types are equally observable, which was not a bad feature, after all.

Before we move on to our last section, let us mention that the main ideas of sequential algorithms have been recast in more abstract terms by Bucciarelli and Ehrhard [BE1]. They develop a framework where cells are linear opens (i.e. linear maps into $\{\perp, \top\}$) of suitable domains, which gives some more mathematical status to the theory. They obtain a non-extensional model of PCF where arrows are pairs (function, computation strategy), as shown above. It is not clear how close this "abstract" model is to the "concrete" model of [BC1,Curien 86], but it ties up with recent developments in denotational semantics, triggered by linear logic.

5 One, two, hundred errors

Recently, R. Cartwright and M. Felleisen have contributed new ideas in the study of sequentiality by taking an explicit account of errors [CF]. This comes of course from practical programming languages, which have errors and error handlers. We first look at the semantic aspect of this. Consider the type T, and its elements $\emptyset, \{(?,\underline{t})\}, \{(?,\underline{f})\}$, which we shall write more conventionally $\perp, \underline{t}, \underline{f}$. If this set of elements is augmented with an element err (for error), then we gain the ability of distinguishing left strict and and right strict and as *functions* :

> left strict and .$(\perp,\text{err}) = \perp$
> right strict and .$(\perp,\text{err}) = \text{err}$

Of course we need to justify these equalities, i.e. give formal definitions of the input-output behaviour of sequential algorithms on states containing error values. The intuition, however, is simple: an algorithm should *propagate* errors. Hence right strict and, which needs its second argument, when fed with an error in this place, propagates it to the output. In contrast, left strict and waits for its first argument, whence the first equation.

The quite magic consequence is that the difference between left strict and and right strict and is observable *extensionally* and at *basic* types.

In other words, sequential algorithms can be turned into an *extensional* model, just by extending inputs and outputs to be able to carry error information. Not quite though: the algorithms themselves are states, and possible inputs or outputs for other algorithms; they too must be able to carry error informations. Hence we do not only extend the input-output observation of algorithms, but we also allow for *more* algorithms. We turn to a theory of *observable* sequential algorithms. Observable sequential algorithms may provoke errors.

This should motivate the following definition.

Convention: A *nonempty* set ERR of *error values* (typically written err, err',...) is fixed once for all, and it is assumed that the set of values V of *any* cds is disjoint from ERR.

Definition of Observable State

Let $M=(C,V,E,\vdash)$ be a cds. An observable state of M is a set x of pairs (c,w) s.t. either $(c,w) \in E$ or $w \in ERR$ and which is both consistent and safe (cf. Definition of Concrete Data Structure).

In particular, states are observable states. Implicit in this definition is the fact that enablings are not allowed to contain error values, since the enabling relation is part of the structure of a cds, which we did not change (we shall change the arrows only). In the forest representation of an observable state, error values can occur only at the leaves, as illustrated by the following figure:

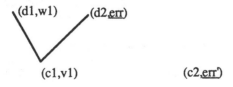

The category which we consider has concrete data strcutures as objects (hence has the *same* objects as the previous category) and observable states of the cds $M \Rightarrow M'$ (which we call *observable sequential algorithms*) as arrows.

The Order-extensionality Theorem

The category of concrete data structures and observable sequential algorithms is cartesian closed and extensional. If ERR (which we remind is a fixed global parameter of the category) has cardinality at least 2, then the category is order-extensional (and the order on arrows coincides with the inclusion of observable states).

Instrumental in the definition of the category and the proof of the Order-extensionality Theorem is an appropriate extension of the input-output function associated with an observable sequential algorithm (and which characterizes it completely, since the model is extensional). We set

$$a.x = \{(c',w')| \ \exists y \leq x \ (yc',w') \in a\},$$
$$\cup \ \{(c',\underline{err}) \ | \ \exists y \leq x \ (yc',c) \in a \ and \ (c, \underline{err}) \in x\}.$$

(where w' ranges over ERR $\cup$ V', and $\underline{err}$ over ERR). The left part of the union is as before, and the right part takes care of error propagation.

The last part of the statement of the Order-extensionality Theorem may seem surprising. The following example shows what's going on. Consider the two following algorithms from **T** to **T**:

- a = {(∅?,?)},
- b = {(∅?,<u>err</u>)}.

Then, clearly a⊄b (remember that states are ordered by inclusion). But we check easily:

a.⊥=⊥≤<u>err</u>=b.⊥ , a.<u>t</u>=⊥≤<u>err</u>=b.<u>t</u> , a.<u>err</u> =<u>err</u>≤<u>err</u>=b.<u>err</u> .

Hence if ERR contains <u>err</u> as its only element, we have no way to witness the inequation pointwise. If instead there is at least one more error, say <u>err'</u>, then we have:

a.<u>err'</u> =<u>err'</u> and b.<u>err'</u>=<u>err</u>, which are *not* comparable.

On the syntactic side, Cartwright and Felleisen extend PCF with the following additional constructs:

(terms) M::= ... | error | catch(M)

with the following typing rules:

$\Gamma \vdash$ error : ι

$$\frac{\Gamma \vdash M : s}{\Gamma \vdash \text{catch}(M) : \iota}$$

Evaluation contexts contain one more clause:

$$E ::= \dots \mid \text{catch}(\lambda x_1 \dots x_m.E)$$

There are two more reduction axioms, and one more reduction rule:

$$\text{catch}(\lambda x_1 \dots x_m.E[x_i M_1 \dots M_l]) \rightarrow \underline{i\text{-}1} \quad (i \le m,\ x_i \text{ free in } E[x_i M_1 \dots M_l])$$

$$\text{catch}(\lambda x_1 \dots x_m.\underline{n}) \rightarrow \underline{m+n}$$

$$E[\text{error}] \mapsto \text{error}$$

Call OS-PCF this extension of PCF. Then the model of observable sequential algorithms is **fully abstract** for OS-PCF. This result is announced in [CF], where the main part of the proof is given: an algorithm is given that constructs defining terms for all isolated elements of the model. Essentially the same algorithm is presented in [Curien 91]. Amazingly enough, the model in [CF] is an independent reconstruction of the (observable) sequential algorithms in the cds's used for the interpretation of PCF. The proof of the Order-extensionality Theorem is given in [Curien 91] (and partly in [CF], in their framework). All these works will be tied up in a forthcoming joint paper.

We end the section (and the paper) with a reincarnation of the GOUTEUR_DE_ET: Define an algorithm taster by the following OS-PCF term:

$$\text{taster} = \lambda z.\ \text{catch } (\lambda xy.\ zxy).$$

Then, applying the rules for catch, we get

taster . left strict and $\rightarrow^* \underline{0}$
taster . right strict and $\rightarrow^* \underline{1}$.

The essential gain w.r.t. CDS is that the observations can now be "projected" to the base type, and there is no need anymore for a shift in the notion of observation. This will not preclude the author from keeping some nostalgie for CDS.

References

[Abramsky1] S. Abramsky, Semantic foundations of applicative multiprogramming, Proc. 10th ICALP, Lecture Notes in Computer Science 154 (1983).

[Abramsky2] S. Abramsky, The lazy lambda-calculus, in Research Topics in Functional Programming, D. Turner ed., Addison-Wesley (1989)

[Abramsky-Ong] S. Abramsky, L. Ong, Full abstraction in the lazy lambda-calculus, to appear in Information and Computation

[Apt-Plotkin] K. Apt, G. Plotkin, Countable non-determinism and random assignment, JACM.

[Asperti] A. Asperti, Stability and computability in coherent domains, Information and Computation 86 (2) (1990).

[Barendregt] H. Barendregt, The Lambda-calculus, its syntax and semantics, North Holland (1984).

[Berry 77] G. Berry, Séquentialité de l'évaluation formelle des Lambda-expressions, in Program Transformations, Proc. 3rd Int. Coll. on Programming, B. Robinet, ed. Dunod (1978).

[Berry 79] G. Berry, Modèles complètement adéquats et stables des lambda-calculs typés, Thèse d'Etat, Université Paris VII (1979).

[BC1] G. Berry, P.-L. Curien, Sequential Algorithms on Concrete Data Structures, Theoretical Computer Science 20 (3) (1982).

[BC2] G. Berry, P.-L. Curien, Theory and practice of sequential algorithms: the kernel of the programming language CDS, in M. Nivat and J. Reynolds eds, Algebraic Methods in Semantics, Cambridge University Press (1985).

[BCL] G. Berry, P.-L. Curien, J.-J. Lévy, Full abstraction for sequential languages: state of the art, in M. Nivat and J. Reynolds eds, Algebraic Methods in Semantics, Cambridge University Press (1985).

[Bloom] B. Bloom, Can LCF be topped? Flat lattice models of typed λ-calculus, Information and Computation 87 (1/2) (1990).

[Bloom-Riecke] B. Bloom, J.G. Riecke, LCF should be lifted, in Proc. Conf. Algebraic Methodology and Software Technology, Department of Computer Science, Iowa (1989).

[Boudol] Lambda-calculi for (strict) parallel functions, to appear in Information and Computation.

[BE1] A. Bucciarelli, T. Ehrhard, A theory of sequentiality, submitted to TCS (1990).

[BE2] A. Bucciarelli, T. Ehrhard, Sequentiality and stong stability, Proc. Logic in Computer Science 1991, Amsterdam.

[BE3] A. Bucciarelli, T. Ehrhard, Extensional embedding of a strongly stable model of PCF, Proc. International Colloquium on Automata, Languages and Programming 91, Lecture Notes in Computer Science 510 (1991).

[CF] R. Cartwright and M. Felleisen, Observable sequentiality and full abstraction, Proc. Principles of Programming Languages 1992, Albuquerque.

[Curien 86] P.-L. Curien, Categorical Combinators, Sequential Algorithms and Functional Programming, Pitman (1986) (out of print, revised edition in preparation).

[Curien 91] P.-L. Curien, Observable sequential algorithms on concrete data structures, Proc. Logic in Computer Science 1992, Santa Cruz.

[Girard 1] J.-Y. Girard, Π^1_2-logic, part 1: dilators, Annals of Mathematical Logic 21 (1981).

[Girard 2] J.-Y. Girard, The system F of variable types, fifteen years later, Theoretical Computer Science 45 (1986).

[Hennessy-Plotkin] M. Hennessy, G. Plotkin, Full abstraction for a simple parallel language, Proc. Mathematical Foundations of Computer Science, Lecture Notes in Computer Science 74 (1979).

[Jim-Meyer] T. Jim, A. Meyer, Full abstraction and the context lemma, preliminary report and invited lecture at Theoretical Aspects of Computer Software, Sendai, Lecture Notes in Computer Science 526 (1991).

[KP] G. Kahn, G. Plotkin, Domaines concrets, Rapport IRIA-LABORIA 336 (1978).

[Lévy 76] J.-J. Lévy, An algebraic interpretation of the $\lambda\beta K$-calculus; and an application of a labelled λ-calculus, Theoretical Computer Science 2 (1) (1976).

[Meyer] A. Meyer, Semantical paradigms, notes for an invited lecture, with two appendices by S. Cosmodakis, in Proc. Logic in Computer Science 1988.

[MeSi] A.Meyer, K. Sieber, Towards fully abstract semantics for local variables: preliminary report, Proc. Principles of Progamming Languages 1988.

[Milner 77] R. Milner, Fully abstract models of the typed lambda-calculus, Theoretical Computer Science 4 (1977).

[Mulmuley] K. Mulmuley, Full abstraction and semantic equivalence, MIT Press (1987).

[Plotkin 77] G.D. Plotkin, LCF considered as a programming language, Theoretical Computer Science 5 (1977).

[Sazonov] V. Sazonov, Expressibility of functions in D. Scott's LCF language, Algebra i Logika 15 (1976) (Russian).

[Scott 76] D. Scott, Data types as lattices, SIAM Journal of Computing 5 (1976).

[Sieber 90] K. Sieber, Relating full abstraction results for different programming languages, Proc. 10th Conf. on Foundations of Software Technology and Theoretical Computer Science, Bangalore, Lecture Notes in Computer Science 472 (1990).

[Sitaram-Felleisen] D. Sitaram and M. Felleisen, Reasoning with continuations II: full abstraction for models of control, Proc. Conf. on Lisp and Functional Porgamming, Nice (1990).

[Sieber 91] K. Sieber, Reasoning about Sequential Functions via Logical Relations, draft (1991).

[Sto1] A. Stoughton, Fully abstract models of programming languages, Pitman (1988).

[Sto2] A. Stoughton, Equationally fully abstract models of PCF, Proc. Mathematical Foundations of Programming Semantics, Lecture Notes in Computer Science 442, Springer (1990).

[Sto3] A. Stoughton, Interdefinability of Parallel Operations in PCF, Theoretical Computer Science (1991).

[Sto4] A. Stoughton, Parallel PCF has a Unique Extensional Model, Proc. Logic in Computer Science 1991, Amsterdam.

Remarks on
Algebraically Compact Categories

Peter Freyd [*]

In Algebraically Complete Categories (in the proceedings of the Category theory conference in Como '90) an **ALGEBRAICALLY COMPLETE CATEGORY** was defined as one for which every covariant endofunctor has an initial algebra. This should be understood to be in a 2-category setting, that is, in a setting in which the phrase "every covariant endofunctor" refers to an understood class of endofunctors.

Given an endofunctor T the category of T- **INVARIANT** objects is best defined as the category whose objects are triples $<A, f, g>$ where $f:TA \to A$, $g:A \to TA$ and fg and gf are both identity maps. T-**Inv** appears as a full subcategory of both T-**Alg** and T-**Coalg**, in each case via a forgetful functor. The Lambek lemma and its dual say that the initial object in T-**Alg** and the final object in T-**Coalg** may be viewed as objects in T-**Inv** wherein they easily remain initial and final. Of course there is a canonical map from the initial to the final. I will say that T is **ALGEBRAICALLY BOUNDED** if this canonical map is an isomorphism, equivalently if T-**Inv** is a punctuated category, that is one with a biterminator, an object that is both initial and final.

An algebraically bicomplete category is **ALGEBRAICALLY COMPACT** if each endofunctor is algebraically bounded. (As with algebraic completeness this should be understood to be in a 2-category setting.) In this context I will use the term **FREE T-ALGEBRA** rather than either initial algebra or final coalgebra.

[*] Departments of Mathematics and Computer Science, University of Pennsylvania. The author was supported by the Office of Naval Research and the National Science foundation and was a guest at St. John's College, Cambridge, when this research was conducted.

1: ON SOME EXAMPLES

At the end of my Como lecture Adamek pointed out that the category of sets has the property that all of its endofunctors have an invariant objects. But even better:

Proposition:
 The category of countable sets is algebraically complete.

Let T be an arbitrary endofunctor. If $T\emptyset = \emptyset$ then $T\emptyset \to \emptyset$ is easily seen to be an initial algebra. Supposing otherwise, let A be an infinite (and, of course, countable) set, and choose an embedding $TA \to A$. Let F be minimal among the subsets of A such that the image of $TF \to TA \to A$ is contained in F. We know that $F \neq \emptyset$. Hence $F \to A$ is a split monic, hence $TF \to TA$ is monic, hence $TF \to F$ is monic and by minimality it is an isomorphism. Now let $b:TB \to B$ be arbitrary. Let K be the partial map from F to B named by the image of $T\emptyset \to TF \times TB \to F \times B$. On the CPO of partial maps from F to B that contain K consider the endofunction that sends a partial map tabulated by $<F' \to F,\ F' \to B>$ to the partial map tabulated by $<TF' \to TF \to F,\ TF' \to TF \to B>$. (By assumption, $F' \neq \emptyset$, therefore $F' \to F$ has a right-inverse, therefore $TF' \to TF$ has a right-inverse and necessarily $TF' \to TF \to F$ is a monomorphism.) It is easily checked that this is an order-preserving map and therefore has a fixed-point.

The minimality of F insures that the domain of this partial map (the thing we just called the fixed-point) is entire, that is, the fixed-point is a map. The fact that it is fixed by the described operation says that it is a map of algebras. Again one may invoke the minimality of F to obtain the uniqueness of this map of algebras by noting that the equalizer of any two such maps would be a proper subalgebra. ∎

The same proof (without needing the exceptional case for the empty set) works for:

Proposition:
 The category of vector spaces of dimension at most countable is algebraically complete. ∎

And both propositions remain true if the word "countable" is replaced with any specific cardinality.

The category whose objects are separable Hilbert spaces and whose maps are linear transformations of norm at most one is, by similar arguments algebraically compact.

2: ON CPO-CATEGORIES

Consider a **CPO-CATEGORY**, that is, a category for which each hom-set is a CPO and in which composition yields maps of CPOs. We have in mind two interpretations of "CPO": in the first case we mean ω-CPOs, that is, partially ordered sets with bottom and with suprema for ascending sequences; in the second case we mean partially ordered sets with bottom and with suprema for all well-ordered chains. In both cases we mean that composition (on either side) preserves bottoms and preserves the stipulated suprema. When we speak of functors between CPO-categories we understand that they preserve the partial ordering. In the first case we will further understand that the functors preserve the suprema of ascending sequences. We will not assume in the second case that the functors preserve any suprema. And in neither case will we assume that functors preserve bottoms. (With the axiom of choice, of course, the existence of suprema for well-ordered chains implies the existence of suprema for arbitrary updirected subsets. As will be seen I here need only well-ordered chains and thus I will not beg the power of the axiom of choice in the definition.)

The motivating example in the first case is the category of extensional PERs that Mulry, Rosolini, Scott and I described in [1]. The motivating example in the second case is the category whose objects are countable sets and whose morphisms are binary relations.

LEMMA FOR CPO-CATEGORIES:
  A T-invariant object is weakly initial as a T-algebra.

Suppose that $f:TA \to A$ is an isomorphism and that $g:TB \to B$ is arbitrary. We seek a commutative diagram:

$$
\begin{array}{ccc}
 & Tx & \\
TA & \to & TB \\
f\downarrow & & \downarrow g \\
A & \to & B. \\
 & x &
\end{array}
$$

Consider the endofunction on the hom-set (A,B) that sends x to $f^{-1};(Tx);g$ and choose a fixed point. (In the second case we are using very little of the hypothesis: only that composition and application of T preserve order, no continuity.) ∎

We may easily infer:

PROPOSITION FOR CPO-CATEGORIES:
  T-Inv, for any endofunctor T, is strongly connected and hence if it has both an initial and final object it is a punctuated category. Thus if the CPO-category is

3

algebraically bicomplete it is necessarily algebraically compact.∎

Recalling that a metric space is compact iff it is complete and totally bounded we might rephrase this last proposition as saying that a CPO-category is **algebraically totally bounded.**

By a **SMALL ENDOMORPHISM** in an ordered category is meant an endomorphism less than or equal to *1*. If $f{:}TA \to A$ is a *T*-invariant object we say that an endomorphism, x, on *A* is invariant if

$$\begin{array}{ccc} & Tx & \\ TA & \to & TA \\ f \downarrow & & \downarrow f \\ A & \to & A. \\ & x & \end{array}$$

We will say that $f{:}TA \to A$ is **SPECIAL INVARIANT** if its only small invariant idempotent is *1*.

Small   idempotents arise as follows: given a CPO-monoid, *M*, and an endomorphism $h{:}M \to M$ such that $h \leq 1$ then the minimal fixed point of *h* is necessarily a small idempotent.

LEMMA FOR CPO-CATEGORIES:
 Special invariant objects are free algebras.

Suppose that $f{:}TA \to A$ is special invariant, and that $g{:}TB \to B$ is arbitrary. We wish to establish the uniqueness of the fixed point of the endofunction on the hom-set *(A,B)* that sends x to $f^{-1}{;}(Tx){;}g$. There seems no way to avoid transfinite induction at this point.

Let $\{x_\alpha\}$ be the transfinite sequence defined by $x_{\alpha+1} = f^{-1}{;}(Tx_\alpha){;}g$ and for limit ordinals β, $x_\beta = sup_{\alpha<\beta} x_\alpha$. In particular, x_0 is bottom.

This ascending sequence stabilizes at the least fixed point. We wish to show that it is the unique fixed point. Let $\{z_\alpha\}$ denote the sequence of endomorphisms that results in the special case that *A = B* and *f = g*. This sequence stabilizes at the least invariant endomorphism of $f{:}TA \to A$. In this case it is the least fixed point of a monoid-endomorphism and, as just observed, therefore a small invariant idempotent. By hypothesis, therefore, the sequence of *z's* stabilizes at *1*. Now suppose that

$$\begin{array}{ccc} & Ty & \\ TA & \to & TB \\ f \downarrow & & \downarrow g \\ A & \to & B. \\ & y & \end{array}$$

An inductive proof establishes that $x_\alpha = z_\alpha{;}y$. Hence the x's stabilize at y. That is, y is the least fixed point.

The argument easily dualizes to yield final coalgebras.∎

There need not be any special invariant objects. In the case of PERs we know that every endofunctor has an invariant object. In the case of relations on countable sets we know that every endofunctor has a **PRE-INVARIANT OBJECT**, that is, a pair of maps $u{:}TA \to A$, and $d{:}A \to TA$ such that $u{;}d = 1$ and $d{;}u \le 1$. (Take A to be infinite, u an inclusion map, d the reciprocal relation of u.) Idempotents in general do not split in categories of relations but small idempotents do.

PROPOSITION FOR CPO-CATEGORIES:
If small idempotents split then an endofunctor with a pre-invariant object has a free algebra.

Given a pre-invariant object as described above consider the function on (A,A) that sends x to $d{;}(Tx){;}u$. This is an example of an endomorphism of a CPO-monoid (A,A) whose least fixed point is necessarily a small idempotent e. If we split e as $d'{:}A \to B$, $u'{:}B \to A$ where $u'{;}d' = 1$, then $(Tu'{;}u{;}d'){:}TB \to B$ is special invariant. (The inverse map is $u'{;}d{;}Td'$. If x is an invariant endomorphism on B then $d'{;}x{;}u'$ would be an invariant endomorphism on A, from which one may conclude that B is special invariant.) ∎

We note here a class of CPO-categories that are algebraically compact in the fullest sense. Suppose that S is a CPO as in the second case and that it is a meet semi-lattice structure in the relevant CPO-sense, that is, the binary operation $\wedge{:}S \times S \to S$ is a map of CPOs. (Note that any finite meet semi-lattice is an example.) If S has a maximal element we could regard S as a one-object CPO-category and consider the Karoubi-closure S. In fact, we may do this whether or not S has a top. The objects of S are the elements of S. The morphisms from x to y are all the elements of S that are less than or equal to $x \wedge y$. The order on the objects of S is the "retract" order, that is $x \le y$ iff x appears as a retract of y, hence any endofunctor of S preserves the order. Since every morphism of S is a split epi followed by a spit mono, the action of a functor on objects determines its action on morphisms (there is at most one way that an object can appear as a retract of another object). Hence every endofunctor preserves order. Every endofunctor has a free algebra, to wit, its least fixed object. (The "impoverished" example in the Como paper is equivalent to the result when S is the two-element CPO.)

3: On Counterexamples

If composition does not preserve suprema it needn't be the case that special invariant objects are free algebras. Consider the CPO obtained by adjoining a top element, ∞, to the poset $2{\times}N$ where 2 is the two-element poset and N is the natural numbers. This is a meet semi-lattice but not in the required sense: $<1,0>{\wedge}sup_n<0,n> = <1,0>$ but $sup_n(<1,0>{\wedge}<0,n>) = <0,0>$. Let S be its Karoubi-closure as described above and let T be the functor that on objects sends $<i,n>$ to $<i,n+1>$. The only T-invariant object is ∞. It is special invariant. It is a weakly initial T- algebra, but not plainly initial: the uniqueness condition fails: for the algebra $<1,0>{:}T<1,0> \rightarrow <1,0>$ there are two maps from $\infty{:}T\infty \rightarrow \infty$, to wit, $<0,0>$ and $<1,0>$.

Composition by a fixed element must preserve suprema whether composing on either the left or right side. Consider a category of bottom-raising maps between CPOs. The identity functor can have an final coalgebra (a one element CPO) which is only weakly initial.

The category of countable sets is not algebraically cocomplete. The endofunctor that sends a set X to $A{\times}X$ does not have a final coalgebra (unless A has fewer than two elements). We do not have a counterexample for the category of sets of cardinality at most the continuum.

4: On the Product Theorems

I used the iterated square theorem to prove:

THE PRODUCT THEOREM FOR ALGEBRAICALLY COMPLETE CATEGORIES:
> If $\mathcal{X}$ and $\mathcal{A}$ are algebraically complete then so is $\mathcal{X} \times \mathcal{A}$

An easier proof is available. It follows the lines of the Bekos proof:

Any endofunctor on the product may be resolved into its coordinate functors. Hence we suppose that we are given $T'{:}\mathcal{X}{\times}\mathcal{A} \rightarrow \mathcal{X}$ and $T{:}\mathcal{X}{\times}\mathcal{A} \rightarrow \mathcal{A}$. We wish to construct an initial algebra for the functor that sends an object $<A',A>$ to $<T'A'A, TA'A>$. (A prime mark will be used for objects and for functors *valued* in $\mathcal{X}$.)

First define a functor $F'{:}\mathcal{A} \rightarrow \mathcal{X}$ as follows: given A in $\mathcal{A}$ let $f'_A{:}T'(F'A)A \rightarrow F'A$ be an initial algebra for the endofunctor $T'(-)A$ on $\mathcal{X}$, that is, the functor that carries X' to $T'X'A$. (In the extensional PERs case a choice-free construction of initial algebras was given, thus obviating the use of the axiom of choice here.) Given a map $x{:}A \rightarrow B$, define $F'x$ to be the unique map of $T'(\text{--})A$-algebras such that

$$
\begin{array}{ccc}
& T'(F'x)1 & \\
T'(F'A)A & \to & T'(F'B)A \\
& & \downarrow\ T'1x \\
f'_A\ \Big\downarrow & & T'(F'B)B \\
& & \downarrow\ f'_B \\
F'A & \xrightarrow{\quad\quad} & F'B. \\
& F'x &
\end{array}
$$

By a little re-diagramming, $F'x$ may be characterized as the unique map that fits in:

$$
\begin{array}{ccc}
& T'(F'x)x & \\
T'(F'A)A & \to & T'(F'B)B \\
f'_A\ \downarrow & & \downarrow\ f'_B \\
F'A & \xrightarrow{\quad\quad} & F'B \\
& F'x &
\end{array}
$$

Now let $G\!:\!\mathcal{A} \to \mathcal{A}$ be defined by $GA = T(F'A)A$. Let $d\!:\!T(F'D)D \to D$ be an initial algebra for G. Define $D' = F'D$ and denote $f'_D\!:\!T'(F'D)D \to F'D$ as $d'\!:\!T'D'D \to D'$. We wish to show that $<D', D>$ is a $<T', T>$-initial algebra with $<d',d>$ as its structure.

Given $<a',a>\!:\!<T'A'A, TA'A> \to <A', A>$ first use the initiallity of $F'A$ to define $y'\!:\!F'A \to A'$ as the unique map such that

$$
\begin{array}{ccc}
& T'y'1 & \\
T'(F'A)A & \to & T'A'A \\
f'_A\ \downarrow & & \downarrow\ a' \\
F'A & \xrightarrow{\quad\quad} & A'. \\
& y' &
\end{array}
$$

Use the initiallity of D to define $x\!:\!D \to A$ as the unique map such that

1:
$$
\begin{array}{ccc}
& T(F'x)x & \\
T(F'D)D & \to & T(F'A)A \\
& & \downarrow\ Ty'1 \\
d\ \Big\downarrow & & TA'A \\
& & \downarrow\ a \\
D & \xrightarrow{\quad\quad} & A. \\
& x &
\end{array}
$$

Define $x' = (F'x);y'$. We may re-diagram the above to obtain

$$
2: \quad
\begin{array}{ccc}
TD'D & \xrightarrow{\ Tx'x\ } & TA'A \\
{\scriptstyle d}\downarrow & & \downarrow{\scriptstyle a} \\
D & \xrightarrow{\ x\ } & A.
\end{array}
$$

Consider the diagram:

$$
\begin{array}{ccccc}
T'(F'D)D & \xrightarrow{\ T'(F'x)x\ } & T'(F'A)A & \xrightarrow{\ T'y'1\ } & T'A'A \\
{\scriptstyle f'_D}\downarrow & & {\scriptstyle f'_A}\downarrow & & \downarrow{\scriptstyle a'} \\
F'D & \xrightarrow{\ F'x\ } & F'A & \xrightarrow{\ y'\ } & A'.
\end{array}
$$

Using the definitions of D' and x' we obtain the other needed diagram:

$$
3: \quad
\begin{array}{ccc}
T'D'D & \xrightarrow{\ T'x'x\ } & T'A'A \\
{\scriptstyle d'}\downarrow & & \downarrow{\scriptstyle a'} \\
D' & \xrightarrow{\ x'\ } & A'.
\end{array}
$$

The uniqueness of $<x',x>$ may be established as follows: suppose that we are given the diagrams **2** and **3**. For any $x:D\to A$ there is a unique x' that makes **3** commute as can be seen from the diagram:

$$
\begin{array}{ccc}
T'(F'D)D & \xrightarrow{\ T'x'1\ } & T'A'D \\
{\scriptstyle f'_D}\Big\downarrow & & \downarrow{\scriptstyle T'1x} \\
 & & T'A'A \\
 & & \downarrow{\scriptstyle a'} \\
F'D & \xrightarrow{\ x'\ } & A'.
\end{array}
$$

On the other hand we know from the first part of the proof that for any x we may take x' as $(F'x);y$ to obtain **3**. Hence we may conclude that $x' = (F'x);y'\pi$.

Thus **2** says that **1** commutes and the initiallity of D says that x is unique.

■

THE PRODUCT THEOREM FOR ALGEBRAICALLY COMPACT CATEGORIES:
 If $\mathcal{X}$ and $\mathcal{A}$ are algebraically compact then so is $\mathcal{X} \times \mathcal{A}$

We may walk through the above proof, checking at each stage that a final coalgebra results when each initial-algebra structure map is inverted. Start with F'. (It is just a bit of a surprise that the definition of $F'x$ remains OK.) ∎

5: ON DINATURALITY

Using the iterated square theorem and the product theorem for algebraically closed categories I gave a proof of the "dinaturality" of the two-functor that delivers initial algebras. That is, given a functor $F:\mathcal{A} \to \mathcal{B}$ the induced diagram of categories commutes in the relevant sense:

$$(\mathcal{B},\mathcal{A}) \nearrow \begin{array}{ccc} (\mathcal{A},\mathcal{A}) & \longrightarrow & \mathcal{A} \\ & & \downarrow \\ (\mathcal{B},\mathcal{B}) & \longrightarrow & \mathcal{B}. \end{array}$$

The commutativity at the object level says:

Given $F:\mathcal{A} \to \mathcal{B}$ and $G:\mathcal{B} \to \mathcal{A}$, F applied to the initial GF-algebra is the initial FG-algebra.

A much more primitive proof is available: The categories *GF-Inv* and *FG-Inv* are equivalent categories. If $g:GFA \to A$ is in *GF-Inv* then $Fg:FGFA \to FA$ is in *FG-Inv*, that is, F yields a functor from *GF-Inv* to *FG-Inv* and G yields one from *FG-Inv* to *GF-Inv* and together they establish an equivalence of categories. Clearly, then F carries an initial object of *GF-Inv* to an initial object of *FG-Inv*. ∎

The iterated square theorem is still used for the proof that $Y(x^2) = Yx$ for the fixed-point operator that we constructed using algebraic compactness and that, in turn, yields the dinaturality of Y assuming the existence of products:

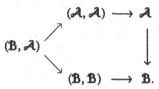

$$(B,A) \nearrow \begin{array}{ccc} (A,A) & \xrightarrow{Y} & (1,A) \\ & & \downarrow \\ (B,B) & \xrightarrow[Y]{} & (1,B). \end{array}$$

Given $g:B \to A$ we wish to show that $Y(f;g);f = Y(g;f)$. Consider the endomorphism x on $A \times B$ which would be described using elements as

$x<a,b> = <gb, fa>$. x^2 is the map $(f;g)\times(g;f)$. The Plotkin axiom easily
yields that $Y((f;g)\times(g;f)) = (Y(f;g))\times(Y(g;f))$. From $Y(x^2) = Yx$ we obtain the
desired identity (and the other one: $Y(g;f);g = Y(f;g)$.)

6: ON MULTI-COREFLECTIVITY

We saw that if $\mathcal{A}$ is a reflective subcategory of a category $\mathfrak{D}$ and if $T:\mathfrak{D} \to \mathfrak{D}$
is a functor that has $\mathcal{A}$ as an invariant subcategory, then a final T-coalgebra as
defined on $\mathcal{A}$ remains a final T-coalgebra for all of $\mathfrak{D}$. We suppose now that
$\mathcal{A}$ is the lluf subcategory of $\mathfrak{D}$ consisting of strict maps. We suppose that it is
multi-coreflective in the sense of Diers, that is, for every object A there is a collection of maps $\{A_i \to A\}_i$ such that every map to A uniquely factors as a strict
map through a unique $A_i \to A$. In the cases of domains in the usual sense, the
indexing set may be taken to be the elements of A with $A_i \to A$ the inclusion
map of the updeal of all elements greater than or equal to i. (In the extensional
PERs case some work must be done.) In all cases there is a one-to-one correspondence between the morphisms of the multi-coreflection and the points of the
object, that is, for every $x:1\to A$ there is a unique $A_i \to A$ such that
$1 \rightarrowtail A_i \to A= x$. Put another way, we may in all cases take the points of an object as the indexing set of its multi-coreflection. For arbitrary $b:B \to A$ define
$x:1\to A$ as $1 \rightarrowtail B \to A$. If $A_x \to A$ is the map in the multi-coreflection of A
such that $1 \rightarrowtail A_x \to A$ $= x$ then there is unique $B \rightarrowtail A_x$ such that
$B \rightarrowtail A_x \to A = b$.

Suppose that $f:TF \to F$ is a free T-algebra and that $a:TA \to A$ is an
arbitrary T-algebra. For each T-algebra map g from F to A we have a commutative diagram:

$$
\begin{array}{ccccc}
 & \perp & T\perp & Tg & \\
1 & \rightarrowtail T1 & \rightarrowtail TF & \to & TA \\
\downarrow & & \downarrow & & \downarrow a \\
1 & \xrightarrow{\;\;\;\perp\;\;\;} & F & \xrightarrow{g} & A.
\end{array}
$$

By defining $x = \perp ;g$ we obtain the diagram

$$
\begin{array}{ccc}
 & \perp & Tx \\
1 & \rightarrowtail T1 & \to TA \\
\downarrow & & \downarrow a \\
1 & \xrightarrow{\;\;\;x\;\;\;} & A.
\end{array}
$$

In other words, every T-algebra map from F to A yields a fixed point for the
endofunction $\lambda x. \perp;Tx;a$ on the points of A. The multi-coreflectivity of strict
maps in all maps forces this to be a one-to-one correspondence. That is, given
such $x:1\to A$ let $A_x\to A$ be the map in the multi-coreflection such that

$1 \rightarrowtail A_x \to A = x$. The condition on x says that
$1 \rightarrowtail TA_x \to TA \to A = 1 \rightarrowtail T1 \rightarrowtail TA_x \to TA \to A = x$. Hence there is a
unique strict map $TA_x \rightarrowtail A_x$ such that

$$
\begin{array}{ccc}
TA_x & \to & TA \\
\downarrow & & \downarrow \\
A_x & \to & A.
\end{array}
$$

We may now use the initiality of $f{:}TF \to F$ to obtain

$$
\begin{array}{ccccc}
TF & \rightarrowtail & TA_x & \to & TA \\
\downarrow & & \downarrow & & \downarrow \\
F & \rightarrowtail & A_x & \to & A.
\end{array}
$$

This T-algebra map $F \to A$ has the required property that $1 \rightarrowtail F \to A = x$. It is routine that it is the only T-algebra map with that property.

In the special case that T has all of its values in the lluf subcategory of strict maps (e.g. for $T = \Sigma$) there is just one x such that

$$
\begin{array}{ccccc}
 & \perp & & Tx & \\
1 & \rightarrowtail & T1 & \rightarrowtail & TA \\
\downarrow & & & & \downarrow \quad a \\
1 & & \xrightarrow{\hspace{2cm}} & & A \\
 & & x & &
\end{array}
$$

and we may conclude that the free T-algebra remains initial in the larger category of all bottom raising maps.

7: ON RELATION CATEGORIES

As observed, the category $\mathcal{R}$ whose objects are countable sets and whose morphisms are binary relations is algebraically compact with respect to endofunctors that preserve the order on relations. Note that if R and S are relations such that $1 \leq R;S$ and $S;R \leq 1$ then necessarily $S = R^\circ$ and that R is a map. Any order-preserving endofunctor, therefore, preserves maps and their reciprocals. Since every relation can be obtained as the composition of the reciprocal of a map with a map, we have that any order-preserving endofunctor preserves reciprocation. R is a partial map iff $R^\circ;R \leq 1$, hence order-preserving endofunctors preserve partial maps.

The lluf subcategory of partial maps is also a CPO category (in the second sense) and the characterization of free algebras as special invariant objects easily proves that for any endofunctor, T, on $\mathcal{R}$ it is the case that the free T-algebra remains free when viewed as an endofunctor on the subcategory of partial maps.

Let $\mathcal{M}$ denote the lluf subcategory composed of maps. As observed above every relevant endofunctor is $\mathcal{M}$-invariant. Let T be such an endofunctor and let $f{:}TF \to F$ be a free T-algebra. f is an isomorphism and all isomorphisms in $\mathcal{R}$ are automatically in $\mathcal{M}$. It is a pleasant fact that $f{:}TF \to F$ remains an initial T-algebra when T is understood to be an endofunctor on $\mathcal{M}$. To prove it, let $a{:}TA \to A$ be a T-algebra in $\mathcal{M}$. We have just argued that there is a unique partial map $h{:}F \to A$ that is a morphism of T-algebras. It remains only to show that the domain, D, of h is the identity map of F. D may be characterized by the conditions $D \leq 1$, $D \leq h;h^\circ$ and $h = D;h$, therefore TD is the domain of Th. But the domain of Th is the same as the domain of $Th;a$, the same as the domain of $f;h$, the same as the domain of $f;D$, from which one may infer $TD;f = f;D$, that is, D is an invariant endomorphism of the T-algebra F. Of course the only invariant endomorphism of an initial T-algebra is the identity.

If the algebraically compact category $\mathcal{R}$ is a category of strict maps, what is the ambient category of bottom raising maps? In the Como paper I used the term "restrict" map to define a lluf subcategory of $\mathcal{R}$. Carl Gunter pointed out a better term. Given any punctuated category $\mathcal{R}$ define the lluf subcategory of **TOTALLY STRICT** morphisms, $\mathcal{T}$, by $t{:}A \to B \in \mathcal{T}$ iff whenever $X \to A \to B$ factors through the biterminator it is already the case that $X \to A$ does so. In the case at hand, $\mathcal{T}$ is the category of **ENTIRE RELATIONS**, that is, those relations "everywhere defined." If $\mathcal{T}$ is a coreflective subcategory the coreflector functor is necessarily an embedding and we may view $\mathcal{T}$ not as a subcategory but as an ambient category of "predomains".

In the case at hand, however, $\mathcal{T}$ is not coreflective. It is only weakly coreflective: for any set A let $u{:}A \to (1+A)$ be the inclusion of A into the coproduct of A with a one-element set. Then $u^\circ{:}(1+A) \to A$ is a weak coreflection. The uniqueness condition fails.

But we are in a category in which not all idempotents split. I will describe here just the case that we split all the idempotents (which, in fact, is a case of overkill). The resulting category can be redescribed as the category whose objects are countably generated completely distributive lattices, the maps being the join-preserving functions. The original category of sets and relations is the full subcategory whose objects are atomic boolean algebras. $\mathcal{T}$ may then be described as the category whose objects are countably generated *conditionally* complete join-semilattices, that is those join-semilattices in which every *nonempty* subset has a lub.

At a later time I will describe the wealth of computationaly interesting categories that result when one makes more judicious choices of classes of idempotents to be split.

Dinaturality for free

Peter J. Freyd [*]   *Edmund P. Robinson* [†]
Giuseppe Rosolini [‡]

The first aim of this paper is to attack a problem posed in [1] about uniform families of maps between realizable functors on PER's.

To put this into context, suppose that we are given a category C to serve as our category of types. The authors of [1] observe that the types representable in the second-order lambda calculus and most extensions thereof can be regarded as being obtained from functors $(C^{op} \times C)^n \longrightarrow C$ by diagonalisation of corresponding contra and covariant arguments. Terms in the calculus give rise to dinatural transformations. This suggests a general structure in which parametrised types are interpreted by arbitrary functors $(C^{op} \times C)^n \longrightarrow C$, and their elements by dinatural transformations. Unfortunately as the authors of the original paper point out, this interpretation can not be carried out in general since dinaturals do not necessarily compose.

However, suppose we are in the extraordinary position that all families of maps which are of the correct form to be a dinatural transformation between functors $(C^{op} \times C)^n \longrightarrow C$ are in fact dinatural, a situation in which we have, so to speak, the dinaturality for free. In this situation dinaturals compose. The result is a structure for a system in which types can be parametrised by types (second-order lambda calculus without the polymorphic types). Suppose, in addition, the category in question is complete, then we can perform the necessary quantification (which is in fact a simple product), and obtain a model for the second-order lambda calculus. Such a model can justifiably be described as inherently parametric, since the dinaturality, which is a parametricity condition, is forced on us by the choice of our category of types, and is not influenced by any further choices about which parametrised types, or which elements, we allow.

The question asked in [1] is, essentially, whether the PER model, suitably interpreted, is such a model. In other words, if we interpret all the above in the

---
[*]Department of Mathematics, University of Pennsylvania

[†]School of Cognitive and Computing Sciences, University of Sussex

[‡]Dipartimento di Matematica, Università di Parma. This author was partially supported by the ESPRIT BRA Programme of the EC no. 3003, and M.P.I. 40%.

internal logic of the effective topos, is the internal category M of modest sets an inherently parametric model for polymorphism? Following long-standing mathematical tradition, we shall not answer this precise question. Rather we shall instead produce a whole family of related categories which are inherently parametric models in the sense above. Thus we have yet another example of a situation in which we might wish to abandon the simple PER model, for one which is closely related and in some sense better-behaved, but is also more complex (c.f. [3, 6]). For those upset by this betrayal of an old friend we have some consolation. We shall show that the models are sufficiently closely related for the parametricity of certain types in the new model to imply its parametricity in the old. In particular the results given here subsume those in [5].

1.   The problem

Recall that the external view of the category of modest sets M of the effective topos is that of partial equivalence relations (=symmetric transitive relations) on the natural numbers where a map $[n]: X \longrightarrow Y$ is a[n equivalence] class of those numbers n such that the n-th partial recursive function ϕ_n satisfies

$$\mathrm{dom}\,(X) \subset \mathrm{dom}\,(\phi_n) \text{ and } (\phi_n \times \phi_n)[X] \subset Y$$

with respect to the equivalence relation

$$(\phi_n \times \phi_m)[X] \subset Y.$$

We shall call this category Per. It is important to notice for future reference that the inclusion order on partial equivalence relations is a subcategory of Per since

$$X \subset Y \text{ if and only if } [i]: X \longrightarrow Y$$

where i is any code of the identity recursive function. To improve readability, we may write $x =_X x'$ in place of the usual $x \, X \, x'$.

The "external" picture of an internal functor M $\longrightarrow$ M is a functor $F:$ Per $\longrightarrow$ Per which comes endowed with a recursive function f such that

$$F([n]: X \longrightarrow Y) = [f(n)]: FX \longrightarrow FY.$$

We refer the reader who is not familiar with all this to [2, 4, 7] for further information.

The question asked in [1] is the following: let $F, G: (\mathsf{Per^{op}})^m \times \mathsf{Per}^n \longrightarrow \mathsf{Per}$ be two (internal) functors and $([s]: F(X, \ldots; X, \ldots) \longrightarrow G(X, \ldots; X, \ldots))_{X \in \mathsf{Per}}$ any uniform family of maps indexed over the PERs themselves—the semicolon separates contravariant and covariant entries. Is $([s])_X$ necessarily a dinatural transformation? In other words, given any $[n]: X \longrightarrow Y$ in Per, does the following diagram

$$F(X,\ldots;X,\ldots) \xrightarrow{\;[s]\;} G(X,\ldots;X,\ldots)$$

$$F([n],\ldots;\mathrm{id}_X,\ldots) \qquad\qquad G(\mathrm{id}_X,\ldots;[n],\ldots)$$

$$F(Y,\ldots;X,\ldots) \qquad\qquad\qquad\qquad G(X,\ldots;Y,\ldots)$$

$$F(\mathrm{id}_Y,\ldots;[n],\ldots) \qquad\qquad G([n],\ldots;\mathrm{id}_Y,\ldots)$$

$$F(Y,\ldots;Y,\ldots) \xrightarrow{\;[s]\;} G(Y,\ldots;Y,\ldots)$$

commute?

The intuition about the possibly positive answer rests on the effective, constructive nature of the logic of $\mathcal{E}\!f\!f$ which allows sometimes very few elements in certain derived sets. Internal products over PER's are so uniform that an internal product $\prod_{X \in \mathbf{Per}} A_X$ gives nothing more than a modest set. For instance, the products of families like $[[X \to X] \to [X \to X]]$ or $[[A \to X] \to X]$ (for a fixed A) consist only of λ-definable functions, thus their elements yield dinatural transformations.

2.   Some properties of an internal functor

In this section we recall some general properties of internal functors and of dinatural transformations, most of which are to be found in [1]. These will be used to reduce the problem to checking commutativity of the hexagon for a particular collection of maps.

For the purpose of the exposition in this section we consider PER's on an arbitrary partial combinatory algebra $(P, \cdot)$ where we have fixed elements $\mathbf{k}, \mathbf{s} \in P$ such that

$$(\mathbf{k}x)y = x \quad \text{and} \quad ((\mathbf{s}x)y)z = (xz)(yz)$$

for all $x, y, z \in P$. A map $[f]: X \longrightarrow Y$ between partial equivalence relations is a[n equivalence class] of those elements $f \in P$ such that

$$\forall x, x'[x =_X x' \implies fx =_Y fx']$$

with respect to the equivalence relation

$$f =_{YX} f' \quad \text{iff} \quad \forall x \in \mathrm{dom}\,(X).fx =_Y f'x.$$

Composition of maps $[f]$ and $[g]$ is defined by the map $[g \circ f]$ where $g \circ f = (\mathbf{s}(\mathbf{k}g))f \in P$ and identities are always defined by the element $\mathbf{i} = (\mathbf{s}\mathbf{k})\mathbf{k} \in P$

as $ix = x$ for all $x \in P$. The category of PER's on $(P, \cdot)$ is still denoted by Per. An internal functor is (again) a functor $F\colon \mathsf{Per} \longrightarrow \mathsf{Per}$ which comes endowed with an element $f \in S$ such that

$$F([n]\colon X \longrightarrow Y) = [fn]\colon FX \longrightarrow FY.$$

The first lemma holds for an arbitrary category.

Lemma 2.1 *Suppose* $F, G\colon \mathsf{C}^{\mathrm{op}} \times \mathsf{C} \longrightarrow \mathsf{C}$ *are functors and* $([s]\colon F(X; X) \longrightarrow G(X; X))_{X \in \mathsf{C}}$ *a family of maps in* C. *Then the maps for which the hexagon commute form a subcategory of* C *and its inclusion in* C *reflects isos, in the sense that*

  (i) *the hexagon for* $\mathrm{id}\colon X \longrightarrow X$ *commutes;*

  (ii) *if the hexagons commute for* $f\colon X \longrightarrow Y$ *and* $g\colon Y \longrightarrow Z$, *then the hexagon for* $g \circ f\colon X \longrightarrow Z$ *commutes;*

  (iii) *if* $f\colon X \longrightarrow Y$ *is iso and the hexagon for it commutes, then the hexagon for* $f^{-1}\colon Y \longrightarrow X$ *commutes.*

Proof : Easy diagram chasing.   □

Lemma 2.2 *Let* $F, G\colon \mathsf{Per}^{\mathrm{op}} \times \mathsf{Per} \longrightarrow \mathsf{Per}$ *be internal functors on PER's over an arbitrary partial combinatory algebra* $(P, \cdot)$. *Suppose* $\mathsf{i} \in P$ *is the identity combinator and* $X \subset Y$ *are PER's. Then the hexagon for* $[\mathsf{i}]\colon X \longrightarrow Y$ *commutes.*

Proof : By hypothesis, the values $F([\mathsf{i}], \mathrm{id}_X)$, $F([\mathsf{i}], [\mathsf{i}])$, and $F(\mathrm{id}_Y, [\mathsf{i}])$ are all determined by the function $(fi)i$. Similarly for G. Then the two paths are computed evaluating the same element $((gi)i)\circ(s\circ((fi)i))$.   □

Suppose now that elements $\mathsf{t}, \mathsf{f}, \mathsf{w} \in P$ are given which allow *definition by cases*, in the sense that

$$((\mathsf{wf})x)y = x \text{ and } ((\mathsf{wt})x)y = y$$

for every $x, y \in P$. Such a property does not determine the three elements uniquely, rather it allows freedom of choice: for instance, $\mathsf{t} = \mathsf{k}, \mathsf{f} = \mathsf{ki}, \mathsf{w} = \mathsf{i}$ allow definition by cases. In particular algebras, there can be other triples enjoying further desirable properties. Say that $X, Y \subset P$ are *separable* (w.r.t. t, f, w) if there is $a \in P$ such that

$$ax = \begin{cases} \mathsf{f} & \text{if } x \in X \\ \mathsf{t} & \text{if } x \in Y \end{cases}$$

Lemma 2.3 *Suppose that elements* t, f, w $\in P$ *allow definition by cases, and suppose* $X, Y \in$ Per *have separable domains. Then the hexagon for any map in* Per *between them commutes.*

Proof :   Consider $[f]: X \longrightarrow Y$, and take the partial equivalence relation K which is the kernel of $[f]$: thus $x \, K \, x'$ if and only if $x, x' \in \text{dom}\,(X)$ and $fx \, Y \, fx'$. Then extend the PER $K \cup Y$ to Z by letting $x \, Z \, y$ when $fx \, Y \, y$ for $x \in \text{dom}\,(X)$ and $y \in \text{dom}\,(Y)$. Thus $X \subset Z \supset Y$. Since K and Y are separated, $[i]: Y \longrightarrow Z$ is iso. The inverse is given by any element $j \in P$ such that $jx = \mathsf{w}(ax)(fx)x$, for instance $j = (\mathsf{s}(\mathsf{s}(\mathsf{s}(\mathsf{kw})a)f))i$ where a separates X and Y. Moreover the composition $X \xrightarrow{[i]} Z \xrightarrow{[i]^{-1}} Y = X \xrightarrow{[n]} Y$. The result now follows from 2.2.   □

In fact any function can be factored as composition of particular isomorphisms and maps satisfying the conditions of 2.3.

Lemma 2.4 *Suppose* $q, r \in P$ *define a retraction pair on* P, *i.e.* $((r \circ q)x = x$ *for every* $x \in P$. *Then there is an internal functor* $(-)^q: $ Per $\longrightarrow$ Per *definable from the retraction pair, which is naturally isomorphic to the identity.*

Proof :   For X a PER, let $y \, X^q \, y'$ if there are $x \, X \, x'$ such that $y = qx$ and $y' = qx'$. This defines a partial equivalence relation as q is left $\circ$-cancellable, and the maps $X \xrightarrow{[q]} X^q \xrightarrow{[r]} X$ are inverse to each other. The function $(-)^q$ is extended to a functor by sending a map $[f]: X \longrightarrow Y$ to the composite $X^q \xrightarrow{[r]} X \xrightarrow{[f]} Y \xrightarrow{[q]} Y^q$. The conclusion follows.   □

Proposition 2.5 *Let* $q, r \in P$ *and* $A \subset P$. *Suppose that* q, r *define a retraction pair on* P, *and that the sets* $\{qx : x \in P\}$ *and* A *are separable. For any PER's* X *and* Y *such that* $\text{dom}\,(X), \text{dom}\,(Y) \subset A$, *the hexagon for any map between them commutes.*

Proof :   Apply 2.4 and 2.3 to $X \xrightarrow{[f]} Y = X \xrightarrow{[f]^q \circ [q]} Y^q \xrightarrow{[r]} Y$.   □

Consider now the case $P = \mathsf{N}$ with Kleene application and with the standard definition by cases where $\mathsf{t} = 1$, $\mathsf{f} = 0$. Consider the retraction $n \mapsto \lfloor n/2 \rfloor : \mathsf{N} \longrightarrow \mathsf{N}$ which has two disjoint sections $n \mapsto 2n : \mathsf{N} \longrightarrow \mathsf{N}$ (say e is a code for this) and $n \mapsto 2n + 1 : \mathsf{N} \longrightarrow \mathsf{N}$. Then any map $[n] : X \longrightarrow Y$ can be factored as a composite $X \longrightarrow X^e \longrightarrow Y^e \longrightarrow Y$. It follows from the previous results that a solution to the problem depends precisely on a solution of the problem for that particular instance. To wit, a positive answer to the question follows from the commutativity of the hexagon

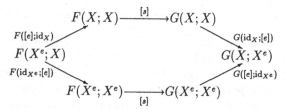

for every PER X.

3.   When the hexagon commutes

From now on, the functors F and G will be simply from $\mathsf{Per}^{\mathrm{op}} \times \mathsf{Per}$ on N. We shall prove that the previous hexagon commutes for any PER X such that the PER $G(X; X)$ satisfies a certain discernability condition: there is a PER B such that inclusion $B \subset \mathrm{dom}\,(B)^2$ either a Σ-subset or a Π-subset, and there is $G(X; X) \rightarrowtail [A \to B]$ for some suitable PER A. In fact, the condition is satisfied for some A if and only if it is when A is $[G(X; X) \to B]$, see [9].

Proposition 3.1 *Let* $F, G : \mathsf{Per}^{\mathrm{op}} \times \mathsf{Per} \longrightarrow \mathsf{Per}$. *Suppose* X *is a PER and there is* $G(X; X) \rightarrowtail [A \to B]$ *for some PER* A *and* B *where* $B \subset \mathrm{dom}\,(B)^2$ *is either r.e. or a complement of an r.e. set of pairs. Then the hexagon above commutes for* X.

Proof :   The argument is by contradiction: if the hexagon does not commute for such an X, then we can solve the halting problem. Consider the following auxiliary recursive functions

$$\phi_{r_m}(n) = \begin{cases} 2n & \text{if } \phi_m(m) \text{ converges in } < n \text{ steps} \\ n & \text{otherwise.} \end{cases}$$

In case $\phi_m(m)$ does not converge, we have $\phi_{r_m} = $ id. Otherwise, ϕ_{r_m} agrees with ϕ_e after a certain value n_0. Consider then the hexagon drawn as follows

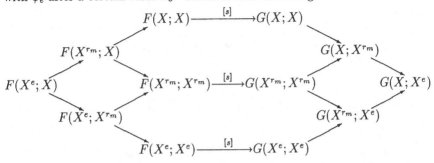

In the first case the upper hexagon commutes trivially, in the second case the lower hexagon commutes by applying 2.5, taking advantage of the infinitely many odd numbers after n_0. Let

$$[u], [c(m)], [l]\colon F(X^e; X) \longrightarrow G(X; X^e)$$

be the composites along the uppermost path, (either) central, and the lowest respectively. Since the two lateral squares always commute by bifunctoriality, we have that

$$\phi_m(m)\downarrow \Longrightarrow \forall X \in \mathsf{Per}.[c(m)] = [l]\colon F(X^e; X) \longrightarrow G(X; X^e),$$
$$\phi_m(m)\uparrow \Longrightarrow \forall X \in \mathsf{Per}.[c(m)] = [u]\colon F(X^e; X) \longrightarrow G(X; X^e).$$

If the big outer hexagon does not commute, there are $x \in \mathrm{dom}\,(F(X^e; X))$ and $a \in \mathrm{dom}\,(A)$ such that

$$\phi_u(x)(a) \neq_B \phi_l(x)(a).$$

As
$$\phi_m(m)\downarrow \Longrightarrow \phi_{c(m)}(x)(a) =_B \phi_l(x)(a),$$
$$\phi_m(m)\uparrow \Longrightarrow \phi_{c(m)}(x)(a) =_B \phi_u(x)(a);$$

it follows that
$$\phi_m(m)\uparrow \Longleftrightarrow \phi_{c(m)}(x)(a) =_B \phi_u(x)(a),$$
$$\Longleftrightarrow \phi_{c(m)}(x)(a) \neq_B \phi_l(x)(a);$$

presenting the complement of the halting set as an r.e. set when $B \subset (\mathrm{dom}\,(B))^2$ is r.e. or the complement of an r.e. set, respectively.   □

Casting back to the discussion before 3.1, consider a fixed PER B and, given any PER Z, consider the canonical map $p\colon Z \longrightarrow [[Z \to B] \to B]$. Then there is a monic $Z \rightarrowtail [A \to B]$ if and only if p is already monic. Therefore the inclusion of the full subcategory H_B on all those Z with $p\colon Z \rightarrowtail [[Z \to B] \to B]$ has a left adjoint obtained by taking the image of p. One checks that the left adjoint $L_B\colon \mathsf{Per} \longrightarrow \mathsf{H}_B$ is internal, see [9] for more details.

Theorem 3.2 *Suppose B is such that $B \subset \mathrm{dom}\,(B)^2$ is either r.e. or the complement of an r.e. set of pairs. Let $F, G: (\mathsf{H}_B^{\mathrm{op}})^m \times \mathsf{H}_B^n \longrightarrow \mathsf{H}_B$ be any two functors. Then*

$$\prod_{X \in \mathsf{H}_B} [F(X,\ldots;X,\ldots) \to G(X,\ldots;X,\ldots)] = \mathrm{Din}(F,G).$$

Proof :   Apply 3.1 to the composite functors $F(L_B^{\mathrm{op}} \times L_B)$ and $G(L_B^{\mathrm{op}} \times L_B)$.
□

It is rather interesting to notice that the internal category corresponding to H_B inherits completeness and closure properties from M. This result therefore tells us that every uniform family of maps between functors defined on the category is dinatural. It follows that any of these categories provides us with a model for functorial polymorphism, in fact an inherently parametric model as in the introduction.

Remark 3.3 The argument above can also be internalized to give a property of the internal categories of functors. Also, no notion of completeness has been used in the proof of 3.1, and the proof can be reworked for any subcategory of H_B closed under isomorphic objects and with a representing set of objects which is uniform.

4.   The algebraic case

It is well-known that the study of closed algebraic types such as the Church integers $\prod_{X \in \mathsf{M}} [[X \to X] \times X \to X]$ can be reduced to PERs $X \rightarrowtail \mathsf{N}$, thus to total PERs. But it is interesting to see that the characterisation of the elements in the closed type can be deduced from their dinaturality. We reproduce a proof of this in the sequel to test the method of dinaturality; in fact we can make it follow from more general properties derived from the automatic dinaturality of uniform families.

Proposition 4.1 *Suppose B is such that $B \subset \mathrm{dom}\,(B)^2$ is either r.e. or the complement of an r.e. set of pairs. Let $F: \mathsf{H}_B \longrightarrow \mathsf{H}_B$ be an internal functor, $\varepsilon: FI \longrightarrow I$ an initial F-algebra, and $[s]: [FX \to X] \longrightarrow X$ a uniform family of maps as X varies in H_B. Suppose moreover that*

(i) *$\mathrm{dom}\,(I)$ is disjoint from a recursive infinite subset of N;*

(ii) *there is a recursive function f which tracks the values of F on maps with the property that $\phi_{f(i)}|_{\mathrm{dom}(FI)} = \mathrm{id}$ for some code i of the identity function on I; and*

(iii) $\mathrm{dom}(FI) \subset \mathrm{dom}(FX)$ *is recursive in* $\mathrm{dom}(FX)$ *whenever the PER X contains I.*

Then for every F-algebra $\alpha\colon FX \longrightarrow X$

$$[s]_X(\alpha) =_X f_\alpha([s]_Y(\varepsilon))$$

where $f_\alpha\colon I \longrightarrow X$ is the unique F-homomorphism.

Proof :   Consider a factorization $f_\alpha = I \xrightarrow{[i]} X' \overset{\sim}{\longrightarrow} X$; this is always possible by (i). Notice that there is an induced F-algebra structure $\alpha'\colon FX' \longrightarrow X'$. Moreover a code of α' can be chosen so that it also codes ε. Indeed, by (ii) the diagram

$$
\begin{array}{ccc}
FI & \xrightarrow{[f(i)]} & FX' \\
{\scriptstyle \varepsilon}\Big\downarrow{\scriptstyle l} & & \Big\downarrow{\scriptstyle \alpha'} \\
I & \xrightarrow{\;[i]\;} & X'
\end{array}
$$

commute (so $f_{\alpha'} = [i]$). Use (iii) to define a partial recursive function such that

$$\alpha''(n) = \begin{cases} \varepsilon(n) & \text{if } n \in \mathrm{dom}(FI) \\ \alpha'(n) & \text{if } n \in \mathrm{dom}(FX) \setminus \mathrm{dom}(FI) \end{cases}$$

and check that $\alpha'' =_{[FI \to I]} \varepsilon$ and $\alpha'' =_{[FX' \to X']} \alpha'$. Because $[s]$ acts on codes and is extensional, we have that $[s]_I(\varepsilon) =_I [s]_I(\alpha')$, thus $[s]_{X'}(\varepsilon) =_I [s]_{X'}(\alpha')$ as $I \subset X'$. Therefore

$$f_{\alpha'}([s]_{X'}(\varepsilon)) =_{X'} [i]([s]_I(\varepsilon)) =_{X'} [s]_{X'}(\alpha').$$

To get the result we must now recall 3.2 asserting that the diagram

$$
\begin{array}{ccc}
[FX' \to X'] & \xrightarrow{\;[s]_{X'}\;} & X' \\
{\scriptstyle l}\Big\downarrow & & \Big\downarrow{\scriptstyle l} \\
[FX \to X] & \xrightarrow{\;[s]_X\;} & X
\end{array}
$$

commutes.   □

In case the functor F is obtained as the interpretation of a polymorphic type, then condition (ii) is automatically satisfied, and (i) and (iii) are satisfied if one can produce an effective presentation of the F-initial algebra.

Remark 4.2 Although condition (ii) in 4.1 is satisfied in all the standard cases, one cannot expect to find a good tracking function in general. Take for instance the following functor $F: \mathsf{Per} \longrightarrow \mathsf{Per}$ which results isomorphic to the functor "coproduct with 1": it sends a PER X to the PER

$$FX = \{\langle\langle i, x\rangle, \langle j, x'\rangle\rangle : x - 1 =_X x' - 1 \text{ or } x = x' = 0; i, j \in \mathsf{N}\}$$

and a map $[f]: X \longrightarrow Y$ to the map coded by the recursive function

$$\langle i, x\rangle \mapsto \langle i + 1, x\rangle.$$

It is obvious that the function described above does not satisfy condition (ii), but that a careful search finds a better behaved function.

Suppose B is as in 4.1; let $P = \prod_{X \in \mathsf{H}_B} [[FX \to X] \to X]$. Recall that it is possible to define a structure of weakly initial F-algebra $\chi: FP \longrightarrow P$, see for instance [8].

Corollary 4.3 *Let* $F: \mathsf{H}_B \longrightarrow \mathsf{H}_B$ *be an internal functor and* $\varepsilon: FI \longrightarrow I$ *an initial F-algebra. Suppose the hypotheses of 4.1 are satisfied; then* $\chi: FP \longrightarrow P$ *is isomorphic to* ε.

We now apply this corollary in the particular case that B is $2 = 1 + 1$.

Corollary 4.4 *Let* $F: \mathsf{Per} \longrightarrow \mathsf{Per}$ *be a polynomial functor, i.e. built from $+$ and $\times$ with no parameters. Then*

$$\prod_{X \in \mathsf{Per}} [[FX \to X] \to X] \;\overset{\sim}{\longrightarrow}\; \prod_{X \in \mathsf{H}_2} [[FX \to X] \to X]$$

and it bears a structure of initial F-algebra.

Proof :   The functor F is clearly internal and it is easy to find a map tracking the values on maps which fixes codes of the identity. Moreover, for every PER X it is

$$\mathrm{dom}\,(FX) = F(\mathrm{dom}\,(X))$$

by construction, therefore every F-algebra $[a]: FX \longrightarrow X$ on X determines an algebra $[a]: F(\mathrm{dom}\,(X)) \longrightarrow \mathrm{dom}\,(X)$ —it depends on the code. It follows that an element $[s] \in \prod_{X \in \mathsf{Per}} [[FX \to X] \to X]$ is totally determined by its components on the subsets of N, hence the isomorphism with the product over H_2. Note now that 2 has decidable equality. Moreover, it is well-known that an initial F-algebra of terms is recursively presentable. One can thus choose a representation of the initial algebra consisting of only even numbers ensuring conditions (i) and (iii) of 4.1. Now apply 4.3 with $B = 2$. □

Another corollary shows that the product $\prod_{X \in \mathsf{Per}} [[A \to X] \to X]$ is isomorphic to A when A is in H_2.

Corollary 4.5 *Given any A in H_2, we have that*

$$A \xrightarrow{\sim} \prod_{X \in \mathsf{Per}} [[A \to X] \to X].$$

Proof : Consider the constant functor of value A. If we let $\phi_e \colon n \mapsto 2n \colon \mathsf{N} \longrightarrow \mathsf{N}$ as before, then $[e] \colon A \longrightarrow A^e$ is initial for the constant functor, and apply 4.3 with $B = 2$. □

The reader who has followed so far can easily imagine how to extend 4.4 to allow parameters from the subcategory H_2 and to algebraic types for signatures with more than one sort and more than one operation!

References

[1] S. Bainbridge, P. Freyd, A. Scedrov, and P. Scott. Functorial polymorphism. Theoretical Comput. Sci., 70:35–64, 1990.

[2] P. Freyd. Structural Polymorphism. Theoretical Comput. Sci., 1991. To appear.

[3] P. Freyd, P. Mulry, G. Rosolini, and D.S. Scott. Extensional PERs. In J. Mitchell, editor, Proc. 5th Symposium in Logic in Computer Science, pages 346–354, Philadelphia. I.E.E.E. Computer Society, 1990.

[4] J. Hyland. The effective topos. In A. Troelstra and D. van Dalen, editors, The L.E.J. Brouwer Centenary Symposium, pages 165–216. North Holland Publishing Company, 1982.

[5] J. Hyland, E. Robinson, and G. Rosolini. Algebraic types in PER models. In M. Main, A. Melton, M. Mislove, and D.Schmidt, editors, Mathematical Foundations of Programming Semantics, Lectures Notes in Computer Science vol. 442, pages 333–350. Springer-Verlag, 1990.

[6] W. Phoa. Effective domains and intrinsic structure. In J. Mitchell, editor, Proc. 5th Symposium in Logic in Computer Science, pages 366–377, Philadelphia. I.E.E.E. Computer Society, 1990.

[7] E. Robinson. How complete is PER? In A. Meyer, editor, Proc. 4th Symposium in Logic in Computer Science, pages 106–111, Asilomar. I.E.E.E. Computer Society, 1989.

[8] E. Robinson and G. Rosolini. Polymorphism, set theory and call by value. In J. Mitchell, editor, Proc. 5th Symposium in Logic in Computer Science, pages 12–18, Philadelphia. I.E.E.E. Computer Society, 1990.

[9] G. Rosolini. About modest sets. Int. J. Found. Comp. Sci., 1:341–353, 1990.

Simply typed and untyped Lambda Calculus revisited

Bart Jacobs *

Abstract

In [6] one finds a general method to describe various (typed) λ-calculi categorically. Here we give an elementary formulation in terms of indexed categories of the outcome of applying this method to the simply typed λ-calculus. It yields a categorical structure in which one can describe exponent types without assuming cartesian product types. Specializing to the "monoid" case where one has only one type yields a categorical description of the untyped λ-calculus.

In the literature there are two categorical notions for the untyped λ-calculus: one by Obtulowicz and one by Scott & Koymans. The notion we arrive at subsumes both of these; it can be seen as a mild generalization of the first one.

1. Introduction

The straightforward way to describe the simply typed λ-calculus (denoted here by λ1) categorically is in terms of cartesian closed categories (CCC's), see [10]. On the type theoretic side this caused some discomfort, because one commonly uses only exponent types without assuming cartesian product types — let alone unit (*i.e.* terminal) types. The typical reply from category theory is that one needs cartesian products in order to define exponents. Below we give a categorical description of exponent types without assuming cartesian product types. We do use cartesian products of contexts; these always exist by concatenation. Thus both sides can be satisfied by carefully distinguishing between cartesian products of types and cartesian products of contexts. We introduce an appropriate categorical structure for doing so.

In [6] one can find a general method to describe typed λ-calculi categorically. One of the main organizing principles used there is formulated below. It deserves the status of a slogan.

---

*During '91 - '92 at Department of Pure Mathematics, University of Cambridge, UK.

A context is an index for the category of types and terms derivable in that context.

This work can be read as the result of applying the slogan to the simply typed and untyped λ-calculus. It naturally leads us to the definition of a λ1-category in 3.1. In [6] these matters can be found in terms of so-called "constant comprehension categories". Here we reformulate them in terms of the more elementary indexed categories. We like to mention a few advantages of this approach.

- It is conceptually satisfying in the sense that it follows the above slogan.

- It allows a description of exponent types without assuming cartesian product types. The exponent types are described by *right* adjoints and the cartesian product types by *left* adjoints to so-called weakening functors (see lemmas 3.3 and 3.7).

- Non-extensional versions are obtained by simply replacing the above adjoints by semi-adjoints. This follows a general categorical understanding of non-extensionality, see [5].

- The untyped λ-calculus is covered by the "monoid" subcase where one has only one type. This follows Dana Scott's insistance that the untyped λ-calculus should be considered as a special case of the typed λ-calculus, *viz.* as a calculus with a single type. Again this is in line with a general categorical understanding of untyped as described by the monoid case.

The force of this last point is that in our approach the untyped λ-calculus can be treated in a corollary. The outcome appears to be (by lemma 4.4) essentially the same notion as introduced by Obtulowicz, see [11, 12]. There is a slight difference though, which concerns the fact that where we use an arbitrary cartesian category of contexts, Obtulowicz restricts himself to an algebraic theory. In the fourth section on the untyped λ-calculus one finds theorem 4.10 which yields an adjunction between set-theoretical λ-algebras (as described in [1]) and our categorical λ-algebras. The main ingredients of this theorem already occur in [2, constructions V and VI in 1.5]. Hence with respect to this fourth section, we only have a very moderate claim to originality. The best way to see it is probably as providing an *a posteriori* justification for the early definition of Obtulowicz.

It is worthwhile noticing that the abovementioned theorem 4.10 could not be obtained in the Scott-Koymans approach to the untyped λ-calculus (see example 4.8 (iii)): turning a CCC with a reflexive object first into a λ-algebra and then again into a category yields incomparable results. This is due to the use of the Karoubi-envelope: it introduces too many new objects and arrows, see [9, 1]. In the approach described here, this Karoubi-construction can be avoided altogether.

Finally we mention some general points. Since the categories we use are close to syntax we describe categorical structure "on-the-nose", *i.e.* up-to-equality and not up-to-isomorphism. For example, **CCC** denotes the category of categories with distinguished terminals, cartesian products and exponents; the morphisms in this category are functors which preserve this structure on-the-nose. Similarly an indexed category is really a functor (and not just a pseudo functor). Further, aspects of size will be ignored. In an adjoint situation the transpose operations will often be denoted by $(-)^{\vee}$ and $(-)^{\wedge}$.

2. Preliminaries

Let C be a non-empty set of "constants". The simply typed λ-calculus $\lambda 1(C)$ has as set of types the smallest collection $Type_C$ satisfying $C \subseteq Type_C$ and $\sigma, \tau \in Type_C \Rightarrow (\sigma \to \tau) \in Type_C$. Hence there are no type variables and no cartesian product types. Terms are then formed by the usual context projection rule and the rules for application and abstraction:

$$\frac{\Gamma \vdash M : \sigma \to \tau \quad \Gamma \vdash N : \sigma}{MN : \tau} \qquad \frac{\Gamma, x : \sigma \vdash M : \tau}{\Gamma \vdash \lambda x : \sigma. M : \sigma \to \tau}$$

We mention the (β)- and (η)-conversions

$$(\lambda x : \sigma. M)N = M[x := N] \qquad \lambda y : \sigma. My = M,$$

where in the latter case the variable y is supposed not to be free in M.

An *indexed category* is a functor of the form $\Psi : \mathbf{B}^{op} \to \mathbf{Cat}$. The category $\mathbf{B}$ is then called the *base*; a category ΨA (for $A \in \mathbf{B}$) is called a *fibre* (above A). For every morphism $u : A \to B$ in $\mathbf{B}$ one has a so-called *reindexing* or *subsitution* functor $\Psi u : \Psi B \to \Psi A$; notice the reverse direction. If no confusion arises one often writes u^* for Ψu. An elementary example is given by the functor $I \mapsto \mathbf{C}^I$, which maps a set I to the category of I-indexed collections of objects and arrows of an arbitrary category $\mathbf{C}$. A *morphism of indexed categories* from $\Phi : \mathbf{A}^{op} \to \mathbf{Cat}$ to $\Psi : \mathbf{B}^{op} \to \mathbf{Cat}$ consists of a pair (H, α), where H is a functor $\mathbf{A} \to \mathbf{B}$ and α is a natural transformation $\Phi \overset{\cdot}{\to} \Psi H^{op}$. In this way one obtains a category **IC**.

The categorical basis for the whole of this work is what we like to call a *CT-structure* — where 'C' stands for 'context' and 'T' for 'type'. It is given by an arbitrary cartesian category $\mathbf{B}$ of "contexts" (*i.e.* a category with distinguished finite products) together with a collection of "types"

$$T \subseteq Obj\mathbf{B}.$$

The inclusion of types into contexts can be understood as an identification of types with the corresponding singleton contexts. A *morphism of CT-structures* from $(\mathbf{A}, S)$ to $(\mathbf{B}, T)$ consists of a functor $H: \mathbf{A} \to \mathbf{B}$ which preserves the distinguished finite products on-the-nose and maps types to types, *i.e.* satisfies $H[S] \subseteq T$. The resulting category will be denoted by **CT**.
We mention some basic matters concerning CT-structures.

(1) A CT-structure $(\mathbf{B}, T)$ will be called *non-trivial* if there is at least one $X \in T$ such that the hom-set $\mathbf{B}(t, X)$ is non-empty — where t is the terminal in $\mathbf{B}$. This gives a non-degeneracy condition.

(2) A *monoid* CT-structure is one of the form $(\mathbf{A}, \{\Omega\})$. It describes a type-free setting.

(3) For a CT-structure $(\mathbf{B}, T)$, let $\overline{T} \subseteq Obj\,\mathbf{B}$ be the smallest collection satisfying

$$
\begin{array}{rcl}
T & \subseteq & \overline{T} \\
t & \in & \overline{T} \\
A, B \in \overline{T} & \Rightarrow & (A \times B) \in \overline{T}
\end{array}
$$

i.e. $\overline{T}$ is obtained from T by adding finite products. One can then form the full subcategory of $\mathbf{B}$ determined by $\overline{T}$; it will also be denoted by $\overline{T}$. A CT-structure $(\mathbf{B}, T)$ will now be called *full* if there is an equivalence of categories $\overline{T} \simeq \mathbf{B}$. Then every context is a finite product of types.

(4) For a CT-structure $(\mathbf{B}, T)$, one can define a so-called *constant* indexed category of the form $\mathcal{K}_T: \mathbf{B}^{op} \to \mathbf{Cat}$ as follows. For $A \in \mathbf{B}$ the fibre category $\mathcal{K}_T(A)$ has elements $X \in T$ as objects. A morphism $f: X \to Y$ in $\mathcal{K}_T(A)$ is a map $f: A \times X \to Y$ in $\mathbf{B}$. Composition of $f: X \to Y$ and $g: Y \to Z$ in $\mathcal{K}_T(A)$ — *i.e.* of $f: A \times X \to Y$ and $g: A \times Y \to Z$ in $\mathbf{B}$ — is given by $g \bullet f = g \circ \langle \pi, f \rangle : A \times X \to A \times Y \to Z$. The identity map on X is $\pi': A \times X \to X$. Thus one obtains a category. For $u: B \to A$ in $\mathbf{B}$ there is a reindexing functor $u^* = \mathcal{K}_T(u) : \mathcal{K}_T(A) \to \mathcal{K}_T(B)$ given by $X \mapsto X$ and $f \mapsto f \circ u \times id$.
This indexed category $\mathcal{K}_T: \mathbf{B}^{op} \to \mathbf{Cat}$ is called "constant" because the objects in the fibres $\mathcal{K}_T(A)$ do not depend on the index A.
Recall that $T \subseteq Obj\,\mathbf{B}$ is a parameter in this construction. In two extreme cases, we adapt the notation a bit: if $T = Obj\,\mathbf{B}$ we write $\mathcal{K}_\mathbf{B}$ for $\mathcal{K}_{Obj\,\mathbf{B}}$ and if T is a singleton, say $T = \{\Omega\}$, we write $\mathcal{K}_\Omega$ for $\mathcal{K}_{\{\Omega\}}$.
This construction extends to a functor $\mathbf{CT} \to \mathbf{IC}$: for a morphism $H : (\mathbf{A}, S) \to (\mathbf{B}, T)$ of CT-structures one obtains a morphism $(H, \widetilde{H}): \mathcal{K}_S \to \mathcal{K}_T$. The natural transformation $\widetilde{H}: \mathcal{K}_S \xrightarrow{\cdot} \mathcal{K}_T H^{op}$ has component functors $(\widetilde{H})_A : \mathcal{K}_S(A) \to \mathcal{K}_T(HA)$ given by $X \mapsto HX$ and $f \mapsto Hf$.

(5) There is an elementary form of quantification for indexed categories associated with a CT-structure $(\mathbf{B}, T)$. We say that an indexed category $\Psi : \mathbf{B}^{op} \to \mathbf{Cat}$ has T-products (resp. T-sums) if both

- for each $A \in \mathbf{B}$ and $X \in T$, the reindexing functor

$$\pi^*_{A,X} \; : \; \Psi(A) \;\; \to \;\; \Psi(A \times X)$$

has a right adjoint $\Pi_{(A,X)}$ (resp. a left adjoint $\Sigma_{(A,X)}$);

- the "Beck-Chevalley" condition holds, i.e. for each $u : B \to A$ in $\mathbf{B}$ and $X \in T$, the pair $(u^*, (u \times id)^*)$ is a morphism of adjunctions from $\pi^*_{A,X} \dashv \Pi_{(A,X)}$ to $\pi^*_{B,X} \dashv \Pi_{(B,X)}$ in

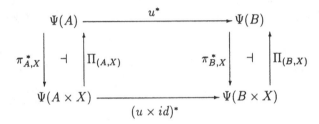

(resp. the pair $((u \times id)^*, u^*)$ is a morphism of adjunctions from $\Sigma_{(A,X)} \dashv \pi^*_{A,X}$ to $\Sigma_{(B,X)} \dashv \pi^*_{B,X}$).

The corresponding notion of morphism is as follows. Let $\Phi : \mathbf{A}^{op} \to \mathbf{Cat}$ have S-products (resp. sums) and $\Psi : \mathbf{B}^{op} \to \mathbf{Cat}$ have T-products (resp. sums). A morphism (H, α) of indexed categories from Φ to Ψ is said to preserve products (resp. sums) if

- $H : (\mathbf{A}, S) \to (\mathbf{B}, T)$ is a morphism of CT-structures;

- for every $A \in \mathbf{A}$ and $X \in S$, the pair $\langle \alpha_{A \times X} : \Phi(A \times X) \to \Psi(HA \times HX), \; \alpha_A : \Phi(A) \to \Psi(HA) \rangle$ is a map of adjunctions from $\pi^*_{A,X} \dashv \Pi_{(A,X)}$ to $\pi^*_{HA,HX} \dashv \Pi_{(HA,HX)}$
  (resp. $\langle \alpha_A, \alpha_{A \times X} \rangle$ from $\Sigma_{(A,X)} \dashv \pi^*_{A,X}$ to $\Sigma_{HA,HX} \dashv \pi^*_{HA,HX}$).

(6) In order to handle non-extensionality one uses the "semi-adjunctions" from [5]. To obtain suitable notions of semi-product/sum and of morphisms of these, one simply replaces in (5) adjunction by semi-adjunction and map of adjunction by map of semi-adjunction.

These non-extensional concepts will only be used in the fourth section on the untyped λ-calculus.

Convention. In case the above collection T consists of a singleton, say $T = \{\Omega\}$, we speak about Ω-products instead of $\{\Omega\}$-products.

We often loosely say that a constant indexed category $\mathcal{K}_T : \mathbf{B}^{op} \to \mathbf{Cat}$ has products/sums when we mean T-products/sums.

One standardly calls right adjoints to the "weakening" functors π^* *products*; left adjoints are called *sums*. In the constant case (without type dependency) the Π becomes $\to$ and the Σ becomes $\times$. This will be shown in the next section. The Beck-Chevalley requirement guarantees the proper distribution of substitution over Π's and Σ's and over the associated operations.

3.   Simply typed λ-calculus

We start with a categorical structure which is appropriate for the simply typed λ-calculus. It will be investigated in the rest of this section.

Definition 3.1 (i) *A λ1-category is a non-trivial CT-structure* $(\mathbf{B}, T)$ *such that the corresponding constant indexed category* $\mathcal{K}_T : \mathbf{B}^{op} \to \mathbf{Cat}$ *has products.*

(ii) *A morphism of λ1-categories from* $(\mathbf{A}, S)$ *to* $(\mathbf{B}, T)$ *is given by a morphism* $H : (\mathbf{A}, S) \to (\mathbf{B}, T)$ *of CT-structures such that the corresponding morphism of indexed categories* $(H, \widetilde{H}) : \mathcal{K}_S \to \mathcal{K}_T$ *preserves products.*

Before submitting the notion of a λ1-category to a close analysis, we present a term model example (*i.e.* a generalized Lindenbaum - Tarski construction); it is particularly suitable to convey the type theoretic intuition.

Example 3.2 Let's consider the calculus $\lambda1(C)$ for a given non-empty set of constants C. A base category $\mathbf{C}$ is formed with contexts $\Gamma = x_1 : \sigma_1 \ldots x_n : \sigma_n$ as objects. A morphism $\Delta \to \Gamma$ in $\mathbf{C}$ — with Γ as before — consists of an n-tuple of $\beta\eta$-equivalence classes of terms $\langle [M_1], \ldots, [M_n] \rangle$ where $\Delta \vdash M_i : \sigma_i$. Composition is done by substitution: if $\langle [N_1], \ldots, [N_m] \rangle : \Gamma \to \Theta$, then $\langle [N_1], \ldots, [N_m] \rangle \circ \langle [M_1], \ldots, [M_n] \rangle = \langle [L_1], \ldots, [L_m] \rangle$ with $L_i = N_i[\vec{x} := \vec{M}]$. The identity on Γ is $\langle [x_1], \ldots, [x_n] \rangle$.

The category $\mathbf{C}$ has finite products: the empty context is terminal and concatenation of contexts yields cartesian products.

The collection of types in $\lambda1(C)$ is denoted here by $T = Type_C$. By identifying a type σ with the context $x : \sigma$, one has $T \subseteq Obj\,\mathbf{C}$. Notice that the resulting CT-structure $(\mathbf{C}, T)$ is non-trivial because the set of constants is non-empty. As a result we have an indexed category $\mathcal{K}_T : \mathbf{C}^{op} \to \mathbf{Cat}$. The fibre $\mathcal{K}_T(\Gamma)$ above context Γ has types and terms in context Γ as objects and morphisms: objects are types $\sigma \in T$ and morphisms $\sigma \to \tau$ in

$\mathcal{K}_T(\Gamma)$ are terms $[P(y)]$ with $\Gamma, y : \sigma \vdash P : \tau$. Substitution functors between the fibres are obtained by substitution: if $\langle [M_1], \ldots, [M_n] \rangle : \Delta \to \Gamma$ in $\mathbf{C}$ as above, then $\langle [M_1], \ldots, [M_n] \rangle^* : \mathcal{K}_T(\Gamma) \to \mathcal{K}_T(\Delta)$ is defined by $\sigma \mapsto \sigma$ and $[P] \mapsto [P[\vec{x} := \vec{M}]]$. All this follows from the construction of constant indexed categories described in (4) in the preliminary section.

Finally we have to show that $\mathcal{K}_T : \mathbf{C}^{op} \to \mathbf{Cat}$ has products. For a context Γ and a type σ we have to define a functor $\Pi_{(\Gamma, \sigma)} : \mathcal{K}_T(\Gamma, y : \sigma) \to \mathcal{K}_T(\Gamma)$. This is done on objects by $\tau \mapsto \sigma \to \tau$. By abstraction and application one easily obtains an isomorphism

$$\mathcal{K}_T(\Gamma, y : \sigma) \Big(\pi^*_{\Gamma, y : \sigma}(\rho), \ \tau \Big) \ \cong \ \mathcal{K}_T(\Gamma) \Big(\rho, \ \Pi_{(\Gamma, \sigma)}(\tau) \Big).$$

Remaining verifications (including the Beck-Chevalley condition) are left to the reader.

The next result shows that products for a constant indexed category can be described simply in terms of exponents. The result (a) $\Rightarrow$ (b) follows from lemma 5.4 in [8]. The formalism used there is somewhat different; therefore we give the details here.

Lemma 3.3 *Let $(\mathbf{B}, T)$ be a non-trivial CT-structure. The following two statements are equivalent.*

(a) *$(\mathbf{B}, T)$ is a λ1-category.*

(b) *The collection T is closed under exponents in the following sense. For "types" $X, Y \in T$ there is an exponent type $X \Rightarrow Y \in T$ together with a map $ev : (X \Rightarrow Y) \times X \to Y$ such that for each $A \in \mathbf{B}$ and $f : A \times X \to Y$ in $\mathbf{B}$ there is a unique map $\Lambda(f) : A \to X \Rightarrow Y$ with $ev \circ \Lambda(f) \times id = f$.*

Proof : (b) $\Rightarrow$ (a) For $A \in \mathbf{B}$ and $X \in T$ one defines $\Pi_{(A,X)} : \mathcal{K}_\mathbf{B}(A \times X) \to \mathcal{K}_\mathbf{B}(A)$ by $Y \mapsto X \Rightarrow Y$. Then indeed,

$$
\begin{aligned}
\mathcal{K}_T(A \times X) \Big(\pi^*_{A,X}(Z), \ Y \Big) &= \mathcal{K}_T(A \times X) \Big(Z, \ Y \Big) \\
&= \mathbf{B} \Big((A \times X) \times Z, \ Y \Big) \\
&\cong \mathbf{B} \Big((A \times Z) \times X, \ Y \Big) \\
&\cong \mathbf{B} \Big(A \times Z, \ X \Rightarrow Y \Big) \\
&= \mathcal{K}_T(A) \Big(Z, \ X \Rightarrow Y \Big) \\
&= \mathcal{K}_T(A) \Big(Z, \ \Pi_{(A,X)}(Y) \Big).
\end{aligned}
$$

Hence one has an adjunction $\pi^*_{A,X} \dashv \Pi_{(A,X)}$.

(a) $\Rightarrow$ (b) For types $X, Y \in T$ one has $Y \in \mathcal{K}_T(t \times X)$ and thus we define

$$X \Rightarrow Y \; = \; \Pi_{(t,X)}(Y) \; \in \; T.$$

We notice that for an object $A \in \mathbf{B}$ one has $!_A : A \to t$ in $\mathbf{B}$ and thus

$$
\begin{aligned}
X \Rightarrow Y \; &= \; !_A^*(X \Rightarrow Y) \; = \; !_A^*\big(\Pi_{(t,X)}(Y)\big) \\
&= \; \Pi_{(A,X)}\big((!_A \times id)^*(Y)\big) \qquad \text{by Beck-Chevalley} \\
&= \; \Pi_{(A,X)}(Y).
\end{aligned}
$$

The unit and counit of the adjunction $\pi_{A,X}^* \dashv \Pi_{(A,X)}$ are maps $\eta_Z^{(A,X)} : A \times Z \to \Pi_{(A,X)}(Z)$ and $\varepsilon_Y^{(A,X)} : (A \times X) \times \Pi_{(A,X)}(Y) \to Y$ in $\mathbf{B}$. Hence in $\mathcal{K}_T(t \times X)$ one has $\varepsilon_Y^{(t,X)} : (t \times X) \times \Pi_{(t,X)}(Y) \to Y$. We define

$$ev_{X,Y} \; = \; \varepsilon_Y^{(t,X)} \; \circ \; \langle\langle !, \pi' \rangle, \pi \rangle \; : \; (X \Rightarrow Y) \times X \; \to \; Y.$$

The definition of abstraction is a bit tricky. Since $(\mathbf{B}, T)$ is by assumption non-trivial, we may assume an object $Z \in T$ with an arrow $z_0 : t \to Z$. For a map $f : A \times X \to Y$ in $\mathbf{B}$ (with $A \in \mathbf{B}$ and $X, Y \in T$) one has $f \circ \pi : (A \times X) \times Z \to Y$ in $\mathbf{B}$ and thus $f \circ \pi : \pi_{A,X}^*(Z) \to Y$ in $\mathcal{K}_T(A \times X)$. Taking the transpose accross the adjunction $\pi_{A,X}^* \dashv \Pi_{(A,X)}$ yields $(f \circ \pi)^{\vee} : Z \to \Pi_{(A,X)}(Y)$ in $\mathcal{K}_T(A)$. As noticed above, $X \Rightarrow Y = \Pi_{(A,X)}(Y)$ and thus $(f \circ \pi)^{\vee} : A \times Z \to X \Rightarrow Y$ in $\mathbf{B}$. Hence we take

$$\Lambda(f) \; = \; (f \circ \pi)^{\vee} \; \circ \; \langle id, z_0 \circ ! \rangle \; : \; A \; \to \; A \times Z \; \to \; X \Rightarrow Y.$$

(The auxiliary type Z is first used to introduce a dummy variable which is later removed by substituting z_0.)

The validity of the categorical (β)- and (η)-equations follows from computations in the fibres. We shall do (β).

$$
\begin{aligned}
ev &\circ \Lambda(f) \times id \\
&= \; \varepsilon_Y^{(t,X)} \; \circ \; \langle\langle !, \pi' \rangle, \pi \rangle \; \circ \; \Lambda(f) \times id \\
&= \; \varepsilon_Y^{(t,X)} \; \circ \; (!_A \times id) \times id \; \circ \; \langle id, \Lambda(f) \circ \pi \rangle \\
&= \; (!_A \times id)^*(\varepsilon_Y^{(t,X)}) \; \circ \; \langle id, \Lambda(f) \circ \pi \rangle \\
&= \; \varepsilon_Y^{(A,X)} \; \circ \; \langle id, \Lambda(f) \circ \pi \rangle \\
&= \; \varepsilon_Y^{(A,X)} \; \circ \; \langle id, \Pi_{(A,X)}(f \circ \pi) \; \circ \; \langle \pi, \eta_Z^{(A,X)} \rangle \; \circ \; \langle id, z_0 \circ ! \rangle \circ \pi \rangle \\
&= \; \varepsilon_Y^{(A,X)} \; \circ \; \langle \pi, \Pi_{(A,X)}(f \circ \pi) \circ \pi \times id \rangle \; \circ \; \langle \pi, \eta_Z^{(A,X)} \circ \pi \times id \rangle \; \circ \; \langle id, z_0 \circ ! \rangle \\
&= \; \varepsilon_Y^{(A,X)} \; \bullet \; \pi_{A,X}^* \Pi_{(A,X)}(f \circ \pi) \; \bullet \; \pi_{A,X}^*(\eta_Z^{(A,X)}) \; \circ \; \langle id, z_0 \circ ! \rangle \\
&= \; (f \circ \pi) \; \bullet \; \varepsilon_{\pi_{A,X}^*(Z)}^{(A,X)} \; \bullet \; \pi_{A,X}^*(\eta_Z^{(A,X)}) \; \circ \; \langle id, z_0 \circ ! \rangle \\
&= \; (f \circ \pi) \; \bullet \; id \; \circ \; \langle id, z_0 \circ ! \rangle \\
&= \; f \; \circ \; \pi \; \circ \; \langle id, z_0 \circ ! \rangle \\
&= \; f. \qquad \square
\end{aligned}
$$

Remark 3.4 The previous lemma gives a reasonably clear picture of what a λ1-category is about. It gives rise to some further points.

(i) One obtains an alternative characterization of CCC's: a category **B** with finite products is a CCC if and only if $\mathcal{K}_\mathbf{B}$ is a λ1-category.

(ii) If $(\mathbf{B}, T)$ is a λ1-category, then T is enriched over **B**.

(iii) We describe the often used set theoretic modelling (see [4]) of a simply typed λ-calculus $\lambda1(C)$ in terms of λ1-categories. Suppose that for each type constant $c \in C$ one has a set D_c. Put $D_{\sigma \to \tau} = D_\tau^{D_\sigma}$, the set of functions from D_σ to D_τ. One forms a base category **B** with sequences of types as objects. A morphism $\langle \sigma_1, \ldots, \sigma_n \rangle \to \langle \tau_1, \ldots, \tau_m \rangle$ in **B** is a sequence $\langle f_1, \ldots, f_m \rangle$ of functions $f_i : D_{\sigma_1} \times \cdots \times D_{\sigma_n} \to D_{\tau_i}$. Composition is done in the obvious way. The empty sequence is terminal and concatenation yields cartesian products. As types take $Type_C \subseteq Obj\,\mathbf{B}$. It obviously yields a λ1-category with evaluation $\langle \sigma \to \tau, \sigma \rangle \to \tau$ defined by $(\varphi, a) \mapsto \varphi(a)$ and for $f : \langle \vec{\mu}, \sigma \rangle \to \tau$ the abstraction map $\Lambda(f) : \vec{\mu} \to (\sigma \to \tau)$ described by $\Lambda(f)(\vec{x}) = \lambda y \in D_\sigma . f(\vec{x}, y)$.

The above elementary description of λ1-categories will be extended to morphisms of these.

Lemma 3.5 *Let* $(\mathbf{A}, S)$ *and* $(\mathbf{B}, T)$ *be λ1-categories and* $H : (\mathbf{A}, S) \to (\mathbf{B}, T)$ *be a morphism of CT-structures. One has that H is a morphism of λ1-categories iff H preserves the exponents in S (i.e. for $X, Y \in S$ one has $H(X \Rightarrow Y) = HX \Rightarrow HY$ and $H(ev_{X,Y}) = ev_{HX,HY}$).*

Proof : Straightforward. □

Corollary 3.6 *The assignment* $\mathbf{B} \mapsto \mathcal{K}_\mathbf{B}$ *extends to a functor* **CCC** → **λ1-cat**. □

The next result is similar to lemma 3.3. It shows that — quite independently from exponent types — cartesian product types are described by left adjoints to weakening functors.

Lemma 3.7 *Let* $(\mathbf{B}, T)$ *be a CT-structure. Then $\mathcal{K}_T$ has sums if and only if T is closed under cartesian products.*

9

Proof : The if-part is easy: for $A \in \mathbf{B}$ and $X \in T$ define $\Sigma_{(A,X)} : \mathcal{K}_T(A \times X) \to \mathcal{K}_T(A)$ by $Y \mapsto X \times Y$. Then

$$
\begin{aligned}
\mathcal{K}_T(A)\big(\Sigma_{(A,X)}(Y),\ Z\big) &= \mathbf{B}\big(A \times (X \times Y),\ Z\big) \\
&\cong \mathbf{B}\big((A \times X) \times Y,\ Z\big) \\
&= \mathcal{K}_T(A \times X)\big(Y,\ \pi^*_{A,X}(Z)\big)
\end{aligned}
$$

The only-if-part requires some computations. For $X, Y \in T$, put $X \& Y = \Sigma_{(t,X)}(Y)$. It suffices to show that the unit $\eta = \eta_Y^{(t,X)} : (t \times X) \times Y \to X \& Y$ in $\mathbf{B}$ is an isomorphism. Therefore we define two new projections $p : X \& Y \to X$ and $p' : X \& Y \to Y$. The counit $\varepsilon = \varepsilon_Y^{(t,X)} : t \times (X \& Y) \to Y$ yields the second projection. The first one is obtained by transposing the obvious arrow $(t \times X) \times Y \to X$. Then

$$
\begin{aligned}
p' \circ \eta &= \varepsilon \circ \langle !, id \rangle \circ \eta \\
&= \varepsilon \circ \langle !, \eta \rangle \\
&= \pi' \qquad \text{by a triangular identity;} \\
p \circ \eta &= (\pi' \circ \pi)^\wedge \circ \langle !, id \rangle \circ \eta \\
&= (\pi' \circ \pi)^\wedge \circ \pi \times id \circ \langle \pi, \eta \rangle \\
&= \pi^*\big((\pi' \circ \pi)^\wedge\big) \bullet \eta \\
&= (\pi' \circ \pi)^{\wedge\vee} \\
&= \pi' \circ \pi.
\end{aligned}
$$

Let's put $\alpha = \langle\langle \pi, p \circ \pi' \rangle, p' \circ \pi' \rangle : t \times (X \& Y) \to (t \times X) \times Y$. In order to show that $\eta \circ \langle\langle !, p \rangle, p' \rangle = id$, it suffices to establish that $\eta \circ \alpha = \pi' : t \times (X \& Y) \to X \& Y$, i.e. that $\eta \circ \alpha$ is the identity in the fibre. The latter follows from $(\eta \circ \alpha)^\vee = \eta$. Well,

$$
\begin{aligned}
(\eta \circ \alpha)^\vee &= \pi^*(\eta \circ \alpha) \bullet \eta \\
&= \eta \circ \alpha \circ \pi \times id \circ \langle \pi, \eta \rangle \\
&= \eta \circ \langle\langle \pi \circ \pi, p \circ \eta \rangle, p' \circ \eta \rangle \\
&= \eta \circ \langle\langle \pi \circ \pi, \pi' \circ \pi \rangle, \pi' \rangle \qquad \text{as proved above} \\
&= \eta. \qquad \square
\end{aligned}
$$

Remark 3.8 Given the functor $\mathbf{CCC} \to \lambda\mathbf{1}\text{-cat}$ in 3.6, one naturally asks for an operation working in the reverse direction. Below, two constructions are described; the second one is based on a remark of Peter Johnstone.

(i) Let $(\mathbf{B}, T)$ be a $\lambda\mathbf{1}$-category. By a "brute force" method, one takes $\hat{\mathbf{B}}$ to be the free CCC generated by $\mathbf{B}$, see [10, I.4]; it comes equipped with an inclusion functor $\mathbf{B} \to \hat{\mathbf{B}}$ which preserves the exponents which exist in T up-to-isomorphism (by a uniqueness argument). Since morphisms of $\lambda\mathbf{1}$-categories are required to preserve such exponents on-the-nose (see lemma 3.5), one does not have an adjoint situation. Thus, one might want to adapt the definition of morphisms of λ1-categories in such a way that one does obtain an adjointness here.

(ii) Assume $(\mathbf{B}, T)$ is a λ1-category again. In the preliminary section we described the category $\overline{T}$ obtained by closing T under finite products. It is a CCC since one can put for the newly added objects

$$A \Rightarrow t \;=\; t$$
$$t \Rightarrow A \;=\; A$$
$$C \Rightarrow (A \times B) \;=\; (C \Rightarrow A) \times (C \Rightarrow B)$$
$$(A \times B) \Rightarrow C \;=\; A \Rightarrow (B \Rightarrow C).$$

Like before one does not have an adjoint situation, this time due to the fact that the inclusion $\overline{T} \to \mathbf{B}$ needs not map $\overline{T}$ into T.

After the next example one finds a further discussion of this construction.

Example 3.9 Let D be a cpo. Non-empty closed subsets $I \subseteq D$ (*wrt.* the Scott topology) can be described by the following three requirements. (i) $\bot \in I$; (ii) $x \leq y \in I \;\Rightarrow\; x \in I$; (iii) directed $X \subseteq I \;\Rightarrow\; \bigsqcup X \in I$. We form a base category $\mathbf{B}$ with finite sequences $\langle I_1, \ldots, I_n \rangle$ (for some $n \in \mathbb{N}$) of such non-empty closed subsets as objects. A morphism $\langle I_1, \ldots, I_n \rangle \to \langle J_1, \ldots, J_m \rangle$ in $\mathbf{B}$ is given by a sequence $\langle f_1, \ldots, f_m \rangle$ of continuous functions $f_i : D^n \to D$ satisfying $f_i[\vec{I}] \subseteq J_i$. The empty sequence in $\mathbf{B}$ is then terminal and concatenation yields cartesian products.

Now let's assume that D is isomorphic to its own space of continuous functions $[D \to D]$, via maps $F : D \to [D \to D]$ and $G : [D \to D] \to D$ satisfying $F \circ G = id$ and $G \circ F = id$. As usal, we write $x \cdot y$ for $F(x)(y)$ and $\lambda x.-$ for $G(\lambda x.-)$. An example of such a cpo is Dana Scott's D_∞. In a standard way one forms an exponent of $I, J \subseteq D$ by $I \Rightarrow J = \{x \in D \mid \forall y \in I.\; x \cdot y \in J\}$. One easily verifies that $I \Rightarrow J$ is non-empty and closed in case I, J are.

Let $T \subseteq Obj\,\mathbf{B}$ be the collection of non-empty closed subsets (*i.e.* the sequences consisting of one set). One obtains a full CT-structure $(\mathbf{B}, T)$. We claim that it is a λ1-category; using lemma 3.3 this is readily established: one has $ev : \langle I \Rightarrow J, I \rangle \to J$ by $ev = \lambda xy.\; x \cdot y$. For $f : \langle \vec{K}, I \rangle \to J$ one takes $\Lambda(f)(\vec{x}) = \lambda y.\; f(\vec{x}, y)$.

Having seen the the notion of a λ1-category, Peter Johnstone suggested the construction in 3.8 (ii) above and argued that one is thus still working with a CCC but one considers only some of its exponents. Our position is that the λ1-category structure is the more fundamental one: for example, in the above example 3.9 the construction from 3.8 (ii) yields the following exponent in $\overline{T} = \mathbf{B}$.

$$\langle I_1, I_2 \rangle \Rightarrow \langle J_1, J_2, J_3 \rangle \;\; = \;\; \langle I_1 \Rightarrow (I_2 \Rightarrow J_1), I_1 \Rightarrow (I_2 \Rightarrow J_2), I_1 \Rightarrow (I_2 \Rightarrow J_3) \rangle.$$

We think that it is not obvious that this category $\mathbf{B}$ is a CCC (and that exponents are given in such a way) unless one knows the underlying λ1-category structure. In a similar way one has that the category of contexts in the term model example 3.2 is a CCC; there is an exponent of contexts which is obtained from the exponent on types.

The interpretation to be described next proceeds in a completely standard way. It forms a preparation for the subsequent theorem — which is as one would expect.

Suppose $(\mathbf{B}, T)$ is a λ1-algebra. We use $\Rightarrow$, ev and $\Lambda(-)$ as described in lemma 3.3. For a non-empty collection of constants C together with a function $C \to T$ one can write out an interpretation of the calculus λ1(C) in $(\mathbf{B}, T)$. This is done as follows. Let's denote the function $C \to T$ by $[\![-]\!]$. It can be extended to all types by putting

$$[\![\sigma \to \tau]\!] \;\; = \;\; [\![\sigma]\!] \Rightarrow [\![\tau]\!].$$

Hence types are interpreted as elements of T and thus as objects of the fibres of $\mathcal{K}_T$. For a context $\Gamma = x_1 \colon \sigma_1, \ldots, x_n \colon \sigma_n$ we put

$$[\![\Gamma]\!] \;\; = \;\; [\![\sigma_1]\!] \times \cdots \times [\![\sigma_n]\!]$$

A certain order is fixed implicitly. The empty context will be interpreted as the terminal object. A term $\Gamma \vdash M : \sigma$ will be interpreted as an arrow $[\![M]\!]_\Gamma : [\![\Gamma]\!] \to [\![\sigma]\!]$. This is done in three steps.

- Variables. For $\Gamma = x_1 \colon \sigma_1, \ldots, x_n \colon \sigma_n$ one uses appropriate projections

$$[\![x_i]\!]_\Gamma \;\; = \;\; \pi_i \; : \; [\![\sigma_1]\!] \times \cdots \times [\![\sigma_n]\!] \;\; \to \;\; [\![\sigma_i]\!].$$

- Application. For terms $\Gamma \vdash M : \sigma \to \tau$ and $\Gamma \vdash N : \sigma$ one has $[\![M]\!]_\Gamma : [\![\Gamma]\!] \to [\![\sigma]\!] \Rightarrow [\![\tau]\!]$ and $[\![N]\!]_\Gamma : [\![\Gamma]\!] \to [\![\sigma]\!]$. Hence we put

$$[\![MN]\!]_\Gamma \;\; = \;\; ev \circ \langle [\![M]\!]_\Gamma, [\![N]\!]_\Gamma \rangle \; : \; [\![\Gamma]\!] \;\; \to \;\; [\![\tau]\!].$$

- Abstraction. For $\Gamma, x \colon \sigma \vdash M : \tau$ one has $[\![M]\!]_{\Gamma, x \colon \sigma} : [\![\Gamma]\!] \times [\![\sigma]\!] \to [\![\tau]\!]$. Thus one takes

$$[\![\lambda x \colon \sigma . M]\!]_\Gamma \;\; = \;\; \Lambda([\![M]\!]_{\Gamma, x \colon \sigma}) \; : \; [\![\Gamma]\!] \;\; \to \;\; [\![\sigma]\!] \Rightarrow [\![\tau]\!].$$

Notice that a closed term $M:\sigma$ becomes an arrow $[\![\,M\,]\!]: t \to [\![\,\sigma\,]\!]$.
In order to show that everything works well, one needs the following weakening and substitution lemmas.

(a) For $\Gamma \vdash M:\sigma$ with $y:\tau \notin \Gamma$ one has

$$[\![\,M\,]\!]_{\Gamma,y:\tau} = [\![\,M\,]\!]_{\Gamma} \circ \pi.$$

(b) For $\Gamma, y:\tau \vdash M:\sigma$ and $\Gamma \vdash N:\tau$ one has

$$[\![\,M[x := N]\,]\!]_{\Gamma} = [\![\,M\,]\!]_{\Gamma,y:\tau} \circ \langle id, [\![\,N\,]\!]_{\Gamma}\rangle.$$

Proofs are by induction on the structure of M. Finally we come to the validity of the (β)- and (η)-conversions.

$$
\begin{aligned}
[\![\,(\lambda x:\sigma.M)N\,]\!]_{\Gamma} &= ev \circ \langle [\![\,\lambda x:\sigma.M\,]\!]_{\Gamma}, [\![\,N\,]\!]_{\Gamma}\rangle \\
&= ev \circ \langle \Lambda([\![\,M\,]\!]_{\Gamma,x:\sigma}), [\![\,N\,]\!]_{\Gamma}\rangle \\
&= ev \circ \Lambda([\![\,M\,]\!]_{\Gamma,x:\sigma}) \times id \circ \langle id, [\![\,N\,]\!]_{\Gamma}\rangle \\
&= [\![\,M\,]\!]_{\Gamma,x:\sigma} \circ \langle id, [\![\,N\,]\!]_{\Gamma}\rangle \\
&= [\![\,M[x := N]\,]\!]_{\Gamma} \quad \text{by the above substitution result.} \\
[\![\,\lambda x:\sigma.Mx\,]\!] &= \Lambda([\![\,Mx\,]\!]_{\Gamma,x:\sigma}) \\
&= \Lambda(ev \circ \langle [\![\,M\,]\!]_{\Gamma,x:\sigma}, [\![\,x\,]\!]_{\Gamma,x:\sigma}\rangle) \\
&= \Lambda(ev \circ \langle [\![\,M\,]\!]_{\Gamma} \circ \pi, \pi'\rangle) \quad \text{by the above weakening result} \\
&= \Lambda(ev \circ [\![\,M\,]\!]_{\Gamma} \times id) \\
&= [\![\,M\,]\!]_{\Gamma}.
\end{aligned}
$$

Theorem 3.10 *Let* $\mathbf{Sets}_{\neq\emptyset}$ *be the category of non-empty sets. There is a forgetful functor* $\lambda\mathbf{1\text{-}cat} \to \mathbf{Sets}_{\neq\emptyset}$ *given by* $(\mathbf{B},T) \mapsto T$; *it maps a* $\lambda 1$-*category to its collection of types.*
The assignment $C \mapsto$ [*the term model of* $\lambda 1(C)$] *from example 3.2 yields a left adjoint. Hence the term model of* $\lambda 1(C)$ *is the free* $\lambda 1$-*category generated by* C.

Proof : We have to establish a bijective correpondence

$$
\frac{C \longrightarrow T \quad \text{in } \mathbf{Sets}_{\neq\emptyset}}{[\text{term model of } \lambda 1(C)] \longrightarrow (\mathbf{B},T) \quad \text{in } \lambda\mathbf{1\text{-}cat}}
$$

The passage downwards is essentially given by the interpretation described above. The substitution lemma mentioned there yields the required functoriality. Upwards, one obtains a function $C \to T$ by restricting the interpretation of types to constants.
Notice that the unit of the adjunction inserts a constant $c \in C$ in the set of types of the calculus $\lambda 1(C)$.   □

Remark 3.11 In the proof above we cheated a bit. The functor from the term model of $\lambda 1(C)$ to $(\mathbf{B}, T)$ does not preserve cartesian products in the bases "on-the-nose": there is only an isomorphism $[\![\Gamma \times \Delta]\!] \cong [\![\Gamma]\!] \times [\![\Delta]\!]$ instead of an identity. By viewing contexts as appropriate trees, this problem can be avoided. We chose not to do so because it would make the notation rather heavy and distract from the essentials.

In the rest of this section we consider an alternative notion of model without finite product types.

A *multicategory* is given by a set of objects and a set of morphisms. Unlike in ordinary category theory, one allows domains of morphisms to be (possibly empty) sequences of objects. One writes

$$A_1, \ldots, A_n \xrightarrow{f} B.$$

Special morphisms of the following form are required.

$$A \xrightarrow{id} A \qquad \text{identity}$$
$$A_1, \ldots, A_n \xrightarrow{\pi_i} A_i \qquad \text{projection}$$

Composition is done in the obvious way: for $f : \langle A_1, \ldots, A_n \rangle \to B$ and $g_i : \vec{C} \to A_i$ for $1 \le i \le n$ one has $f \circ (g_1, \ldots, g_n) : \vec{C} \to B$. Straightforward identity and associativity laws should be satisfied. Further one has

$$\pi_i \circ (g_1, \ldots, g_n) = g_i$$
$$f \circ (\pi_1, \ldots, \pi_n) = f.$$

A functor $H : \mathbf{M}_1 \to \mathbf{M}_2$ between multicategories is given by two maps $Obj\mathbf{M}_1 \to Obj\mathbf{M}_2$ and $Mor\mathbf{M}_1 \to Mor\mathbf{M}_2$, both denoted by H, such that domain, codomain, identities, composition and projections are preserved. For example, one requires

$$A_1, \ldots, A_n \xrightarrow{f} B \text{ in } \mathbf{M}_1 \quad \Rightarrow \quad HA_1, \ldots, HA_n \xrightarrow{Hf} HB \text{ in } \mathbf{M}_2.$$

In [3], a multicategory is called *closed* if for each pair of objects A, B an exponent object $A \Rightarrow B$ exists together with an evaluation map $ev : A \Rightarrow B, A \to B$ such that for each $f : A_1, \ldots, A_n, A \to B$ there is a unique $\Lambda(f) : A_1, \ldots, A_n \to A \Rightarrow B$ such that

$$ev \circ (\Lambda(f) \circ (\pi_1, \ldots \pi_n), \pi_{n+1}) = f.$$

Let $\mathbf{CMC}$ denote the category of closed multicategories; a morphism $H : \mathbf{M}_1 \to \mathbf{M}_2$ in $\mathbf{CMC}$ is a functor of multicategories which preserves exponents, *i.e.* satisfies $H(A \Rightarrow B) = HA \Rightarrow HB$ and $H(ev_{A,B}) = ev_{HA,HB}$.

Theorem 3.12 *There is a "forgetful" functor λ1-cat $\rightarrow$ CMC; it has a left adjoint.*

Proof :   The functor $U : \lambda$1-cat $\rightarrow$ CMC will be described first. A λ1-category $(\mathbf{B}, T)$ determines a multicategory with $X \in T$ as objects; a morphism f from $X_1, \ldots, X_n$ to Y is simply a map $f : X_1 \times \cdots \times X_n \rightarrow Y$ in $\mathbf{B}$. The resulting multicategory is obviously closed, see lemma **3.3**. A morphism H of λ1-categories yields a morphism UH of closed multicategories since it preserves cartesian products and exponents on-the-nose.

A functor $F : \mathbf{CMC} \rightarrow \lambda$1-cat is obtained by assigning to a closed multicategory $\mathbf{M}$ a category having sequences $A_1, \ldots, A_n$ of objects from $\mathbf{M}$. A morphism $A_1, \ldots, A_n \rightarrow B_1, \ldots, B_m$ between such sequences is given by an m-tuple $(f_1, \ldots, f_m)$ of maps $f_i : A_1, \ldots, A_n \rightarrow B_i$ in $\mathbf{M}$. The resulting category has finite products: the empty sequence is terminal and concatenation yields cartesian products. As collection of types we take $Obj\mathbf{M}$. By definition of closedness for multicategories, it is closed under exponents. Hence one obtains a λ1-category.

There is a natural bijective correspondence

$$\cfrac{F\mathbf{M} \xrightarrow{\ H\ } (\mathbf{B}, T) \quad \text{in } \lambda\text{1-cat}}{\mathbf{M} \xrightarrow{\ K\ } U(\mathbf{B}, T) \quad \text{in } \mathbf{CMC}.}$$

One takes $H^\vee(A) = HA$ and $H^\vee(f : \vec{A} \rightarrow B) = Hf : \vec{H}A \rightarrow HB$ using that H preserves finite products. Further, $K^\wedge(A_1, \ldots, A_n) = KA_1 \times \cdots \times KA_n$ and $K^\wedge(f_1, \ldots, f_m) = \langle Kf_1, \ldots, Kf_m \rangle$.   □

4.   The untyped λ-calculus

As mentioned in the preliminary section we seek to capture non-extensional models of the type-free λ-calculus. This will be done by using "semi-notions". Restriction to the monoid case gives appropriate untyped settings.

Definition 4.1 *A* categorical λ-algebra *is given by a cartesian category* $\mathbf{B}$ *containing a distinguished object* Ω *such that*
    (a) Ω *is a non-empty object, i.e.* $\mathbf{B}(t, \Omega) \neq \emptyset$;
    (b) $\mathcal{K}_\Omega : \mathbf{B}^{op} \rightarrow \mathbf{Cat}$ *has semi-products.*

In point (b) we mean semi-Ω-products, see the convention at the end of the preliminary section. The first point expresses that the monoid CT-structure $(\mathbf{B}, \{\Omega\})$ is non-trivial. A categorical λ-algebra will be called full if this underlying CT-structure is full.

In order to avoid problems like in remark 3.11 we shall require that appropriate morphisms preserve only specific finite products of the form $A \leftarrow A \times \Omega \rightarrow \Omega$ and $A \rightarrow t$. This matter will be of relevance in the proof of theorem 4.10.

Definition 4.2 *A morphism of categorical λ-algebras from* $(\mathbf{B}, \Omega)$ *to* $(\mathbf{B}', \Omega')$ *is a functor* $H : \mathbf{B} \rightarrow \mathbf{B}'$ *satisfying*

- $H\Omega = \Omega'$ *and* $Ht = t'$, *the terminal object in* $\mathbf{B}'$; *moreover,* $H(!_A) = !_{HA}$;

- *for every* $A \in \mathbf{B}$ *one has* $H(A \times \Omega) = (HA) \times \Omega'$ *with* $H(\pi_{A,\Omega}) = \pi_{HA,\Omega'}$ *and* $H(\pi'_{A,\Omega}) = \pi'_{HA,\Omega'}$;

- *the pair* $(H, \widetilde{H})$ *is a morphism* $\mathcal{K}_\Omega \rightarrow \mathcal{K}_{\Omega'}$ *of semi-products.*

This yields a category **Cat-λ-Alg**.

Finally we introduce the usual ramifications.

Definition 4.3 *Let* $(\mathbf{B}, \Omega)$ *be a categorical λ-algebra. It will be called*
  (a) *a categorical λ-model if* Ω *has enough points, i.e. if for* $f, g : \Omega \rightarrow A$,

$$\forall x : t \rightarrow \Omega. \ f \circ x = g \circ x \ \Rightarrow \ f = g;$$

  (b) *a categorical λη-algebra if* $\mathcal{K}_\Omega$ *has ordinary products;*
  (c) *a categorical λη-model if it is both a categorical λ-model and a λη-algebra.*

Instead of describing examples in terms of semi-adjunctions, we first introduce the following two lemmas which give a more elementary description. The proofs are as in the previous section; they can also be found in [6].

Lemma 4.4 *Let* $\mathbf{B}$ *be a category with finite products and* $\Omega \in \mathbf{B}$ *be non-empty. Then*
  (a) $(\mathbf{B}, \Omega)$ *is a categorical λ-algebra if and only if there is a map*

$$app \ : \ \Omega \times \Omega \ \rightarrow \ \Omega$$

together with an operation

$$\lambda(-) \ : \ \mathbf{B}(A \times \Omega, \Omega) \ \rightarrow \ \mathbf{B}(A, \Omega)$$

such that

$$app \circ \lambda(f) \times id = f$$
$$\lambda(f \circ g \times id) = \lambda(f) \circ g.$$

(b) $(\mathbf{B}, \Omega)$ *is a categorical λη-algebra if and only if there are app and λ as in* (a) *which additionally satisfy*

$$\lambda(app) = id. \qquad \square$$

Lemma 4.5 *Let* $(\mathbf{B}, \Omega)$ *and* $(\mathbf{B}', \Omega')$ *be categorical λ-algebras. A functor* $H : \mathbf{B} \to \mathbf{B}'$ *is a morphism of categorical λ-algebras if and only if*

- $H\Omega = \Omega'$ *and* $H(!_A) =!_{HA}$*;*

- $H(\pi_{A,\Omega}) = \pi_{HA,\Omega'}$ *and* $H(\pi'_{A,\Omega}) = \pi'_{HA,\Omega'}$*;*

- $H(app) = app'$ *and* $H(\lambda(f)) = \lambda'(Hf).$ $\qquad \square$

Lemma 4.4 shows that the above notion of categorical λ-algebra is slightly more general then the notion used by Obtulowicz. We recall the latter in the present terminology.

Definition 4.6 (see [11, 12]) *A* Church algebraic theory *is a categorical λ-algebra* $(\mathbf{B}, \Omega)$ *where* $\mathbf{B}$ *is an algebraic theory.*
In particular, a Church algebraic theory is a full categorical λ-algebra.

The next result is essentially theorem 5.7 in [12].

Proposition 4.7 *If* $(\mathbf{B}, \Omega)$ *is a full categorical λη-algebra, then* $\mathbf{B}$ *is a CCC.*

Proof : By the construction in 3.8 (ii), the category $\overline{\{\Omega\}}$ is a CCC. Hence $\mathbf{B} \simeq \overline{\{\Omega\}}$ as well. $\qquad \square$

Next we turn to some examples.

Examples 4.8 (i) Let D be a reflexive cpo via maps $F : D \to [D \to D]$ and $G : [D \to D] \to D$ with $F \circ G = id$, see [1], 5.4. As before we write $a \cdot b$ for $F(a)(b)$ and $\lambda x.-$ for $G(\lambda x.-)$.
A base category $\mathbf{D}$ is formed with $n \in \mathbb{N}$ as objects; n can be considered as the context containing the first n variables from an enumeration $\{x_n \mid n \in \mathbb{N}\}$. Morphism $n \to m$ are sequences $(f_1, \ldots, f_m)$ where each f_i is a continuous

function $D^n \to D$, i.e. $f_i \in [D^n \to D]$. Composition in $\mathbf{D}$ is done in the obvious way and identities are sequences of projections. The object $0 \in \mathbf{D}$ is terminal and $n + m$ is a product. Thus $\mathbf{D}$ is an algebraic theory. As distinguished object ("Ω") we take $1 \in \mathbf{D}$. Notice that 1 is a non-empty object iff the set D is non-empty. We show that $(\mathbf{D}, 1)$ is a Church algebraic theory.

One has $app : 1 + 1 \to 1$ as a continuous function $D \times D \to D$ described by $(x, y) \mapsto x \cdot y$. For $f : n + 1 \to 1$ in $\mathbf{D}$ one has $\lambda(f)(\vec{x}) = \lambda y. f(\vec{x}, y)$. Then

$$\begin{aligned}
\left(app \circ \lambda(f) \times id\right)(\vec{x}, z) &= \left(\lambda y. f(\vec{x}, y)\right) \cdot z \\
&= f(\vec{x}, z).
\end{aligned}$$

The other equation from lemma 4.4 is left to the reader.

One easily verifies that $(\mathbf{D}, 1)$ is a categorical λ-model. In case $G \circ F = id$ — i.e. $D \cong [D \to D]$ — it becomes a categorical $\lambda\eta$-model.

(ii) Let $\mathsf{M} = \langle D, \cdot, \mathbf{K}, \mathbf{S} \rangle$ be a (set theoretic) λ-algebra, see [1, 5.2]. One writes $\mathbf{1}_n = \lambda x_0 \ldots x_n. x_0 \ldots x_n$; inductively, one can define $\mathbf{1}_0 = I = \mathbf{SKK}$ and $\mathbf{1}_{n+1} = \mathbf{S}(\mathbf{K}\mathbf{1}_n)$, see *loc. cit.* 5.6. Let's put $(D^n \to D) = \{a \in D \mid \mathbf{1}_n \cdot a = a\}$. Then $(D^0 \to D) = D$; we write $\mathbf{1}$ for $\mathbf{1}_1$ and $(D \to D)$ for $(D^1 \to D)$.

Let $\mathbf{D}$ be a base category, once again with $n \in \mathbb{N}$ as objects, but with m-tuples $(a_1, \ldots, a_m)$ with $a_i \in (D^n \to D)$ as morphisms $n \to m$. Then $(b_1, \ldots, b_k) \circ (a_1, \ldots, a_m) = (c_1, \ldots, c_k)$ where c_i is the term $\lambda x_1 \ldots x_n. b_i(a_1 x_1 \ldots x_n) \ldots (a_m x_1 \ldots x_n)$. The identity on n is $(\lambda x_1 \ldots x_n. x_1, \ldots, \lambda x_1 \ldots x_n. x_n)$. The category $\mathbf{D}$ has terminal 0 and products $n + m$ as before. Hence it is an algebraic theory again. We take $1 \in \mathbf{D}$ as distinguished object.

Lemma 4.4 will be used to show that $(\mathbf{D}, 1)$ is a categorical λ-algebra. One has $app = \lambda xy. xy \in (D^2 \to D)$. For $a \in (D^{n+1} \to D)$ one takes $\lambda(a) = a$, which is in $(D^n \to D)$. Then

$$\begin{aligned}
ev \circ \lambda(a) \times id &= \lambda x_1 \ldots x_{n+1}. (a \cdot x_1 \cdot \ldots \cdot x_n) \cdot x_{n+1} \\
&= \lambda x_1 \ldots x_{n+1}. a \cdot x_1 \cdot \ldots \cdot x_n \cdot x_{n+1} \\
&= \mathbf{1}_{n+1} \cdot a \\
&= a.
\end{aligned}$$

In case M is a λ-model, i.e. $\forall x \in D. \; a \cdot x = b \cdot x \Rightarrow \mathbf{1} \cdot a = \mathbf{1} \cdot b$, one obtains a categorical λ-model: suppose morphisms $(a_1, \ldots, a_m), (b_1, \ldots, b_m) : 1 \to m$ in $\mathbf{D}$ are given with $\forall x : 0 \to 1. \; (a_1, \ldots, a_m) \circ x = (b_1, \ldots, b_m) \circ x$. Then $a_i, b_i \in (D \to D)$ satisfy $\forall x \in D. \; a_i \cdot x = b_i \cdot x$. Hence $a_i = \mathbf{1} \cdot a_i = \mathbf{1} \cdot b_i = b_i$. Thus the object $1 \in \mathbf{D}$ has enough points.

(iii) Suppose $\mathbf{B}$ is a CCC which has a reflexive object Ω. The latter means that there are maps $F : \Omega \to (\Omega \Rightarrow \Omega)$ and $G : (\Omega \Rightarrow \Omega) \to \Omega$ with $F \circ G = id$. Such structures are used by Scott and Koymans for the semantics of the untyped λ-calculus, see [13, 9]. Using lemma 4.4 one easily obtains a categorical λ-algebra $(\mathbf{B}, \Omega)$; one defines

$$app \;=\; ev \circ F \times id$$
$$\lambda(f) \;=\; G \circ \Lambda(f).$$

This yields the required equations.

$$
\begin{aligned}
app \circ \lambda(f) \times id &= ev \circ F \times id \circ (G \circ \Lambda(f)) \times id \\
&= ev \circ \Lambda(f) \times id \\
&= f. \\
\lambda(f \circ g \times id) &= G \circ \Lambda(f \circ g \times id) \\
&= G \circ \Lambda(f) \circ g \\
&= \lambda(f) \circ g.
\end{aligned}
$$

Moreover, in case $(\mathbf{B}, \Omega)$ is extensional in the sense of Scott and Koymans — which means that $G \circ F = id$ and thus $(\Omega \Rightarrow \Omega) \cong \Omega$ — then

$$
\begin{aligned}
\lambda(app) &= G \circ \Lambda(ev \circ F \times id) \\
&= G \circ F \\
&= id.
\end{aligned}
$$

Remark 4.9 (i) A categorical λ-algebra as it is used here is "more economical" than the structure used by Scott and Koymans: in our case there is no requirement about exponents in the base category $\mathbf{B}$.

(ii) The reflexive domain D used in 4.8 (i) is a reflexive object in the category of cpo's with continuous functions. It yields a non-full CT-structure and hence one does not have an algebraic theory this time.

(iii) The conditions in lemma 4.4 (i) essentially express that $(\mathbf{B}, \Omega)$ is a categorical λ-algebra if and only if the functor $\mathbf{B}(-, \Omega)$ is a reflexive object in the topos of presheaves $\mathbf{Sets}^{\mathbf{B}^{op}}$. This is because

$$
\begin{aligned}
\big(\mathbf{B}(-, \Omega) \Rightarrow \mathbf{B}(-, \Omega)\big)(A) &= Nat\big(\mathbf{B}(-, A) \times \mathbf{B}(-, \Omega),\, \mathbf{B}(-, \Omega)\big) \\
&\cong Nat\big(\mathbf{B}(-, A \times \Omega),\, \mathbf{B}(-, \Omega)\big) \\
&\cong \mathbf{B}(A \times \Omega,\, \Omega).
\end{aligned}
$$

At the end of this section one finds another way to obtain a CCC with a reflexive object from a categorical λ-algebra.

Let $(\mathbf{B}, \Omega)$ be a categorical λ-algebra. For $a, b \in \mathbf{B}(A, \Omega)$ put $a \cdot b = app \circ \langle a, b \rangle$. We write $\|\Omega\|$ for the (non-empty) collection $\mathbf{B}(t, \Omega)$ and claim that

$(\|\Omega\|, \cdot)$ is a (set theoretic) λ-algebra as described in [1]. Abstraction is done as follows. For a term $a(x) : t \times \Omega \to \Omega$ containing a free variable x one takes

$$\lambda x. a(x) = \lambda(a(x)) : t \to \Omega.$$

Then

$$
\begin{aligned}
\big(\lambda x. a(x)\big) \cdot b &= app \circ \langle \lambda(a(x)), b \rangle \\
&= app \circ \lambda(a(x)) \times id \circ \langle id, b \rangle \\
&= a(x) \circ \langle id, b \rangle \\
&= a(b).
\end{aligned}
$$

Let's write

$$\pi_i^n : t \times \underbrace{\Omega \times \cdots \times \Omega}_{n \text{ times}} \to \Omega$$

for the i-the projection. One has

$$
\begin{aligned}
K &= \lambda(\lambda(\pi_1^2)) \\
S &= \lambda(\lambda(\lambda((\pi_1^3 \cdot \pi_3^3) \cdot (\pi_2^3 \cdot \pi_3^3)))) \\
I &= \lambda(\pi_1^1) \\
1 &= \lambda(\lambda(\pi_1^2 \cdot \pi_2^2))
\end{aligned}
$$

which yields essentially de Bruijn's nameless notation.
Notice that for $a \in \|\Omega\|$ one has $1 \cdot a = \lambda y. a \cdot y = \lambda(app \circ a \times id)$. Hence if $(\mathbf{B}, \Omega)$ is a categorical λ-model, one obtains the (ξ)-rule.

$$
\begin{aligned}
&\forall x \in \|\Omega\|.\ a \cdot x = b \cdot x \\
&\Rightarrow \quad \forall x : t \to \Omega.\ app \circ a \times id \circ \langle id, x \rangle = app \circ b \times id \circ \langle id, x \rangle \\
&\Rightarrow \quad app \circ a \times id = app \circ b \times id, \qquad \text{since } t \times \Omega \cong \Omega \text{ has enough points} \\
&\Rightarrow \quad 1 \cdot a = 1 \cdot b.
\end{aligned}
$$

And if $(\mathbf{B}, \Omega)$ is a categorical $\lambda\eta$-algebra, then (η) holds.

$$
\begin{aligned}
\lambda y. a \cdot y &= \lambda(app \circ a \times id) \\
&= \lambda(app) \circ a \\
&= a.
\end{aligned}
$$

We write **Set-λ-Alg** for the category with (set theoretic) λ-algebras $\langle D, \cdot, K, S \rangle$ as objects; we allow D to be a collection of arbitrary size. Morphisms are maps between the underlying collections preserving application and K, S, see [1, 5.2.2 (ii)].

The assignment $(\mathbf{B}, \Omega) \mapsto \langle \|\Omega\|, \cdot \rangle$ forms the object-part of a "forgetful" functor $U : \mathbf{Cat\text{-}\lambda\text{-}Alg} \longrightarrow \mathbf{Set\text{-}\lambda\text{-}Alg}$: for a morphism $K : (\mathbf{B}, \Omega) \to (\mathbf{B}', \Omega')$ of categorical λ-algebras, one has $UK : \|\Omega\| \to \|\Omega'\|$ defined by $a \mapsto Ka$. By lemma 4.5, K preserves app and λ on-the-nose; hence UK is a morphism of λ-algebras.

Theorem 4.10 *The forgetful functor* $U : \mathbf{Cat\text{-}\lambda\text{-}Alg} \longrightarrow \mathbf{Set\text{-}\lambda\text{-}Alg}$ *has a left adjoint; the unit of the adjunction is an identity.*
The adjointness gives rise to an equivalence between $\mathbf{Set\text{-}\lambda\text{-}Alg}$ *and the full subcategory of* $\mathbf{Cat\text{-}\lambda\text{-}Alg}$ *of Church algebraic theories.*

Proof :   The object-part of a functor $F : \mathbf{Set\text{-}\lambda\text{-}Alg} \longrightarrow \mathbf{Cat\text{-}\lambda\text{-}Alg}$ is described in example 4.8 (ii). For a morphism of λ-algebras $h : \langle D, \cdot \rangle \longrightarrow \langle D', \cdot' \rangle$ one defines $Fh : (\mathbf{D}, 1) \to (\mathbf{D}', 1)$ by $n \mapsto n$ and $(a_1, \ldots, a_m) \mapsto (h(a_1), \ldots, h(a_m))$. By lemma 4.5 and proposition 5.1.14 (i) from [1], h preserves the relevant structure. Notice that the underlying collection of $UF(\langle D, \cdot \rangle)$ is $\|1\| = \mathbf{D}(0, 1) = (D^0 \to D) = D$. One obtains $UF = id$.
A counit $\varepsilon : FU(\mathbf{B}, \Omega) \to (\mathbf{B}, \Omega)$ is defined on objects by $n \mapsto \Omega^n$. To define it on morphisms, we need some notation. For an element $a \in \|\Omega\|$ we define $a^{(n)} : \Omega^n \to \Omega$ by $a^{(n)} = (a \circ !_{\Omega^n}) \cdot \pi_1^n \cdot \ldots \cdot \pi_n^n$ where $\pi_i^n : \Omega^n \to \Omega$ is the i-th projection.
On a morphism $(a_1, \ldots, a_m) : n \to m$ in $FU(\mathbf{B}, \Omega)$ — where $a_i \in (\|\Omega\|^n \to \|\Omega\|)$ — we put $\varepsilon(a_1, \ldots, a_m) = \langle a_1^{(n)}, \ldots, a_m^{(n)} \rangle : \Omega^n \to \Omega^m$. One has

$$\varepsilon(\lambda x_1 \ldots x_n. x_i) \;=\; ((\lambda x_1 \ldots x_n. x_i) \circ !) \cdot \pi_1^n \cdot \ldots \cdot \pi_n^n \;=\; \pi_i^n.$$

Hence ε preserves identities and the projections $n \leftarrow n+1 \to 1$. Composition is preserved since

$$\varepsilon(\lambda x_1 \ldots x_n. b_i(a_1 x_1 \ldots x_n) \ldots (a_m x_1 \ldots x_n)) \;=\; b_i^{(m)} \circ \langle a_1^{(n)}, \ldots, a_m^{(n)} \rangle$$

In order to show that ε is a morphism of categorical λ-algebras it suffices by lemma 4.5 to check

$$
\begin{aligned}
\varepsilon(app) &= (\lambda xy. xy)^{(2)} &&\text{see 4.8 (ii)} \\
&= \pi_1^2 \cdot \pi_2^2 \\
&= app \circ \langle \pi, \pi' \rangle \\
&= app.
\end{aligned}
$$

and for $a \in (\|\Omega\|^{n+1} \to \|\Omega\|)$,

$$\varepsilon(\lambda(a)) \;=\; \varepsilon(a) \qquad \text{see 4.8 (ii)}$$

$$
\begin{aligned}
&= (a \circ !) \cdot \pi_1^n \cdot \ldots \cdot \pi_n^n \\
&= \lambda x. (a \circ !) \cdot \pi_1^n \cdot \ldots \cdot \pi_n^n \cdot x \qquad \text{since } 1_{n+1} \cdot a = a \\
&= \lambda (app \circ ((a \circ !) \cdot \pi_1^n \cdot \ldots \cdot \pi_n^n) \times id) \\
&= \lambda ((a \circ !) \cdot \pi_1^{n+1} \cdot \ldots \cdot \pi_{n+1}^{n+1}) \\
&= \lambda (\varepsilon(a)).
\end{aligned}
$$

The triangular identities boil down to

$$
\varepsilon F = id \qquad \text{and} \qquad U\varepsilon = id.
$$

These are easily verified. Finally the counit $\varepsilon : FU(\mathbf{B}, \Omega) \to (\mathbf{B}, \Omega)$ is an isomorphism if and only if $\mathbf{B}$ is an algebraic theory. $\qquad \square$

The pattern obtained here is the same as established in [7], 7.4.3 for the second order λ-calculus: the functor from categorical to set theoretical models has a left-adjoint-right-inverse.

Next we deal with some categorical properties of categorical λ-algebras.

Theorem 4.11 *Let $(\mathbf{B}, \Omega)$ be a categorical λ-algebra. Then*

(a) *$\mathcal{K}_\Omega$ is an indexed monoid, i.e. all fibre categories are monoids;*

(b) *$\mathcal{K}_\Omega$ is an indexed semi-CCC, i.e. all fibre categories are semi-CCC's and reindexing preserves this structure.*

Proof : (a) Obvious, since one starts from a single type Ω.

(b) The standard (non-surjective) pairing from λ-calculus yields "combinators" $fst, snd : \Omega \to \Omega$ and $pair : \Omega \times \Omega \to \Omega$ satisfying $fst \circ pair = \pi$ and $snd \circ pair = \pi'$. In λ-calculus notation, $fst(z) = z\mathbf{K}, snd(z) = z\mathbf{K}'$ — where $\mathbf{K}' = \lambda xy. y$ — and $pair(x, y) = \lambda z. zxy$. A bit more categorically, $fst = id_\Omega \cdot (\mathbf{K} \circ !_\Omega), snd = id_\Omega \cdot (\mathbf{K}' \circ !_\Omega)$ and $pair = \lambda(\pi' \cdot (\pi \circ \pi) \cdot (\pi' \circ \pi))$. Remember from the preliminary section that composition in the fibre categories $\mathcal{K}_\Omega(A)$ is described by $g \bullet f = g \circ \langle \pi, f \rangle$. We define the semi-CCC structure in the fibre categories, see [5].

(1) $!' = c_0 \circ !_{A \times \Omega} : A \times \Omega \to \Omega$, where $c_0 : t \to \Omega$ is an arbitrary constant; then $!' \bullet f = c_0 \circ !_{A \times \Omega} \circ \langle \pi, f \rangle = c_0 \circ !_{A \times \Omega} = !'$.

(2) $\pi_0 = fst \circ \pi'$, $\pi_1 = snd \circ \pi' : A \times \Omega \to \Omega$. Further, for $f, g : A \times \Omega \to \Omega$ one takes $\langle\langle f, g \rangle\rangle = pair \circ \langle f, g \rangle$. Then $\pi_0 \bullet \langle\langle f, g \rangle\rangle = f$, $\pi_1 \bullet \langle\langle f, g \rangle\rangle = g$ and $\langle\langle f, g \rangle\rangle \bullet h = \langle\langle f \bullet h, g \bullet h \rangle\rangle$.

(3) $ev = app \circ \langle fst, snd \rangle \circ \pi' : A \times \Omega \to \Omega$. For $f : A \times \Omega \to \Omega$ one takes $\Lambda(f) = \lambda(f \circ \langle \pi \circ \pi, pair \circ \pi \times id \rangle)$. Then $ev \bullet \langle\langle \Lambda(f) \bullet g, h \rangle\rangle = \Lambda(f) \bullet \langle\langle g, h \rangle\rangle$, $\Lambda(f \bullet \langle\langle g \bullet \pi_0, \pi_1 \rangle\rangle) = \Lambda(f) \bullet g$ and $ev \bullet \langle\langle \pi_0, \pi_1 \rangle\rangle = ev$. $\qquad \square$

The previous theorem yields a second construction of a "CCC with reflexive object" (see 4.8 (iii)) from a categorical λ-algebra (the first one is in 4.9). The next two facts should be used.

- The Karoubi envelope $K(\mathbf{C})$ of a semi-CCC $\mathbf{C}$ is a CCC, see [5].

- If $X \cong (X \Rightarrow X)$ in a semi-CCC $\mathbf{C}$, then id_X is a reflexive object in $K(\mathbf{C})$. The latter is easily verified.

Obviously, the object, say X, of a monoid semi-CCC satisfies $X \cong (X \Rightarrow X)$. Hence taking the Karoubi envelope of one of the fibre categories of a categorical λ-algebra $\mathcal{K}_\Omega$ yields a CCC with a reflexive object.

References

[1] H.P. Barendregt, *The Lambda Calculus. Its Syntax and Semantics*, 2nd rev. ed. (North Holland, Amsterdam, 1984).

[2] P.-L. Curien, *Categorical Combinators, Sequential Algorithms and Functional Programming* (Pitman, London, 1986).

[3] P.-L. Curien, An abstract framework for environment machines, *Theor. Comp. Sci.* **82** (1991) 389 – 402.

[4] H. Friedman, Equality between Functionals, in: R. Parikh (ed.), *Logic Colloquium, Symposium on Logic held at Boston, 1972 – 1973* (Springer LNM 453, Berlin, 1975) 22 – 37.

[5] S. Hayashi, Adjunction of Semi-functors: Categorical structures in nonextensional lambda-calculus, *Theor. Comp. Sci.* **41** (1985) 95 – 104.

[6] B.P.F. Jacobs, *Categorical Type Theory*, Ph.D. thesis, Univ. Nijmegen, 1991.

[7] B.P.F. Jacobs, Semantics of the second order Lambda Calculus, in: *Math. Struct. in Comp. Sci.* 1991, to appear.

[8] B.P.F. Jacobs, Comprehension Categories and the Semantics of Type Dependency, *Theor. Comp. Sci.*, to appear.

[9] K. Koymans, *Models of the Lambda Calculus*, Ph.D. thesis, Univ. Utrecht. Reprinted as CWI Tract 9 (CWI, Amsterdam, 1984).

[10] J. Lambek & P.J. Scott, *Introduction to Higher Order Categorical Logic* (CUP, Cambridge, 1986).

[11] A. Obtulowicz, Functorial Semantics of the type-free λ-βη calculus, in: Karpinsky, ed. *Fundamentals of Computation Theory* (Springer LNCS 56, Berlin, 1977) 302 – 307.

[12] A. Obtulowicz & A. Wiweger, Categorical, functorial and algebraic aspects of the type-free lambda calculus, *Banach Center Publications* **9** (1982) 399 – 422.

[13] D.S. Scott, Relating Theories of the Lambda Calculus, in: R. Hindley & J.P. Seldin, eds. *To H.B. Curry: Essays on Combinatory Logic, Lambda-Calculus and Formalism* (Academic Press, New York and London, 1980) 403 – 450.

Modelling Reduction in Confluent Categories

C. Barry Jay [*]

Abstract

Rewriting systems can be modelled categorically by representing terms as morphisms and reduction by an order on the homsets. Confluence of the system, and hence of the homset orders, yields a confluent category. Collapsing these hom-orders, so that a term is identified with their reducts, and in particular any normal form it might have, then yields a denotational semantics. Confluent functors, natural transformations and adjunctions can also be defined, so that on passing to the denotations they yield the usual notions.

Two general ways of constructing such models are given. In the first objects are types and morphisms are terms, with substitution represented by equality. The second has contexts for objects and declarations for morphisms, with the call mechanism represented by a reduction. When applied to the simply-typed λ-calculus these constructions yield confluently cartesian closed categories. A third such model, in which the call rule is used to interpret β-reduction, is also given.

1. Introduction

One of the cornerstones of programming language semantics is the correspondence between simply-typed λ-calculi and cartesian closed categories established by Lambek and Scott [LS86]. It provides a clean algebraic semantics for λ-terms, which helps us to reason effectively about the values of programs. Of equal interest to the implementer, however, is the reduction process which computes these values, i.e. the operational semantics. One of the most satisfying approaches to date is Plotkin's structural operational semantics [Plo75, Plo81] which uses a sequent calculus to describe the reduction process and provide a rigorous specification for computation.

[*]Laboratory for Foundations of Computer Science, University of Edinburgh, The King's Buildings, Mayfield Road, Edinburgh, U.K. EH9-3JZ. Research supported by BP/Royal Society of Edinburgh Research Fellowship.

In spite of their separate virtues, it is difficult to relate the logic (of operational semantics) to the categorical algebra (of denotational semantics); the usual approach, using computational adequacy, is rather unwieldy.

The proposal here is to give a *categorical* semantics of reduction with a simple functorial relationship between the operational and denotational semantics. This is achieved by extending the usual mapping of types to objects and terms to morphisms, so that reductions correspond to an order relation on the morphisms. We will impose two conditions on this relation. First, it must be a rewriting system, so that reduction is modular, and the composition of the category preserves the order. Second, the reduction order must be confluent, so that any normal form that exists is unique. This leads to *confluent categories* and the corresponding functors, natural transformations and adjoints.

Confluent functors map objects to objects, and morphisms to morphisms, as usual, and preserve the order on the homsets. Composition and identities, however, are not preserved exactly but only up to compatibility, i.e. having a common reduct. For example, if $g \circ f$ is a composite morphism in C and $F : C \rightarrow D$ is a confluent functor then $F(g \circ f)$ and $Fg \circ Ff$ are compatible. Similarly, a *confluent natural transformation* $\alpha : F \Rightarrow G : C \rightarrow D$ between confluent functors is given by a family of morphisms $\alpha_A : FA \rightarrow GA$ such that, for each $f : A \rightarrow B$ in C there is some $g : FA \rightarrow GB$ such that

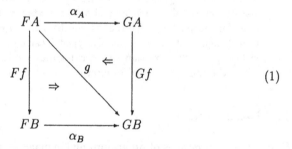

$$(1)$$

The intuition is that since both sides of the square reduce to a common form they have the same denotational semantics. Technically, confluence of the order is necessary to ensure that the horizontal composition of natural transformations and the application of functors to natural transformations are defined. It follows that the usual Godemont rules for functors and natural transformations hold, provided that some of the equalities are relaxed to compatibilities. Confluent adjunctions are defined similarly, and determine the *confluently cartesian closed categories*.

The ease with which the usual calculus of functors and natural transformations is generalised to confluent categories is in stark contrast to the situation for general ordered categories. Without confluence, one is forced to make composition and identities preserved up to a reduction, which may be in one of two possible directions, yielding *lax* and *colax functors* (unlike the symmetric concept of compatibility). Both types of functor arise in practice, but they do not compose. There are similar incompatible choices for natural transformations (namely, the *lax* and *oplax*). And finally, applying lax functors to lax natural transformations does not always yield a lax (or oplax) natural transformation. Since there doesn't appear to be a general framework, one is forced to seek out additional constraints that are satisfied in interesting classes of examples. For example, one may characterise categories of relations in terms of their sub-category of *maps* (internal left adjoints) [CKS84, CKW89] or categories of partial morphisms in terms of their *total morphisms* [Jay90a, Jay90e]. Here confluence of the orders will be used.

Once a reduction system has been modelled in a confluent category C then the corresponding denotational semantics arises by collapsing the order on the homsets, to obtain a set-based category whose homsets are equivalence classes of morphisms of C. Confluent functors, natural transformations and adjunctions are mapped to their more usual counterparts so that, in particular, confluently cartesian closed categories are mapped to their ordinary counterparts.

Some authors [See87, RS87, HR91] seeking categorical models of reduction have chosen to use 2-categories or bicategories, that is, to label their reductions and establish equations and inequations between them. However, without prejudging future developments, there are many interesting phenomena which are obscured by the large conceptual overhead arising from the need to choose, state and verify many equations on 2-cells, and establish the desired coherence theorems. Similar choices are available in modelling refinement, where ordered categories have been used [HHM89].

The construction of a confluently cartesian closed category from a confluent rewriting system can be done in many ways, depending on which features are to be emphasised. Two methods are demonstrated here. The first is chosen for maximum simplicity (objects are types, morphisms are terms, and substitution is an equality), so that the basic ideas can be understood. The second takes contexts as objects and declarations as morphisms, with the call rule (substitution) modelled by a reduction.

Finally, these techniques are used to model the simply-typed λ-calculus in confluently cartesian closed categories. As well as the two canonical models derived from the previous section, there is a third, in which the usual β-reduction is broken into two steps, the creation of a new declaration (corresponding to its interpretation by a **let**), and its call to the body of the

λ-term.

The paper is divided into the following sections: Introduction; Confluent Categories; Confluent Adjunctions; Modelling Reduction; The λ-Calculus; Further Work.

2.   Confluent Categories

Let $(P, \Rightarrow)$ be a pre-ordered set in which the pre-order is called *reduction*. Elements x, y of P are *equivalent*, denoted $x \Leftrightarrow y$ if each reduces to the other, in which case the reductions $x \Rightarrow y$ and $y \Rightarrow x$ are *reversible*. They are *compatible*, denoted $x \sim y$, if they have a common reduct, i.e. there is a z such that $x \Rightarrow z$ and $y \Rightarrow z$. Define P to be *confluently ordered* if the set of reducts of any given element are all pairwise compatible (that is, $\sim$ is an equivalence relation on P).

The ordered category **Conf** is defined to have confluent orders as objects and order-preserving functions as morphisms. The order on the morphisms $f, g : P \rightarrow Q$ between confluent orders is the equivalence relation given by pointwise compatibility, i.e. $f \sim g$ iff for all $x \in P$ we have $f(x) \sim g(x)$. Now P and Q are *equivalent* if there are confluent functions $f : P \rightarrow Q$ and $g : Q \rightarrow P$ such that $g \circ f \sim \mathrm{id}_P$ and $f \circ g \sim \mathrm{id}_Q$.

The interplay between preserving reduction and preserving compatibility is quite delicate. Morphisms in **Conf** will be used to combine reduction systems as modules of a larger system, and so should preserve the actual computation steps, and not merely their denotations, which is all that compatibility ensures. However, in comparing different ways of combining modules, it suffices to compare their denotations, so that compatibility suffices. Note that if the morphisms of **Conf** were functions that merely preserve compatibility then they could not distinguish between $\Rightarrow$ and $\sim$ and the category would be equivalent to its subcategory of equivalence relations.

Recall [Jay90e] that an *ordered category* is a category $\mathcal{C}$ whose homsets have a given pre-order such that composition is order-preserving, i.e. $\mathcal{C}$ is pre-order enriched.

Definition 2.1 A confluently ordered category *is an ordered category whose homsets are confluently ordered. A* confluent functor $F : \mathcal{C} \rightarrow \mathcal{D}$ *between confluent categories* $\mathcal{C}$ *and* $\mathcal{D}$ *is given by a function* $A \mapsto FA$ *from the objects of* $\mathcal{C}$ *to those of* $\mathcal{D}$ *and, for each pair of objects* A, B *of* $\mathcal{C}$ *an order-preserving function* $F_{A,B} : \mathcal{C}(A, B) \rightarrow \mathcal{D}(FA, FB)$ *which together preserve composition and identities, up to compatibility. That is, if* $g \circ f : A \rightarrow C$ *is a composite* *morphism in C then*

$$F(g \circ f) \sim Fg \circ Ff$$
$$F(\mathrm{id}_A) \sim \mathrm{id}_{FA}$$

A confluent (natural) transformation $\alpha : F \Rightarrow G : C \to D$ *between confluent functors is given by a family of components* $\alpha_A : FA \to GA$ *such that for all* $f : A \to B$ *in* C *we have* $Gf \circ \alpha_A \sim \alpha_B \circ Ff$ *(see Diagram 1 above).*

Vertical composition of confluent natural transformations is defined in the usual way: if $\alpha : F \to G : C \to D$ and $\beta : G \to H$ then $\beta \circ \alpha : F \to H$ has components $\beta_A \circ \alpha_A$

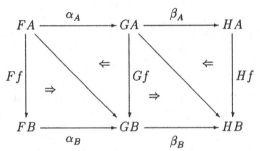

Confluence ensures that the two different reductions of $\beta_B \circ Gf \circ \alpha_A$ have a common reduct, which makes $\beta \circ \alpha$ a confluent natural transformation.

Define confluent natural transformations to be *compatible* if their components at each object are. Also an *equivalence* of the confluent functors $F, G; C \to D$ is given by a pair of confluent natural transformations $\alpha : F \to G$ and $\beta : G \to F$ such that $\beta \circ \alpha$ and $\alpha \circ \beta$ are both equivalent to identity natural transformations.

A diagram in a confluent category is *compatible* if any pair of paths through it are so. Then confluence of the order ensures that a diagram is compatible if all of its cells are. In future, a diagram cell with compatible sides will be labelled with the compatibility symbol $\sim$.

Further, confluent functors preserve compatible diagrams, from which follows several facts. First, composites of confluent functors are confluent. Second, if α is as above and $\alpha' : F' \to G' : D \to E$ is another confluent natural transformation then we can define confluent natural transformations $F'\alpha : F' \circ F \to F' \circ G$ and $\alpha'F : F' \circ F \to G' \circ F$ with components at A given by $F'\alpha_A : F'FA \to F'GA$ and α'_{FA} respectively. Thus we can define the horizontal composition of α and $\alpha' : F' \to G' : D \to E$ to be

$$(\alpha' * \alpha) = \alpha'G \circ F'\alpha$$

A distinct, but compatible, alternative would have the horizontal composite be $G'\alpha \circ \alpha'F$.

Now the usual Godemont rules for functors and natural transformations hold, provided that equality is replaced by compatibility. For example, the interchange law for α, α', β as above and $\beta' : G' \to H'$ asserts

$$(\beta' \circ \beta) * (\alpha' \circ \alpha) \sim (\beta' * \alpha') \circ (\beta * \alpha)$$

which follows from the confluence of G'' and of β' with respect to components of α.

Consequently, key notions such as adjoints, which are defined up to isomorphism in a 2-category, are here defined only up to compatibility. If one must have a 2-category then this can be obtained in at least two different ways, which will not be pursued here. First, one could replace the confluent natural transformations by their equivalence classes under compatibility; this would damage the correspondence with the syntax in the examples. Second, and perhaps more natural, would be to require that the components of confluent natural transformations be normal forms. Of course this will only make sense if composites of normal forms are normalisable.

In the absence of a 2-category, and without wishing to enter ever-higher dimensions, let us consider **ConfCat** to be the 2-graph of confluent categories, functors and natural transformations.

Proposition 2.2 *There is a 2-graph homomorphism* $(-)_r : $ **ConfCat** $\to$ **Cat** *which preserves all horizontal and vertical composites, and adjunctions, and identifies compatible natural transformations.*

Proof   Define the *residual category* C_r of a confluent category C to be the quotient of C by the congruence given by compatibility. The rest is straightforward.   □

Thus by interpreting reduction in a confluent category and then taking the residual category one obtains a denotational semantics of the reduction system. Of course, the problem of interpreting the equivalence classes remains, but the conceptual difficulties of linking the operational and denotational semantics have been removed.

If C is a confluent category then C^{op} is also confluent. The opposite of confluent functors and natural transformations are also defined in the usual way. However, the other dual C^{co} in which only the order on the morphisms is reversed, is not generally confluent. The product of confluent categories is also confluent with the pointwise order.

Conf is a confluent category since the order on its morphisms is actually an equivalence relation. Further, each object A in a confluent category $\mathcal{C}$ yields confluent *hom-functors* $\mathcal{C}(A, -) : \mathcal{C} \to \textbf{Conf}$ and $\mathcal{C}(-, A) : \mathcal{C}^{op} \to \textbf{Conf}$, and each morphism $A \to B$ yields a corresponding pair of confluent natural transformations. Define a confluent functor $\mathcal{C} \to \textbf{Conf}$ (respectively, $\mathcal{C}^{op} \to \textbf{Conf}$) to be *representable* if it is equivalent to a representable functor.

3. Confluent Adjunctions

With the weaker theory that we have chosen it is inevitable, if not desirable, that compatibilities will replace equalities in the definition of adjoints. That said, all the usual ways of representing adjunctions, as natural isomorphisms (or equivalences), via representablility, universal maps or triangle laws are valid.

Definition 3.1 *A confluent adjunction is given by a pair of confluent functors $F : \mathcal{C} \to \mathcal{D}$ and $G : \mathcal{D} \to \mathcal{C}$ and a pair of confluent natural transformations, namely the unit $\eta : \text{id} \to GF$ and the counit $\varepsilon : FG \to \text{id}$ which satisfy the following triangle laws*

We may write $F \dashv G : \mathcal{D} \to \mathcal{C}$ when η and ε are understood.

Of course if $F \dashv G$ then $G^{op} \dashv F^{op}$ and the following results can be dualised.

Lemma 3.2 *Two confluent left adjoints to a confluent functor are equivalent.*

Proof   Let $(F, G, \eta, \varepsilon) : \mathcal{D} \to \mathcal{C}$ and $(F', G, \eta', \varepsilon') : \mathcal{D} \to \mathcal{C}$ yield two left adjoints for G. Then the equivalence is given by $\varepsilon F' \circ F\eta' : F \to F'$ in the usual way.   □

Let $G : \mathcal{D} \to \mathcal{C}$ be a confluent functor and let A be an object of $\mathcal{C}$. A *universal arrow* from A to G is given by an object F_0A in $\mathcal{D}$ and an arrow $\eta_A : A \to GF_0A$ such that for each morphism $f : A \to GB$ there is a chosen arrow $g : F_0A \to B$ such that (i) the diagram is compatible; (ii) if $h : F_0 A \to B$ is such that $Gh \circ \eta_A \sim f$ then $h \sim g$, and; (iii) if $f \Rightarrow f'$ then $g \Rightarrow g'$ where $g' : F_0 A \to B$ is the morphism determined by f'.

Theorem 3.3 *The following statements about a confluent functor $G : \mathcal{D} \to \mathcal{C}$ are equivalent:*

   (i) G has a left adjoint.

   (ii) There is a confluent functor $F : \mathcal{C} \to \mathcal{D}$ such that the confluent functors $\mathcal{D}(F-, ?)$ and $\mathcal{C}(-, G?) : \mathcal{C}^{op} \times \mathcal{D} \to \mathbf{Conf}$ are equivalent, i.e. there are equivalences

$$\mathcal{D}(FA, B) \underset{\psi_{A,B}}{\overset{\phi_{A,B}}{\rightleftarrows}} \mathcal{C}(A, GB)$$

   which are confluently natural in A and B.

   (iii) The functors $\mathcal{C}(A, G-) : \mathcal{C} \to \mathbf{Conf}$ are all representable.

   (iv) For each object A in $\mathcal{C}$ there is a universal arrow from A to G.

Proof   (i) $\Rightarrow$ (ii): define $\phi_{A,B}(g) = Gg \circ \eta_A$ and $\psi_{A,B}(f) = \varepsilon_B \circ Ff$. The compatibility proofs are routine. (ii) $\Rightarrow$ (iii): immediate.

(iii) $\Rightarrow$ (iv): Let $F_0 A$ be the representing object for $\mathcal{C}(A, G-)$ and let $\alpha_B : \mathcal{D}(F_0 A, B) \to \mathcal{C}(A, GB)$ and $\beta_B : \mathcal{C}(A, GB) \to \mathcal{D}(F_0 A, B)$ be the natural equivalences. Define

$$\eta_A = \alpha_{F_0 A}(\mathrm{id}) : A \to GF_0 A$$

Now if there is a g making the triangle above compatible, then

$$\beta_B(f) \sim \beta_B(Gg \circ \eta_A) \sim g \circ \beta_{F_0 A}(\eta_A) = g \circ \beta_{F_0 A}(\alpha_{F_0 A}(\mathrm{id})) \sim g$$

by the naturality of β with respect to B. This establishes the uniqueness of g up to compatibility. The naturality of α with respect to B can be similarly used to prove that $\beta_B(f)$ does indeed make the triangle compatible, and the construction is order-preserving since β is.

(iv) $\Rightarrow$ (i): F_0 can be made the object part of a functor $\mathcal{C} \to \mathcal{D}$ by defining F on arrows as follows. If $f : A \to A'$ in $\mathcal{C}$ then Ff is the chosen map which makes the following square compatible

$$\begin{array}{ccc} A & \xrightarrow{\eta_A} & GFA \\ f \downarrow & \sim & \downarrow GFf \\ A' & \xrightarrow{\eta_{A'}} & GFA' \end{array}$$

The universal property shows that F preserves the order on morphisms, and preserves composition and identities up to compatibility. Thus it is a confluent functor. Also, η is a confluent natural transformation. Define $\varepsilon_B : FGB \to B$ to be the chosen morphism making the following compatible

$$GB \xrightarrow{\eta_{GB}} GFGB$$

with id_{GB} and $G\varepsilon_B$ to GB, $\sim$.

ε_B is natural in B since given $g : B \to B'$ in $\mathcal{D}$ then $\varepsilon_{B'} \circ FGg$ and $g \circ \varepsilon_B :$ $FGB \to B'$ are both equivalent to the induced morphism corresponding to $Gg : GB \to GB'$ and so are equivalent.

One of the triangle laws defines ε. The other follows upon verifying that $\varepsilon_{FA} \circ F\eta_A$ and id_{FA} are both equivalent to the morphism induced by η_A. $\square$

The definitions of product and coproduct functors, and of cartesian closure flow directly from that of adjunction. They will satisfy the usual universal properties except that equalities are replaced by compatibility.

At this point we should also define a *confluent natural numbers object*. Without striving for maximum generality [Rom89, PR89, Coc90, Jay91a], we may define it to be an object N equipped with morphisms $\mathtt{zero} : 1 \to N$ and $\mathtt{succ} : N \to N$ such that for each diagram

$$1 \xrightarrow{a} A \xleftarrow{h} A$$

there is a chosen morphism $f : N \to A$ called the *iterator* of a and h such that (i) the diagram

$$1 \xrightarrow{\mathtt{zero}} N \xleftarrow{\mathtt{succ}} N$$

with a, f, $\sim$, f to $A \xleftarrow{h} A$

is compatible; (ii) if f can be replaced by $g : N \to A$ in the diagram while preserving compatibility then $f \sim g$ and; (iii) if $a \Rightarrow a'$ and $h \Rightarrow h'$ then $f \Rightarrow f'$ where f' is the iterator of a' and h'. If the iterator is exists but without satisfying (ii) above then $(N, \mathtt{zero}, \mathtt{succ})$ is a confluent *weak* natural numbers object.

The temptation to replace the compatibilities in this definition with reductions must be resisted, since these changes would neither be in the spirit of the universal properties above nor be satisfied by the examples in Section 5.3.

4.   Modelling Reduction

A rewriting system is a reduction system R in which arbitrary sub-terms may be reduced. That is, if $t_1 \Rightarrow t_2$ then $t[t_1/x] \Rightarrow t[t_2/x]$ (where $t[t_1/x]$ denotes the expression obtained by replacing each free occurrence of x in t by t_1). We will consider two constructions of ordered categories from R, which will be fixed for the rest of the section. The first will quickly establish the general approach, while the second will pay closer attention to contexts, declarations and substitution.

4.1.   Substitution as Composition

Define an ordered category $\mathcal{C}(R)$ whose objects are the types and whose morphisms

$$X \xrightarrow{\;\;t\;\;} Y$$

are terms $t : Y$ which have at most one free variable, say x, which is of type X (as in [LS86]), that is $FV(t) \subseteq \{x\}$. The order on morphisms is generated by term reduction in R. Composition is given by substitution: if $FV(t') \subseteq \{y\}$ then

commutes.

Composition preserves the order since R is a rewriting system. Thus $\mathcal{C}(R)$ is an ordered category, which is confluent iff R is a confluent rewriting system.

$\mathcal{C}(R)$ suffers from two main operational defects. First, its restriction to terms with at most one free variable only makes sense in the presence of product types, which is perhaps too great a burden for them. Second, treating substitution as an equality ignores its fundamental importance, and, in particular, its irreversibility as a computation step.

4.2.   Contexts and Declarations

The next model will have *contexts* as its objects and *declarations* as its morphisms. The presence of many variables in the context removes the first objection. For the second, composition of declarations will be given freely,

with the call mechanism (replacing substitution) presented as a form of reduction, or computation step. In this way, substitution or calling becomes
more explicit, and thus can be controlled.

A *context* is a finite collection of typed variables of R, i.e. a set of variables
equipped with type assignments. *Declarations* are lists of *atomic declarations*
which in turn have the form $x = t$ where t is an term of the reduction system.
Declarations should be thought of as updating a context or environment, or
as a way of interpreting a let expression, i.e. interpret $y = $ let $x = t_1$ in t_2
by $x = t_1; y = t_2$. Thus atomic declarations bind everything to their right, for
example $x = e$ binds d in $x = e; d$. Any declaration d and context Δ combine
to form a *typed declaration* $d : \Delta$.

The ordered category $\mathcal{D}(R)$ has contexts for objects. Informally, a morphism
$\Gamma \to \Delta$ between contexts is given by a declaration d such that Γ contains the
variables required to produce Δ from d. More precisely, define the *minimum
domain* $\mathrm{dom}(d : \Delta)$ of a *typed declaration* $d : \Delta$ by

$$\mathrm{dom}(nil : \Delta) \;=\; \Delta$$
$$\mathrm{dom}(x = t; d : \Delta) \;=\; FV(t) \cup (\mathrm{dom}(d, \Delta) - \{x\})$$

Now $d : \Gamma \to \Delta$ if $\Gamma \supseteq \mathrm{dom}(d : \Delta)$. Note that a declaration d will be a morphism between many different pairs of contexts, e.g. nil $: \Gamma \to \Delta$ iff $\Gamma \supseteq \Delta$.
Composition of declarations is given by appending of lists and identities are
given by nil.

Reduction of morphisms $d : \Gamma \to \Delta$ are specified by reductions of the typed
declaration $d : \Delta$ (which will always preserve the type). Its *basic reductions*
are given by the following list:

(E)	$x = t$	$>$	$x = t'$	if $t \Rightarrow t'$ in R.
(C)	$x = t_1; y = t_2$	$>$	$y = t_2[t_1/x]; x = t_1$	if $x \neq y$ and $y \notin FV(t_1)$.
(T)	$x = t_1; x = t_2$	$>$	$x = t_2[t_1/x]$	
(R)	$x = t$	$>$	$y = t; x = y$	if $y \notin \Delta$.
(W)	$x = t$	$>$	nil	if $x \notin \Delta$.
(N)	nil	$>$	$x = x$	if $x \in \Delta$.

The first of these is *expression reduction* and is the only rule which refers
to the reduction system R. The rest are *declaration reductions* which are
named, respectively, *call, thinning, renaming of identifiers, weakening and
nil reduction*. In particular, (C) above is a *call* of $x = t_1$ to $y = t_2$.

The *1-step reductions* are given by a decomposition of d as $d = d_1; d_2; d_3$
and a basic reduction $d_2 > d_4 : \mathrm{dom}(d_3, \Delta)$ whence $d \Rightarrow d_1; d_4; d_3 : \Delta$. They
inherit the titles of expression reductions, calls etc. according to the basic

reduction employed. General reduction is the reflexive, transitive closure of 1-step reduction.

Here is an example of reduction of a declaration of type $\{y : N, z : N\}$ where R is any suitable reduction system for the natural numbers N and addition $(+)$.

$$
\begin{aligned}
x = 1; y = x + 2; z = x + 3 \;\; &\Rightarrow \;\; y = 3; x = 1; z = x + 3 \quad &\text{(C,E)} \\
&\Rightarrow \;\; y = 3; z = 4; x = 1 \quad &\text{(C,E)} \\
&\Rightarrow \;\; y = 3; z = 4 \quad &\text{(W)}
\end{aligned}
$$

The fundamental declaration reduction is call. Note that the declaration $x = t_1$ in its definition is not lost, but placed so that it still binds declarations further to the right. When it no longer binds any expressions then it may be weakened out. The two restrictions on call prevent any identifiers from increasing their scope. When $x = y$ then thinning can be applied; the declaration $x = t_1$ disappears since any later occurrences of x are bound by t_2. When y is free in t_1 then renaming, which plays a role comparable to α-conversion, can be used to permit the call to procede. This is illustrated by reduction of $x = y + 1; y = x + 2 : \{y : N\}$. Call cannot be applied, and so without renaming this declaration would be essentially normal, despite the possibility of substituting for x. The difficulty arises since y appears in the minimum domain but is later rebound as an identifier of the declaration. The solution is to separate the two uses of y by introducing another identifier u, as follows:

$$
\begin{aligned}
x = y + 1; y = x + 2 \;\; &\Rightarrow \;\; x = y + 1; u = x + 2; y = u \quad &\text{(R)} \\
&\Rightarrow \;\; u = y + 3; y = u \quad &\text{(C,E,W)} \\
&\Rightarrow \;\; y = y + 3 \quad &\text{(C,W)}
\end{aligned}
$$

In this case the identifier u that was introduced could later be eliminated. However, a second variable in the codomain may prevent this as is shown by

$$
\begin{aligned}
x = y + 1; y = x + z; z = x \\
\Rightarrow \;\; u = y + 1 + z; z = y + 1; y = u : \{y : N, z : N\} \quad (2)
\end{aligned}
$$

whose right-hand side is essentially normal. Here the rebinding $y = u$ must be performed after all uses of the original y are complete. There is, however, a convenient shorthand for this rebinding process.

A *product* of contexts $\Gamma = \{x_i : X_i \mid 1 \leq i \leq m\}$ and $\Delta = \{y_j : Y_j \mid 1 \leq j \leq n\}$ is given by a choice of disjoint union:

$$
\Gamma \times \Delta = \{x_{i1} : X_i \mid 1 \leq i \leq m\} \cup \{y_{j2} : Y_j \mid 1 \leq j \leq n\}
$$

The first projection $\pi_{\Gamma,\Delta}$ is then the translation

$$x_1 = x_{1,1}; x_2 = x_{2,1}; \ldots; x_m = x_{m,1} : \Gamma \times \Delta \to \Gamma$$

with the second projection $\pi'_{\Gamma,\Delta}$ defined similarly. The pairing of declarations $d : \Theta \to \Gamma$ and $d' : \Theta \to \Delta$ is given by $d; \theta_{\Gamma,\Delta}; d'; \theta'_{\Gamma,\Delta} : \Theta \to \Gamma \times \Delta$ where $\theta_{\Gamma,\Delta}$ is defined by $x_{1,1} = x_1; x_{2,1} = x_2; \ldots; x_{m,1} = x_m$ with $\theta'_{\Gamma,\Delta}$ defined similarly. Note that θ is not a morphism $\Gamma \to \Gamma \times \Delta$.

Now if Γ and Δ have no variables in common then we can define the *conjunction* of d and d' by

$$\Theta \xrightarrow{\;d \text{ and } d'\;} \Gamma \cup \Delta$$

by d and $d' = \langle d, d' \rangle; \pi; \pi'$. Then the result of (2) is equivalent to

$$(y = y + 1 + z) \text{ and } (z = y + 1) : \{y : N, z : N\}$$

Aside from any reduction loops that might arise in R there are some new loops built from declaration reductions. For example, in the definition of call above if $y \notin FV(t_1)$ and $x \notin FV(t_2)$ then the call is a *1-step permutation* which can be reversed by another permutation

$$x = t_1; y = t_2 \Leftrightarrow y = t_2; x = t_1$$

Sequences of 1-step permutations are, of course, *permutations*. Also, renamings are reversible since

$$x = t_1 \Rightarrow u = t_1; x = u \Rightarrow x = t_1; u = t_1 \Rightarrow x = t_1$$

by using call and weakening.

These reversible reductions, such as permutations and renamings seem quite harmless (like α-conversion) but force us to modify our notion of normal form accordingly.

A term t is *essentially normal* if any reduction of it is reversible. It is *weakly essentially normalisable* if it reduces to an essentially normal term, and *(strongly) essentially normalisable* if each reduction sequence

$$t = t_0 \Rightarrow t_1 \Rightarrow t_2 \Rightarrow \ldots \Rightarrow t_n \Rightarrow \ldots$$

has a finite number of irreversible steps, called its *essential length*. R is *weakly (respectively, strongly) essentially normalising* if every expression is weakly (respectively, strongly) essentially normalisable.

Let R_0 be the term algebra of R, i.e. the sub-system containing all the terms, but with no non-trivial reduction.

Theorem 4.1 *The system of declarations and declaration reductions is confluent and strongly essentially normalising. A declaration* $d : \Gamma \to \Delta = \{y_j : Y_j\}$ *(for* $j = 1, \ldots n$*) is essentially normal iff it is equivalent to a conjunction*

$$(y_1 = t_1) \text{ and } (y_2 = t_2) \text{ and } \ldots \text{ and } (y_n = t_n)$$

The corresponding confluent category $\mathcal{D}(R_0)$ *has confluent finite products.*

Proof   The full proof is in [Jay91c, Appendix]. It proceeds by progressively stripping out the obstructions to call, and hence the other reduction rules, and then proving that call is essentially normalising by a rank argument. Then confluence follows from a strengthening of weak confluence to account for reversible reductions.   □

Theorem 4.2 *Let* R *be a typed reduction system. Then* $\mathcal{D}(R)$ *is a confluent category (respectively, weakly essentially normalising) iff* R *is a confluent rewriting system (respectively, weakly essentially normalising).*

Proof   The forward direction is trivial since the call rule forces R to be a rewriting system. In the reverse direction, it suffices to prove the confluence of the coarser system whose 1-step reductions are sequences either of expression reductions or of declaration reductions. Now it is straight-forward to show that critical pairs of 1-step reductions can be completed in one step, which yields confluence. For weak essential normalisation, it suffices to reduce to the essential normal form in the sub-system of declaration reductions, and then reduce the expressions to essential normal forms.   □

5.   The λ-Calculus

The models of reduction above will now be instantiated for the simply-typed λ-calculus with pairing, unit or terminal type and iterator. After addressing some issues concerning confluence of expression reduction, the models of the previous section can be built directly and their confluent cartesian closure established. Following this, we will see how to tailor expression reduction to exploit the presence of declarations.

5.1.   The Reduction System

Fix R to be the following variant of the λ-calculus, discussed in detail as the *expansionary reduction system* in [Jay91b]. Its *types* are generated by the type N of natural numbers and the unit type 1 and closed under products $X \times Y$ and arrows $X \to Y$.

Each type X comes equipped with a collection of variables $x : X$. Variables will be denoted by the letters x, y, z and their annotated versions. Other lower case letters will denote arbitrary terms. The *terms* are inductively generated by

(i)   variables

(ii) $\lambda x.b : A{\rightarrow}B$ if $x : A$ and $b : B$

(iii) $fa : B$ if $f : A{\rightarrow}B$ and $a : A$

(iv) $\langle a, b \rangle : A{\times}B$ if $a : A$ and $b : B$

(v) $\pi c : A$ if $c : A{\times}B$

(vi) $\pi' c : B$ if $c : A{\times}B$

(vii) $0 : N$

(viii) $Sn : N$ if $n : N$

(ix) $\mathrm{It}(a, h, n) : A$ if $a : A$ and $h : A{\rightarrow}A$ and $n : N$

(x) $* : 1$

Terms equivalent up to change of bound variable will be identified.

The *basic reductions* are:

$$
\begin{array}{lrcl}
(\beta) & (\lambda x.b)a & > & b[a/x] \\
(\eta') & t & > & \lambda x.tx \\
(\pi) & \pi\langle a, b \rangle & > & a \\
(\pi') & \pi'\langle a, b \rangle & > & b \\
(sp) & c & > & \langle \pi c, \pi' c \rangle \\
(*) & a & > & * \\
(It0) & \mathrm{It}(a, h, 0) & > & a \\
(ItS) & \mathrm{It}(a, h, Sn) & > & h\mathrm{It}(a, h, n)
\end{array}
$$

where $t : A{\rightarrow}B$ and $c : A{\times}B$ and $a : 1$ is not $*$. The (β) rule involves implicit α-conversion whenever substitution threatens to capture free variables of a. The *1-step* reductions of a term are obtained by applying a basic reduction to one of its sub-terms. The transitive closure of the 1-step reductions is the reduction relation, denoted $\Rightarrow$. Reductions built solely from $(\eta'), (sp)$ and $(*)$ are called *expansions*. Conversely, reductions built without any use of expansions are *contractions*.

It is more usual to represent η-conversion and surjective pairing by contractions. The difficulty, observed by Obtulowicz (see [LS86]) is that the presence of the unit type prevents confluence. With expansions, however, we have the following result [Jay91b]

Theorem 5.1 *The expansionary reduction system is a confluent and weakly essentially normalising rewriting system with essential normal forms given by the long $\beta\eta$-normal forms of Huet [Hue 76].*   □

5.2.   A Substitution Model

Theorem 5.2 $C(R)$ *is a confluent cartesian closed category with confluent weak natural numbers object whose hom-orders are essentially normalising.*

Proof   The product of types A and B in $C(R)$ is simply $A \times B$. This yields a confluent right adjoint to the diagonal functor $\Delta : C(R) \to C(R)^2$ whose universal arrow $\Delta(A \times B) \to (A, B)$ in $C(R)^2$ is given by

$$(A \times B, A \times B) \xrightarrow{\ (\pi z, \pi' z)\ } (A, B)$$

Now given $(t, t') : \Delta C \to (A, B)$ (and assuming that t and t' share any free variable they have) the induced morphism is $\langle t, t' \rangle : C \to A \times B$.

$$
\begin{array}{ccccc}
C & & \Delta C & \xrightarrow{\ (t,t')\ } & (A, B) \\
\langle t,t'\rangle \big\downarrow & & \Delta\langle t,t'\rangle \big\downarrow & \overset{\sim}{\diagup} & \\
A \times B & & \Delta(A \times B) & \underset{(\pi z, \pi' z)}{\diagup} &
\end{array}
$$

The compatibility arises because $\pi z \circ \langle t, t' \rangle = \pi \langle t, t' \rangle \Rightarrow t$ in $C(R)$ and similarly for π'. The uniqueness of the induced morphism follows because if $f : C \to A \times B$ satisfies $\pi f \sim t$ and $\pi' f \sim t'$ then $f \sim \langle \pi f, \pi' f \rangle \sim \langle t, t' \rangle$. Finally if $t \Rightarrow s$ and $t' \Rightarrow s'$ then $\langle t, t' \rangle \Rightarrow \langle s, s' \rangle$.

The unit for the adjunction is thus $\langle x, x \rangle : A \to A \times A$ and the product of morphisms $t : A \to B$ and $t' : A' \times B'$ is given by

$$t \times t' = \langle t[\pi z/x], t'[\pi' z/x'] \rangle$$

where x and x' are assumed to be the free variables of t and t' respectively.

The exponential C^B in $C(R)$ is just the function type $B \to C$. The universal arrow is the evaluation $\mathbf{ev}_{B,C} = (\pi z)(\pi' z) : (B \to C) \times B \to C$. Now given $t : A \times B \to C$ with free variable z the induced morphism is $\lambda y.t[\langle x, y \rangle/z] : A \to C^B$ with free variable $x : A$.

$$
\begin{array}{ccccc}
A & & A \times B & \xrightarrow{\ \ t\ \ } & C \\
\lambda y.t[\langle x,y\rangle/z] \big\downarrow & & \langle \lambda y.t[\langle \pi z,y\rangle/z], \pi' z\rangle \big\downarrow & \overset{\sim}{\diagup} & \\
B \to C & & (B \to C) \times B & \underset{(\pi u)(\pi' u)}{\diagup} &
\end{array}
$$

Compatibility follows since the lower composite is

$$
\begin{aligned}
(\pi \langle \lambda y.t[\langle \pi z, y\rangle/z], \pi' z\rangle)(\pi' \langle \lambda y.t[\langle \pi z, y\rangle/z], \pi' z\rangle) \ &\Rightarrow\ (\lambda y.t[\langle \pi z, y\rangle/z])(\pi' z) \\
&\Rightarrow\ t[\langle \pi z, \pi' z\rangle/z] \\
&\Leftarrow\ t
\end{aligned}
$$

Uniqueness of the induced morphism follows by direct computation.

The type N is a weak natural numbers object with zero $= 0 : 1 \rightarrow N$ and succ $= Sn : N \rightarrow N$ where $n : N$ is some variable. The iterator of morphisms $a : 1 \rightarrow A$ and $t : A \rightarrow A$ is given by It$(a, \lambda x.t, n)$ where $x : A$ is the free variable of t.   □

5.3.   A Declaration Model

Theorem 5.3 *Let R' be the sub-system of R without product or unit types. Then $\mathcal{D}(R')$ and $\mathcal{D}(R)$ are confluent cartesian closed categories whose hom-orders are weakly essentially normalising. Further, $\mathcal{D}(R)$ has a confluent weak natural numbers object.*

Proof   Given Theorem 4.2, it remains to prove the existence of the exponential and the natural numbers object. The constructions will be given and the details left to the reader.

If $\Gamma = \{x : X\}$ and $\Delta = \{y : Y\}$ then $\Gamma \rightarrow \Delta = \{u : X \rightarrow Y\}$. More generally, if $\Gamma = \{x_i : X_i\}$ and $\Delta = \{y_j : Y_j\}$ are as above then $\Gamma \rightarrow \Delta = \{u_j : U_j\}$ where

$$U_j = X_1 \rightarrow X_2 \rightarrow \ldots \rightarrow X_m \rightarrow Y_j$$

(with $\rightarrow$ associated to the right). To define the currying of a declaration $d : \Gamma \times \Delta \rightarrow \Theta$ it suffices, since $\Delta \rightarrow (-)$ preserves products, to consider the case where $\Theta = \{z : Z\}$ is a single variable. Then the essential normal form of d in the declaration subsystem is an atomic declaration, say $z = e$, whose curried form is

$$\Gamma \xrightarrow{\quad u = \lambda y_1 y_2 \ldots y_n.e \quad} \{u : Y_1 \rightarrow Y_2 \rightarrow \ldots Y_n \rightarrow Z\}$$

Evaluation is then $z = u y_1 y_2 \ldots y_n : (\Delta \rightarrow \Theta) \times \Delta \rightarrow \Theta$.

Although the existence of products in $\mathcal{D}(R')$ does not depend on the existence of types to represent them, the latter are necessary for the object $\{n : N\}$ to be a natural numbers object, unless the iterator is strengthened to handle multiple recursion. However, in $\mathcal{D}(R)$ every object $\Gamma = \{x_i : X_i\}$ is isomorphic to one with only a single variable $\{x : X\}$ where X is the product of the X_i's, and so we will assume Γ to be of that form. Now reduce all declarations to atomic form. Now we can use the term structure to show that the context $\{n : N\}$ is a weak confluent natural numbers object. Note that the compatibility square for the iterator cannot be strengthened to be a reduction, since both sides must reduce to an atomic declaration for the comparison.   □

5.4. β-Reduction Through Call

The last model of β-reduction illustrated the nature of categories of declarations, but permitted two different ways of realising substitution, namely by β-reduction on expressions, and the call rule on declarations. By modifying the expression reductions we can use call to perform the substitution arising from β-reduction, just as (β) is sometimes interpreted by let.

Define the basic reduction β *through call* or (β') by

$$(\beta') \qquad\qquad y = (\lambda x.t_2)t_1 \Rightarrow x = t_1; y = t_2 : \Delta$$

provided that $x \notin \Delta$. Let R'' be the result of replacing (β) by (β') in R.

Lemma 5.4 *Reductions of R are reductions of R''. Conversely, if $d \Rightarrow d'$ in R'' then $d \sim d'$ in R. Hence R'' is confluent and weakly essentially normalising.*

Proof   Observe that (β) is given by (β') followed by call and weakening. This implies the first statement, and, since compatibility in R is an equivalence relation, implies the next pair of claims too. Finally, weak essential normalisation follows upon observing that β-normal forms are β'-normal too. □

Theorem 5.5 $\mathcal{D}(R'')$ *is confluently cartesian closed with weak natural numbers object and essentially normal hom-orders.*

Proof   The lemma above shows that R'' and R have the same compatibility relation and reversible reductions.   □

6.   Further Work

One line of development is to stop reduction from occurring under λ's. That is, abolish the ξ-rule. Of course this destroys confluence of expression reduction with call as currently described. A possible solution is to introduce closures, so that call defers substitution under λ's as in the $\lambda\sigma$-calculus [ACC90].

Another possible development is to restrict call on $x = t_1; y = t_2$ so that t_1 is required to be a *value* of some kind, e.g. a weak head normal form. This results in something of the flavour of call-by-need semantics (by-need since weakening can always remove unwanted arguments). The expression t_2 could be similarly constrained to capture some aspects of spine reduction (studied by Barendregt and Levy) in which the body of an evaluated λ-term is reduced before the value is passed. These restrictions increase the extent to which parallel reduction of a declaration is enforced. Various reduction strategies, such as call-by-value, call-by-name and call-by-need may perhaps be conveniently expressed as reduction strategies in a category of declarations.

Beyond these considerations lie the problems of extending this work to a polymorphic setting, and to full recursion.

Acknowledgements My warm thanks go to P.-L. Curien, N. Ghani, S. Peyton-Jones, G.D. Plotkin and S. Soloviev for many useful discussions.

References

[ACC90] M. Abadi et al, *Explicit substitutions. INRIA-Rocuencourt Rapports de Recherche 1176 (1990).*

[CKS84] A. Carboni, S. Kasangian and R. Street, *Bicategories of spans and relations. J. Pure and Appl. Alg. 33 (1984) 259 – 267.*

[Coc90] J.R.B. Cockett, *List-arithmetic distributive categories: locoi. J. Pure and Appl. Alg. 66 (1990) 1–29.*

[HR91] B.P. Hilken and D.E. Rydeheard, *Towards a categorical semantics of type classes. Fundamenta Informatica, to appear.*

[HHM89] C.A.R. Hoare, He Jifeng & C.E. Martin, *Pre-adjunctions in Order Enriched Categories. Oxford University Computing Laboratory (1989).*

[Hue76] G. Huet, *Résolution d'équations dans des langages d'ordre $1, 2, \ldots, \omega$. Thèse d'Etat, Université de Paris VII, 1976.*

[CKW89] A. Carboni, G.M. Kelly & R.J. Wood, *A 2-categorical approach to Geometric morphisms, I. Sydney Category Seminar Reports 89–19 (1989).*

[Jay88] C.B. Jay, *Local adjunctions. J. Pure and Appl. Alg. 53 (1988) 227–238.*

[Jay90a] C.B. Jay, *Extendinging properties to categories of partial maps. Tech. Rep. ECS-LFCS 90-107.*

[Jay90e] C.B. Jay, *Partial functions, ordered categories, limits and cartesian closure. in: G. Birtwistle (ed), IV Higher Order Workshop, Banff, 1990. (Springer, 1991).*

[Jay91a] C.B. Jay, *Tail recursion from universal invariants. in: D.H. Pitt et al (eds), Category Theory and Computer Science Paris, France, September 1991 Proceedings. Lecture Notes in Computer Science 530 (Springer, 1991) 151–163.*

[Jay91b] C.B. Jay, *Long βη normal forms and confluence (revised)* preprint.

[Jay91c] C.B. Jay, *Modelling Reduction in Confluent Categories.* Tech. Rep. *ECS–LFCS–91-187.*

[LS86] J. Lambek and P. Scott, *Introduction to higher order categorical logic.* Cambridge Studies in Advanced Mathematics 7 (Cambridge Univ. Press, 1986).

[Min79] G.E. Mints, *Teorija categorii i teoria dokazatelstv.I.* in: Aktualnye problemy logiki i metodologii nauky, Kiev, 1979 252-278.

[PR89] R. Paré and L. Roman, *Monoidal categories with natural numbers object.* Studia Logica 48(3) (1989).

[Plo75] G.D. Plotkin, *Call-by-name, call-by-value and the λ-calculus.* Theoretical Computer Science 1 (1975) 125–159.

[Plo81] G.D. Plotkin, *A structural approach to operational semantics.* Tech. Rep. Aarhus University, Daimi FN-19.

[Rom89] L. Roman, *Cartesian categories with natural numbers object.* J. of Pure and Appl. Alg. 58 (1989) 267–278.

[RS87] D.E. Rydeheard & J.G. Stell, *Foundations of equational deduction: A categorical treatment of equational proofs and unification algorithms.* in: Pitt et al, (eds), Category Theory and Computer Science. Lecture Notes in Computer Science 283 (Springer, 1987) 114 – 139.

[See87] R.A.G. Seely, *Modelling computations: a 2-categorical framework.* in: Proceedings of the Second Annual Symposium on Logic in Computer Science (1987).

On clubs and data-type constructors

G.M. Kelly *

1. Introduction

The diagrams for symmetric monoidal closed categories proved commutative
by Mac Lane and the author in [19] were diagrams of (generalized) natural
transformations. In order to understand the connexion between these results
and free models for the structure, the author introduced in [13] and [14] the
notion of *club*, which was further developed in [15] and applied later to other
coherence problems in [16] and elsewhere.

The club idea seemed to apply to several diverse kinds of structure on a
category, but still to only a restricted number of kinds. In an attempt to understand its natural limits, the author worked out a general notion of "club",
as a monad with certain properties, not necessarily on **Cat** now, but on any
category with finite limits. A brief account of this was included in the 1978
Seminar Report [17] , but was never published; the author doubted that there
were enough examples to make it of general interest.

During 1990 and 1991, however, we were fortunate to have with our research
team at Sydney Robin Cockett, who was engaged in applying category theory
to computer science. In lectures to our seminar he called attention to certain
kinds of monads involved with data types, which have special properties : he
was calling them *shape monads*, but in fact they are precisely examples of
clubs in the abstract sense above.

In these circumstances it seems appropriate to set down those old ideas after
all, and to complete them in various ways, in particular as regards *enriched*
monads.

Moreover there is also in computer science a renewed interest in free cate-
gories-with-structure of various kinds, and in questions of the naturality or
dinaturality of the morphisms in these, in connexion with polymorphism.
Accordingly the account below begins with the original motivating examples

---

*School of Mathematics FO7, University of Sydney, NSW 2006, Australia. The author
gratefully acknowledges the support of the Australian Research Council.

of diagrams of natural transformations and their relation to free structures, before passing on to the abstract notion of club.

Using the techniques of [18], one can give simple proofs that several types of monads on distributive categories arising in computer science are indeed clubs; but only under the hypothesis — perhaps uncomfortable for that subject — of external cocompleteness of the category involved. Much finer results are being prepared for publication by Cockett and Jay, using further developments of their techniques introduced in [4], [11], and [12] — see §5.6 below. The present account may serve as an introduction to their work.

2.   The original club idea

2.1. The commutative diagrams of Mac Lane's coherence theorems [23] were diagrams of natural transformations, involving components like $a_{XYZ}: (X \otimes Y) \otimes Z \to X \otimes (Y \otimes Z)$ and $c_{XY} : X \otimes Y \to Y \otimes X$. So too were the diagrams for symmetric monoidal closed categories proved commutative by Mac Lane and the author in [21]; but the transformations were now natural in the extended sense of Eilenberg and Kelly [8], involving components like $e_{XY} : [X,Y] \otimes X \to Y$ and $d_{XY}: X \to [Y, X \otimes Y]$.

Because our work in [21] was loosely inspired by that of Lambek in [22], we were aware that there was some connexion between our results and the structure of *free* symmetric monoidal closed categories; but we did not see this at all clearly. In fact, we found Lambek's ideas – central though they were for us – hard to follow, and realised later (see [14, p.144]) that his main result here was wrong.

The author elucidated the connexion with free structures in [13] and [14], along the following lines, beginning with structures like a symmetric monoidal category $\mathcal{A}$, described by functors $T : \mathcal{A}^n \to \mathcal{A}$ like $\otimes : \mathcal{A}^2 \to \mathcal{A}$ and $I : \mathcal{A}^0 \to \mathcal{A}$, and by natural transformations f of the form

$$
\begin{array}{ccc}
& \mathcal{A}^n & \\
\mathcal{A}^\xi \Big\uparrow & \overset{T}{\underset{\Downarrow f}{\searrow}} & \mathcal{A} \\
& \mathcal{A}^n & \nearrow_{\ s}
\end{array}
\tag{2.1}
$$

(where ξ is a permutation of n) like a and c above, subject to equational axioms, whether on the objects, like the $(X \otimes Y) \otimes Z = X \otimes (Y \otimes Z)$ of a strict monoidal category, or on the maps, like Mac Lane's pentagon and hexagon conditions – or like the axioms for a monad (T, i, m) on $\mathcal{A}$.

To describe the appropriate formalism we first replace the *codomain* $\mathcal{A}$ in (2.1) by an arbitrary category $\mathcal{B}$, thus introducing a category $\{\mathcal{A}, \mathcal{B}\}$, an object

(n, T) of which is a natural number n together with a functor $T: \mathcal{A}^n \to \mathcal{B}$, and a morphism $(n, T) \to (n, S)$ in which is a permutation ξ of n together with a natural transformation f of the form

$$
\begin{array}{c}
\mathcal{A}^n \\
\mathcal{A}^\xi \Big\uparrow \underset{\Downarrow f}{\overset{T}{\searrow}} \mathcal{B} ~; \\
\mathcal{A}^n \overset{S}{\nearrow}
\end{array}
\qquad (2.2)
$$

there are no morphisms $(n, T) \to (m, S)$ for $m \neq n$. Composition is evident, and we have an obvious forgetful functor $\Gamma: \{\mathcal{A}, \mathcal{B}\} \to \mathbf{P}$, where $\mathbf{P}$ is the category of natural numbers n and permutations ξ; that is, we have an object $\{\mathcal{A}, \mathcal{B}\}$ of $\mathbf{Cat/P}$. One next observes that the 2-functor $\{\mathcal{A}, -\}: \mathbf{Cat} \to \mathbf{Cat/P}$ has a left adjoint $- \circ \mathcal{A}$, so that for $\mathcal{K} = (\mathcal{K}, \Gamma: \mathcal{K} \to \mathbf{P})$ in $\mathbf{Cat/P}$ we have a 2-natural isomorphism

$$
\mathbf{Cat}(\mathcal{K} \circ \mathcal{A}, \mathcal{B}) \quad \cong \quad (\mathbf{Cat/P})(\mathcal{K}, \{\mathcal{A}, \mathcal{B}\}); \qquad (2.3)
$$

here an object of $\mathcal{K} \circ \mathcal{A}$ has the form

$$
P[A_1, \cdots, A_n] \qquad (2.4)
$$

where $P \in \mathcal{K}$ with $\Gamma P = n$ and $A_i \in \mathcal{A}$, while a morphism in $\mathcal{K} \circ \mathcal{A}$ has the form

$$
f[g_1, \cdots, g_n] : P[A_1, \cdots, A_n] \to Q[B_1, \cdots, B_n] \qquad (2.5)
$$

where $f : P \to Q$ in $\mathcal{K}$ with $\Gamma f = \xi$ and $g_i : A_i \to B_{\xi i}$. Note that

$$
\mathcal{K} \circ 1 \quad \cong \quad \mathcal{K}, \qquad (2.6)
$$

where the $\mathcal{K}$ on the left lies in $\mathbf{Cat/P}$ while the $\mathcal{K}$ on the right is seen as a mere category. If $\mathcal{A}$ is itself an object of $\mathbf{Cat/P}$, with "augmentation" $\Gamma: \mathcal{A} \to \mathbf{P}$, we turn $\mathcal{K} \circ \mathcal{A}$ into an object over $\mathbf{P}$, setting $\Gamma(P[A_1, \cdots, A_n]) = \sum \Gamma A_i$, while $\Gamma(f[g_1, \cdots, g_n])$ is the result $\xi(\eta_1, \cdots, \eta_n)$ of "substituting the permutations $\eta_i = \Gamma g_i$ into the permutation $\xi = \Gamma f$". If now $\mathcal{B}$ too is an object of $\mathbf{Cat/P}$, we get a new $\{\mathcal{A}, \mathcal{B}\} \in \mathbf{Cat/P}$, reducing to the original one when both $\Gamma: \mathcal{A} \to \mathbf{P}$ and $\Gamma: \mathcal{B} \to \mathbf{P}$ are constant at 0, for which we have a 2-natural isomorphism

$$
(\mathbf{Cat/P})(\mathcal{K} \circ \mathcal{A}, \mathcal{B}) \quad \cong \quad (\mathbf{Cat/P})(\mathcal{K}, \{\mathcal{A}, \mathcal{B}\}). \qquad (2.7)
$$

Here $\circ$ is coherently associative, with unit $\mathcal{J}$ given by the $\Gamma: 1 \to \mathbf{P}$ naming $1 \in \mathbf{P}$, and so provides on $\mathbf{Cat/P}$ a right-closed(highly non-symmetric) monoidal structure. For the details, see [13].

In [14], the author used the word *club*, or more precisely *club over* $\mathbf{P}$, for a $\circ$−monoid $\mathcal{K}$ in $\mathbf{Cat/P}$. Such a club gives a 2-monad $\mathcal{K} \circ -$ on $\mathbf{Cat}$, an algebra for which, given by a category $\mathcal{A}$ and an action $\mathcal{K}o\mathcal{A} \to \mathcal{A}$, can equally be seen as given by $\mathcal{A}$ and a club-map $\mathcal{K} \to \{\mathcal{A}, \mathcal{A}\}$; of course the endo-object $\{\mathcal{A}, \mathcal{A}\}$ has a canonical club-structure. Now the data describing such a structure as we considered two paragraphs back may be said to constitute a *pre-club* $\mathcal{L}$; it has objects and arrows *naming* the structural functors T and the structural natural transformations f of (2.1), although it knows nothing of a particular carrier $\mathcal{A}$; and it has selected object-pairs and diagrams, corresponding to the axioms. To give such a structure to a category $\mathcal{A}$ is to give , in an evident sense, a *map of pre-clubs* $\mathcal{L} \to \{\mathcal{A}, \mathcal{A}\}$. It is shown in [14] that each pre-club $\mathcal{L}$ generates a club $\mathcal{K}$ — we see $\mathcal{L}$ as consisting of generators and relations for $\mathcal{K}$ — so that such a map $\mathcal{L} \to \{\mathcal{A}, \mathcal{A}\}$ of pre-clubs is the same thing as a map $\mathcal{K} \to \{\mathcal{A}, \mathcal{A}\}$ of clubs. It follows that the category $\mathcal{L}$-Alg of categories $\mathcal{A}$ with the given structure, and functors preserving the structure strictly, is equally the category — canonically in fact a 2-category — $\mathcal{K}$-Alg of algebras for the monad $\mathcal{K} o -$. {What is more, the *non-strict* morphisms of algebras, like the "monoidal functors" of Eilenberg and Kelly [9], have precisely their expected description in terms of the monad, so that the results of Blackwell, Kelly and Power [2] are applicable.}

The connexion between diagrams of natural transformations and free structures now becomes clear, in this very simple case. The commutativity of diagrams made from the structural natural transformations, originally seen by Mac Lane and the author as (in effect) a matter of commutativity in $\{\mathcal{A}, \mathcal{A}\}$, is more cleanly seen as a matter of commutativity in $\mathcal{K}$: for the image under $\mathcal{K} \to \{\mathcal{A}, \mathcal{A}\}$ of a non-commuting diagram may commute for a particular algebra $\mathcal{A}$, and indeed certainly does so for the terminal algebra $\mathcal{A} = 1$.(It cannot, however, commute for all algebras $\mathcal{A}$, since $\mathcal{K} \to \{\mathcal{K}, \mathcal{K}\}$ is faithful — see [14, Lemma 5.1].) Because $\mathcal{K}$ is the quotient $\mathcal{N}/q$ of a certain category $\mathcal{N}$, straightforwardly generated from $\mathcal{L}$, by the congruence q of "the diagrams which commute", determination of this congruence is equivalent to determining $\mathcal{K}$ explicitly. Of course a diagram in $\mathcal{K}$ cannot commute unless its image under $\Gamma : \mathcal{K} \to \mathbf{P}$ does so — we cannot equate natural transformations f, g of different "arities" ξ, η — so a result of the form "*all* diagrams commute" is the assertion that Γ is *faithful*. This is indeed the case where $\mathcal{K}$ is the club for symmetric monoidal categories; in fact Mac Lane's result from [23] may be precisely expressed by saying that, for this club, $\Gamma : \mathcal{K} \to \mathbf{P}$ is an equivalence. $\mathbf{P}$ itself is of course a club: its algebras are *strict* symmetric monoidal categories. The club $\mathcal{K}$ for merely monoidal categories also has $\Gamma : \mathcal{K} \to \mathbf{P}$ faithful; but now $\mathcal{K}$ is equivalent to the discrete category $\mathbf{N}$ on the natural numbers.

On the other hand, $\mathcal{K}o\mathcal{A}$ is the free $\mathcal{K}$-algebra on $\mathcal{A}$; so that to know $\mathcal{K}$ is also to know these free algebras. Indeed $\mathcal{K}$ itself, as a category, is by (2.6) the free $\mathcal{K}$-algebra on 1. Thus these monads arising from clubs have this curious property, that the free algebra on $\mathcal{A}$ is "almost" known when the free algebra on 1 is known. "Almost", because to form $\mathcal{K} \circ \mathcal{A}$ we must know not only the category $\mathcal{K}$ but also the augmentation $\Gamma \colon \mathcal{K} \to \mathbf{P}$; this last, however, in practice is read off directly from the arities of the structural data.

2.2.   We now consider extensions to more general structures on categories. First, everything above (and below) extends directly to many-sorted algebras, where we are dealing with monads not on **Cat** but on some power $\mathbf{Cat}^X$; again the details are given in [13] and [14], and we say no more on this. An extension in a different direction is given by replacing the permutation $\xi : n \to n$ in (2.2) by a function $\phi : n \to m$, so that now Γ takes its values not in **P** but in the skeletal category **S** of finite sets, whose objects are the natural numbers. Natural transformations of this kind have components of the form $T(A_{\phi 1}, \cdots, A_{\phi n}) \to S(A_1, \cdots, A_m)$, and include those describing a category $\mathcal{A}$ with finite coproducts: to wit, $i \colon A \to A + B$, $j \colon B \to A + B$, $f \colon A + A \to A$, and $s \colon 0 \to A$, whose respective arities are the two functions $1 \to 2$, the unique function $2 \to 1$, and the unique function $0 \to 1$. Monoids for the monoidal structure $\circ$ on $\mathbf{Cat/S}$ may be called *clubs over* **S**; and **S** itself is the club for categories with strictly-associative finite coproducts. Another example of a club over **S**, studied in [16], is that describing a category with two symmetric monoidal structures and "distributivity" morphisms $(A_1 \otimes B) \oplus \cdots \oplus (A_n \otimes B) \to (A_1 \oplus \cdots \oplus A_n) \otimes B$. Every club over **P** "is", *a fortiori*, a club over **S**.

However, there is a different extension to *clubs over* $\mathbf{S}^{op}$, where now the arity of $f \colon (n, T) \to (m, S)$ is not a function $\phi \colon n \to m$ as above but a function $\psi \colon m \to n$; such natural transformations have components of the form $T(A_1, \cdots, A_n) \to S(A_{\psi 1}, \cdots, A_{\psi m})$, as typified by the $p \colon A \times B \to A$, $q \colon A \times B \to B$, $d \colon A \to A \times A$, $t \colon A \to 1$ which describe a category with finite products. Again, every club over **P** may be seen as a club over $\mathbf{S}^{op}$. See [13] and [14] again.

The monoids arising from clubs over **S** or $\mathbf{S}^{op}$ have the "curious property" mentioned at the end of §2.1: the free algebra on $\mathcal{A}$ is determined by the free algebra $\mathcal{K}$ on 1 together with the augmentation $\Gamma \colon \mathcal{K} \to \mathbf{S}$ or $\Gamma \colon \mathcal{K} \to \mathbf{S}^{op}$, which in practice is known.

The author speculated in [13] that it might be possible to extend these ideas further, so as to include for instance the monad T for categories with both finite products and finite coproducts. He now strongly doubts that this is the case in any useful sense; it seems most unlikely that TA is determined by $\mathcal{K} = T1$ and some functor Γ from $\mathcal{K}$ to some "category of arities". The matter would be settled by a proof that T is not a club in the abstract sense introduced below.

2.3. It did prove possible in [13] and [14], however, to extend the club formalism in yet another direction, capturing such cases as the monad for symmetric monoidal closed categories, and allowing the elucidation mentioned above of the connexion between free structures and the results of [21] on commutative diagrams of natural transformations. Here the natural transformations, which include the $e_{XY} : [X, Y] \otimes X \to Y$ and $d_{XY} : X \to [Y, X \otimes Y]$ of our first paragraph above, are the generalized ones of Eilenberg and Kelly [8]; as in the $\mathbf{Cat/P}$ case, every variable occurs just twice — but perhaps on the same side of the arrow, and then with opposite variances. $\mathbf{P}$ is now replaced by a certain category $\mathbf{G}$, whose objects are not natural numbers but strings of $+$ and $-$ signs, representing covariance and contravariance; and whose maps, in the first instance, are pairings–off of these signs, representing variables to be linked, and called *graphs*. The arity ΓT of a functor-term like $[-, -] \otimes -$ is the object $(- + +)$ of $\mathbf{G}$, and the arity of a natural-transformation-term like $e : [X, Y] \otimes X \to Y$ is the morphism $\Gamma e : (- + +) \to (+)$ of $\mathbf{G}$ linking the $-$ to the second $+$ and the first $+$ to the third one.

As is made clear in [8], however, these generalized natural transformations are composable only when their graphs are what is called *compatible* – which broadly means that no free loops (or *islands*) appear when one abuts the two graphs at the common object. { A prime example of incompatibility is provided by the canonical maps $\alpha : \mathbf{R} \to A^* \otimes A$ and $\beta : A^* \otimes A \to \mathbf{R}$ in the category of finite-dimensional real vector spaces, with A^* denoting the dual of A; each of α and β is natural, but $\beta\alpha : \mathbf{R} \to \mathbf{R}$ depends on an A hidden in the "island".} As a consequence, we have here for categories $\mathcal{A}$ and $\mathcal{B}$ no analogue of the $\{\mathcal{A}, \mathcal{B}\}$ above. We do not even have, as yet, a category $\mathbf{G}$ — for we have not said how we are to compose incompatible graphs; compatible ones are of course composed as in [8]. The device adopted in [13] and [14] was to compose incompatible graphs by just discarding the islands; but a better device was adopted in the later [15]: the morphisms $P \to Q$ in $\mathbf{G}$ consist of the graphs together with a "zero morphism" $*$, and the composite of two incompatible graphs is $*$. Now $\mathbf{G}$ is a category; and although $\{\mathcal{A}, \mathcal{B}\}$ is lacking, we can easily extend the definition in §2.1 above of $\mathcal{K} \circ \mathcal{A}$ to the case of $\Gamma : \mathcal{K} \to \mathbf{G}$, provided that $*$ is not in the image of Γ. If we write $(\mathbf{Cat/G})'$ for the full subcategory of $\mathbf{Cat/G}$ determined by such Γ, we have $\circ : (\mathbf{Cat/G})' \times \mathbf{Cat} \to \mathbf{Cat}$, extending to $\circ : (\mathbf{Cat/G})' \times (\mathbf{Cat/G})' \to (\mathbf{Cat/G})'$, and providing a monoidal structure — not a right-closed one here — on $(\mathbf{Cat/G})'$.

A $\circ$-monoid $\mathcal{K}$ in $(\mathbf{Cat/G})'$ may be called a *club over* $\mathbf{G}$(although here $\mathbf{G}$ itself is not in $(\mathbf{Cat/G})'$, and is not a club). It gives rise to a monad $\mathcal{K} \circ -$ on $\mathbf{Cat}$, although (unlike the clubs over $\mathbf{P}$, $\mathbf{S}$ or $\mathbf{S}^{op}$) not in general a *2-monad* on $\mathbf{Cat}$; more precisely, it gives a 2-monad on the 2-category $\mathbf{Cat}_g$

of categories, functors, and natural *isomorphisms*. { Indeed, it is shown that in [2, §6.2] that , if $\mathcal{K}$ is the club for symmetric monoidal closed categories, the functor $\mathcal{K} \circ - :$ **Cat** $\to$ **Cat** underlies no 2-functor **Cat** $\to$ **Cat** whatsoever.} There is again the notion of a pre-club $\mathcal{L}$, describing the basic functors and natural transformations of the structure, and the equational axioms these are to satisfy; and it generates a club $\mathcal{K}$, unless during the generation process we come to a point where incompatibles would have to be composed. This can well happen, even if $\mathcal{L}$ itself has no incompatibles; an example is the monad for *compact closed categories*, studied by Laplaza and Kelly [20]; with the abstract notion of club introduced in the present paper, we are able to show in §5.5 below that this monad does not arise from any club. It is otherwise for structures like symmetric monoidal closed categories, or monoidal closed or biclosed categories, where we begin with a structure like symmetric monoidal categories or monoidal categories, describable by a club over **P**, and then postulate the existence of right adjoints to some of its functors, such as $- \otimes B$ $\dashv [B, -]$; the author showed in [15] that in such cases $\mathcal{L}$ always generates a club $\mathcal{K}$. Although $\{\mathcal{A}, \mathcal{A}\}$ is lacking, one can make sense, for a $\mathcal{K}$-algebra $\mathcal{A}$, of "that part of $\{\mathcal{A}, \mathcal{A}\}$ which would be the image of $\mathcal{K} \to \{\mathcal{A}, \mathcal{A}\}$", and so once again clarify the relation of free structures to commuting diagrams of natural transformations. See [14] for the details. Since here too the free algebra $\mathcal{K} \circ 1$ on 1 is the category $\mathcal{K}$, we have again the "curious property" that the free algebra $\mathcal{K} \circ \mathcal{A}$ on $\mathcal{A}$ is "almost" determined by that on 1.

2.4.   This seems to be about as far as the club notion can usefully take us, as far as categories with structure are concerned — except that, now that the place of *braided* monoidal categories has been recognized, there are obvious extensions of the above with **P** replaced by the braid category **B**. Here too it is possible to use the analogue of $(\mathbf{Cat}/\mathbf{G})'$, and so consider such structures as braided monoidal closed categories. However such structures as *cartesian* closed categories, where we start with a cartesian category, described by a club in $\mathbf{Cat}/\mathbf{S}^{op}$, and postulate a right adjoint for each $- \times B$, would seem to elude the club formulation. Indeed it is notorious that the morphisms in the free cartesian closed category on some category $\mathcal{A}$ (possibly discrete) do not appear as natural transformations; and much recent effort has gone into establishing for them such weaker properties as *dinaturality*.

Such comments as the above are necessarily vague. After all, motivated by the desire to explicate the results of [21], we merely introduced *ad hoc* clubs over **P** and over **G**, and then observed that we could extend the first of these to clubs over **S**, $\mathbf{S}^{op}$, or **B**, while seeing no way to include categories with both finite products and finite coproducts, or cartesian closed categories. To go beyond vagueness and make rational statements on the limits of the club formalism, we need an abstract notion of a club, as a monad with certain properties.

The "curious property" mentioned at the end of §2.1 for clubs over $\mathbf{P}$, and reappearing for clubs over $\mathbf{S}, \mathbf{G}$, and so on, plays a role in suggesting the appropriate abstract notion; but not quite a direct one, since this "curious property" is rather relative than absolute. For a club $\mathcal{K}$ over $\mathbf{P}$, the map $\Gamma \colon \mathcal{K} \to \mathbf{P}$ of o-monoids gives a monad-map $\alpha \colon T \to S$, where $T = \mathcal{K} \circ -$ and $S = \mathbf{P} \circ -$. When we say that the free T-algebra $T\mathcal{A} = \mathcal{K} \circ \mathcal{A}$ on any $\mathcal{A}$ is known when we know (besides $\mathcal{A}$) the free algebra $T1 = \mathcal{K}$ and the $\alpha 1 \colon T1 \to S1$ given by $\Gamma \colon \mathcal{K} \to \mathbf{P}$, it is understood that S is already fully known to us. The description above of the objects and morphisms of $\mathcal{K} \circ \mathcal{A}$ shows that $T\mathcal{A}$ is in fact given by the pullback

$$
\begin{array}{ccc}
T\mathcal{A} & \xrightarrow{\;T!\;} & T1 \\
\alpha\mathcal{A}\downarrow & & \downarrow\alpha 1 \\
S\mathcal{A} & \xrightarrow[\;S!\;]{} & S1 \ ,
\end{array}
\tag{2.8}
$$

and the monoidal structure on $\mathbf{Cat}/\mathbf{P} = \mathbf{Cat}/S1$ translates under (2.8) to a monoidal structure on the full subcategory $\mathcal{M}/S$ of $[\mathbf{Cat}, \mathbf{Cat}]/S$ determined by those endofunctors $\alpha \colon T \to S$ for which (2.8) is a pullback. Moreover this is just the restriction of the obvious monoidal structure on $[\mathbf{Cat}, \mathbf{Cat}]/S$ given by composition. This leads to the abstract definition: a monad S on $\mathbf{Cat}$ is called a *club* if this composition-monoidal-structure on $[\mathbf{Cat}, \mathbf{Cat}]/S$ does indeed restrict to $\mathcal{M}/S$. In this definition, $\mathbf{Cat}$ can clearly be replaced by any category with finite limits; when we do this, we find many more examples of clubs that occur in computer science. We turn now to the details.

3.   The abstract notion of club

3.1. Recall from [10] the notion of a *factorization system* $(\mathcal{E}, \mathcal{M})$ on a category $\mathcal{B}$; and call the factorization system *local* if (i) $fg \in \mathcal{E}$ and $f \in \mathcal{E}$ imply $g \in \mathcal{E}$, and (ii) $\mathcal{E}$ is stable by pullbacks. For a finitely-complete $\mathcal{B}$, it is shown in §4 of Cassidy, Hébert, and Kelly [3] that local factorization systems on $\mathcal{B}$ are in bijection with *localizations* of $\mathcal{B}$. By such a localization we mean a full, replete, reflective subcategory $\mathcal{A}$ of $\mathcal{B}$ for which the reflexion $r \colon \mathcal{B} \to \mathcal{A}$ is left exact. In terms of the localization, $\mathcal{E}$ consists of the morphisms inverted by r, and $\mathcal{M}$ of those morphisms $m : B \to C$ for which

$$
\begin{array}{ccc}
B & \xrightarrow{\;\rho B\;} & rB \\
m\downarrow & & \downarrow r(m) \\
C & \xrightarrow[\;\rho C\;]{} & rC
\end{array}
\tag{3.1}
$$

is a pullback, where ρ is the unit of the reflexion. The $(\mathcal{E}, \mathcal{M})$– factorization $f = me$ of an arbitrary $f : B \to C$ is given by taking for m the pullback along ρC of $r(f)$. In terms of the factorization system, $\mathcal{A}$ consists of those $A \in \mathcal{B}$ for which the unique $A \to 1$ lies in $\mathcal{M}$, and the reflexion $\rho B : B \to rB$ of B into $\mathcal{A}$ is the $\mathcal{E}$-part of the $(\mathcal{E}, \mathcal{M})$-factorization of $B \to 1$.

For any $C \in \mathcal{B}$, write $\mathcal{M}/C$ for the full subcategory of $\mathcal{B}/C$ determined by those $m : B \to C$ which lie in $\mathcal{M}$. Clearly, $\mathcal{M}/C$ is reflective in $\mathcal{B}/C$, the reflexion of $f : B \to C$ again being given by the $(\mathcal{E}, \mathcal{M})$–factorization of f.

In fact $\mathcal{M}/C$ is equivalent to $\mathcal{A}/rC$: the functor $\mathcal{M}/C \to \mathcal{A}/rC$ sends $m : B \to C$ to $r(m) : rB \to rC$, while the functor $\mathcal{A}/rC \to \mathcal{M}/C$ sends $g : A \to rC$ to its pullback m in

$$
\begin{array}{ccc}
B & \xrightarrow{\;h\;} & A \\
{\scriptstyle m}\downarrow & & \downarrow{\scriptstyle g} \\
C & \xrightarrow[\rho C]{} & rC \quad ;
\end{array}
\tag{3.2}
$$

note that $g \in \mathcal{M}$ since it lies in $\mathcal{A}$, so that its pullback m lies in $\mathcal{M}$. That $\mathcal{M}/C \to \mathcal{A}/rC \to \mathcal{M}/C$ is isomorphic to the identity is clear; while $\mathcal{A}/rC \to \mathcal{M}/C \to \mathcal{A}/rC$ is isomorphic to the identity because h in (3.2) is essentially $\rho B : B \to rB$: indeed $h = k.\rho B$ for some $k : rB \to A$, but h, as the pullback of ρC, lies in $\mathcal{E}$, so that $k \in \mathcal{E}$, whence k, which is in $\mathcal{M}$ since it lies in $\mathcal{A}$, is invertible. Clearly, under the equivalence $\mathcal{M}/C \simeq \mathcal{A}/rC$, the reflexion of $\mathcal{B}/C$ onto $\mathcal{M}/C$ becomes just the reflexion of $\mathcal{B}/C$ onto $\mathcal{A}/rC$ sending $f : B \to C$ to $r(f) : rB \to rC$.

3.2. We apply the above to the case where $\mathcal{A}$ is a finitely-complete category and $\mathcal{B}$ is a functor category $[\mathcal{C}, \mathcal{A}]$ where $\mathcal{C}$ has a terminal object 1. The diagonal $\Delta : \mathcal{A} \to [\mathcal{C}, \mathcal{A}]$, sending A to the functor $\Delta A : \mathcal{C} \to \mathcal{A}$ constant at A, embeds $\mathcal{A}$ as a full subcategory of $[\mathcal{C}, \mathcal{A}]$; and Δ has the left adjoint $E : [\mathcal{C}, \mathcal{A}] \to \mathcal{A}$ given by evaluation at 1, the unit of the adjunction having components $T! : TC \to T1$, where $! : C \to 1$ is the unique map in $\mathcal{C}$. Since E is of course left exact, $\mathcal{A}$ is indeed a localization of $[\mathcal{C}, \mathcal{A}]$ (except that it is not replete — a matter of no importance here), and we have on $[\mathcal{C}, \mathcal{A}]$ a local factorization system $(\mathcal{E}, \mathcal{M})$ as above.

Here $\alpha : T \to S$ belongs to $\mathcal{E}$ if and only if $\alpha 1 : T1 \to S1$ is invertible, and α belongs to $\mathcal{M}$ if and only if each

$$
\begin{array}{ccc}
TC & \xrightarrow{\;T!\;} & T1 \\
{\scriptstyle \alpha C}\downarrow & & \downarrow{\scriptstyle \alpha 1} \\
SC & \xrightarrow[S!]{} & S1
\end{array}
\tag{3.3}
$$

is a pullback; *cf.* (2.8). Observe that, by the elementary properties of pull-backs, α belongs to $\mathcal{M}$ if and only if

$$
\begin{array}{ccc}
TC & \xrightarrow{\;Tf\;} & TD \\
{\scriptstyle \alpha C}\downarrow & & \downarrow{\scriptstyle \alpha D} \\
SC & \xrightarrow[\;Sf\;]{} & SD
\end{array}
\tag{3.4}
$$

is a pullback for each $f : C \to D$ in $\mathcal{C}$. Such transformations, all of whose naturality squares are pullbacks, have recently been observed in various computer-science contexts; some writers have called them *cartesian* natural transformations, but this name does not seem particularly apt.

Now for $S \in [\mathcal{C}, \mathcal{A}]$ we have as above the reflexion, given by the factorization, of $[\mathcal{C}, \mathcal{A}]/S$ onto its full subcategory $\mathcal{M}/S$; and the equivalence $\mathcal{M}/S \simeq \mathcal{A}/S1$, which in the forward direction sends $\alpha : T \to S$ in $\mathcal{M}$ to $\alpha 1 : T1 \to S1$ and in the other sends $g : K \to S1$ in $\mathcal{A}$ to $\alpha : T \to S$ given by the pullbacks

$$
\begin{array}{ccc}
TC & \xrightarrow{\;hC\;} & K \\
{\scriptstyle \alpha C}\downarrow & & \downarrow{\scriptstyle g} \\
SC & \xrightarrow[\;S!\;]{} & S1 \quad ;
\end{array}
\tag{3.5}
$$

of course there is a unique way of defining $Tf : TC \to TD$ that makes T a functor and $\alpha : T \to S$ and $h : T \to \Delta K$ natural, and then we have $T1 = K$ and $T! = hC$ – provided that we choose the trivial pullback (3.5) when $C = 1$.

Note that, if $\mathcal{C}$ like $\mathcal{A}$ is only locally small, the category $[\mathcal{C}, \mathcal{A}]/S$ is not in general locally small; but its reflective subcategory $\mathcal{M}/S \simeq \mathcal{A}/S1$ is so.

3.3.   The case of central interest to us is that where $\mathcal{C} = \mathcal{A}$. The category $[\mathcal{A}, \mathcal{A}]$ has a monoidal structure, whose tensor product is composition and whose unit is the identity endofunctor $I : \mathcal{A} \to \mathcal{A}$. Suppose now that S is a monad on $\mathcal{A}$, with unit $j : I \to S$ and multiplication $n : SS \to S$. Then $[\mathcal{A}, \mathcal{A}]/S$ has an induced monoidal structure, the value $\alpha \otimes \alpha'$ of whose tensor product on objects $\alpha : T \to S$ and $\alpha' : T' \to S$ is the composite

$$
TT' \xrightarrow{\;\alpha\alpha'\;} SS \xrightarrow{\;n\;} S,
\tag{3.6}
$$

while its value $\rho \otimes \rho' : \alpha \otimes \alpha' \to \beta \otimes \beta'$ on the morphisms

$$
\begin{array}{ccccc}
T & \xrightarrow{\;\rho\;} & R & \qquad & T' \xrightarrow{\;\rho'\;} R' \\
 {\scriptstyle \alpha}\searrow & & \swarrow{\scriptstyle \beta} & & {\scriptstyle \alpha'}\searrow \quad \swarrow{\scriptstyle \beta'} \\
 & S & & , & \qquad S
\end{array}
\tag{3.7}
$$

is given by

$$\rho \otimes \rho' = \rho\rho' : TT' \to RR'; \tag{3.8}$$

the unit object for $\otimes$ is $j : I \to S$.

We shall call the monad S on $\mathcal{A}$ a *club* if the full subcategory $\mathcal{M}/S$ is closed under the monoidal structure $(\otimes, j)$ on $[\mathcal{A}, \mathcal{A}]/S$.

Proposition 3.1. *The monad (S, j, n) on $\mathcal{A}$ is a club if and only if j and n lie in $\mathcal{M}$, and $S\mathcal{M} \subset \mathcal{M}$; this last is equivalent to the preservation by S of all pullbacks of the form (3.5).*

Proof. It is of course necessary that j be in $\mathcal{M}$; that it is necessary for n to be in $\mathcal{M}$ is seen by taking $\alpha = \alpha' = 1_S$ in (3.6). We can write (3.6) as $n.S\alpha'.\alpha T'$, and clearly $\alpha T' \in \mathcal{M}$ if $\alpha \in \mathcal{M}$; so given $j, n \in \mathcal{M}$, it suffices that $S\alpha'$ be in $\mathcal{M}$ for $\alpha' \in \mathcal{M}$; and this is necessary since α could be 1_S. Thus the condition is that S of (3.3) be a pullback whenever $\alpha \in \mathcal{M}$; which, by the preceding discussion, is to say that S preserves pullbacks of the form (3.5).

Remark 3.2. In most, if not all, of the examples we shall give, S preserves *all* pullbacks.

Remark 3.3. One might envisage an alternative way of obtaining from the monoidal structure $\otimes$ on $[\mathcal{A}, \mathcal{A}]/S$ a monoidal structure on the reflective $\mathcal{M}/S$: not by requiring $\otimes$ to restrict, but by asking whether there is some (necessarily unique) monoidal structure on $\mathcal{M}/S$ for which the reflexion adjunction between $\mathcal{M}/S$ and $[\mathcal{A}, \mathcal{A}]/S$ becomes a monoidal adjunction. For this we have the criterion provided by Day [7]; and it turns out never to be satisfied except in the trivial cases $\mathcal{A} = 1$ and $\mathcal{A} = (0 \to 1)$, with $S = I$.

A monoid in the monoidal $[\mathcal{A}, \mathcal{A}]/S$ is of course an $\alpha : T \to S$ where T is a monoid in $[\mathcal{A}, \mathcal{A}]$ — that is, a monad (T, i, m) — and α is a map of monads. So a monoid in $\mathcal{M}/S$ is just such an $\alpha : T \to S$ with $\alpha \in \mathcal{M}$.

Proposition 3.4. *If S is a club and $\alpha : T \to S$ is a $\otimes$-monoid in $\mathcal{M}/S$ – that is, a monad-map with $\alpha \in \mathcal{M}$ – then T too is a club.*

Proof. The map i is in $\mathcal{M}$ since j and α are so and $\alpha i = j$. Moreover if $\beta : R \to T$ and $\beta' : R' \to T$ are in $\mathcal{M}$ so is $m.\beta\beta'$; for α is in $\mathcal{M}$ and $\alpha.m.\beta\beta' = n.\alpha\alpha.\beta\beta' = n.(\alpha\beta)(\alpha\beta')$ is in $\mathcal{M}$ since S is a club.

We can define a category $\mathbf{Club}\mathcal{A}$ to consist of those monads on $\mathcal{A}$ which are clubs, along with those monad-maps that lie in $\mathcal{M}$. Then, for a club S, the category $(\mathbf{Club}\mathcal{A})/S$ of "clubs over S" is, by the above, the category of

$\otimes$-monoids in the monoidal $\mathcal{M}/S$, which is equivalent to $\mathcal{A}/S1$, and thus locally small when $\mathcal{A}$ is so. We now consider the corresponding monoidal structure on $\mathcal{A}/S1$.

3.4. Fixing therefore on a particular club $S = (S, j, n)$ on $\mathcal{A}$, set

$$G = S1,\ k = j1 : 1 \to G,\ \nu = n1 : SG \to G,\ \sigma A = S!: SA \to S1 = G. \quad (3.9)$$

The monoidal structure $\otimes$ on $\mathcal{M}/S$ translates, via the equivalence $\mathcal{M}/S \simeq \mathcal{A}/S1$, to a monoidal structure on $\mathcal{A}/G$, with the tensor product say $\circ : \mathcal{A}/G \times \mathcal{A}/G \to \mathcal{A}/G$. Moreover the monoidal $\mathcal{A}/G$ has an *action* $* : \mathcal{A}/G \times \mathcal{A} \to \mathcal{A}$ on $\mathcal{A}$, corresponding via the strong monoidal functor

$$\mathcal{A}/G \simeq \mathcal{M}/S \to [\mathcal{A}, \mathcal{A}]/S \to [\mathcal{A}, \mathcal{A}] \quad (3.10)$$

to the evaluation action $[\mathcal{A}, \mathcal{A}] \times \mathcal{A} \to \mathcal{A}$. { Note that here, in contrast to our usage in §2 above, we use different symbols for $\circ$ and $*$; this is because, in the present generality, it may not be possible to see $*$ as a restriction of $\circ$; we return to this point in §3.5 below.} We now give explicit expressions for $*$ and $\circ$, which are of course automatically coherently associative.

We write an object $g : K \to G$ of $\mathcal{A}/G$ as (K, g), sometimes suppressing from the notation the *augmentation* g; compare the $\Gamma : \mathcal{K} \to \mathbf{P}$ and so on of §2. The corresponding object $\alpha : T \to S$ of $\mathcal{M}/S$ is given by the pullback (3.5), which we may re-write to give the value TA of $(K, g) * A$ as the pullback

$$
\begin{array}{ccc}
(K, g) * A & \xrightarrow{\partial_1} & SA \\
{\scriptstyle \partial_0}\downarrow & & \downarrow{\scriptstyle \sigma A} \\
K & \xrightarrow{\hspace{1em} g \hspace{1em}} & G \ ;
\end{array}
\quad (3.11)
$$

in other words, we have

$$(K, g) * A = (K, g) \times_G (SA, \sigma A). \quad (3.12)$$

On the other hand we can express S in terms of $*$, for clearly

$$(G, 1_G) * A = SA, \qquad (G, 1_G) * f = Sf; \quad (3.13)$$

if we add to this the observation that

$$(K, g) * 1 = K, \quad (3.14)$$

which is clear since $\sigma 1 = 1_G$, we can re-write (3.11) yet again as the pullback

$$
\begin{array}{ccc}
(K,g) * A & \xrightarrow{\ g*A\ } & (G,1_G) * A \\
{\scriptstyle (K,g)*!}\downarrow & & \downarrow{\scriptstyle (G,1_G)*!} \\
(K,g) * 1 & & (G,1_G) * 1 \\
{\scriptstyle \cong}\downarrow & & \downarrow{\scriptstyle \cong} \\
K & \xrightarrow[\ g\ ]{} & G.
\end{array}
\qquad (3.15)
$$

Note that the ν of (3.9) may be written as

$$
\nu : (G,1_G) * G \to G. \qquad (3.16)
$$

We turn now to the monoidal structure $\circ$ on $\mathcal{A}/G$. Clearly its unit object, corresponding to $j : I \to S$, is $k : 1 \to G$. As for its tensor product, let (A,h) in $\mathcal{A}/G$ correspond to $\beta : R \to S$ as (K,g) corresponds to $\alpha : T \to S$. Then $\alpha \otimes \beta$ is, as in (3.6), the composite $n.\alpha\beta : TR \to SS \to S$, and $(K,g) \circ (A,h)$ is the result $TR1 \to SS1 \to S1$ of evaluating this at 1. It follows at once that:

Proposition 3.5. $(K,g) \circ (A,h)$ is the object $(K,g) * A$ of $\mathcal{A}$ with the augmentation

$$
(K,g) * A \xrightarrow{\ g*h\ } (G,1_G) * G \xrightarrow{\ \nu\ } G. \qquad (3.17)
$$

Of course the $\circ$-monoids (K,g) in $\mathcal{A}/G$ correspond to the clubs T over S, with T given explicitly as the monad $(K,g) * -$; here — see (3.13) — the club S itself corresponds to the $\circ$-monoid $((G,1_G), k, \nu)$. If we use an alternative language, calling the $\circ$-monoid $K = (K,g)$ itself *a club over* $G = (G,1_G)$, we recapture the terminology of §2; and we shall show in §5 that the clubs of §2 are indeed clubs in the present abstract sense. Note what has become of the "curious property" at the end of §2.1: by (3.12), the free T-algebra $(K,g) * A$ on A is determined by K, which by (3.14) is the free T-algebra on 1, along with the augmentation $g : K \to G$. This, however, supposes that S is known; and is thus a property, not of clubs abstractly, but of clubs over S, as discussed in §2.4 above.

3.5. Sometimes we can embed $\mathcal{A}$ fully in $\mathcal{A}/G$, by assigning to $A \in \mathcal{A}$ a "trivial augmentation" $A \to G$, in such a way that $*$ becomes the restriction of $\circ$; this was so in the examples of §2.

It is so whenever there is an S-algebra map $\rho : 1 \to G$ from the terminal S-algebra 1 to the free S-algebra $S1$; and *a fortiori* when the category S-Alg is pointed.

Indeed, writing G for $(G, 1_G)$ and identifying $G * 1$ with G, to say that ρ is a map of S-algebras is to demand commutativity in

$$
\begin{array}{ccc}
G & \xrightarrow{\;!\;} & 1 \\
{\scriptstyle G*\rho}\big\downarrow & & \big\downarrow{\scriptstyle \rho} \\
G * G & \xrightarrow[\nu]{} & G
\end{array}
\qquad (3.18)
$$

The functor $\mathcal{A} \to \mathcal{A}/G$ sending A to itself with the "trivial augmentation"

$$
A \xrightarrow{\;!\;} 1 \xrightarrow{\;\rho\;} G \qquad (3.19)
$$

is clearly fully faithful, and by (3.17) the augmentation of $(K, g) \circ (A, \rho!)$ is $\nu(g * \rho!) = \nu(G * \rho)(G*!)(g * A)$, which by (3.18) is $\rho!.(G*!)(g * A) = \rho!$, again a trivial augmentation.

4.   The enriched case

4.1. In §2.1 we mentioned, but did not particularly emphasize, that the "functor" $\circ : \mathbf{Cat/P} \times \mathbf{Cat/P} \to \mathbf{Cat/P}$ is in fact a 2-functor, as is its right adjoint $\{-,-\}$, while (2.3) is a 2-adjunction; so that a club $\mathcal{K}$ in $\mathbf{Cat/P}$ gives not just a mere monad $\mathcal{K}\circ-$ on the category $\mathbf{Cat}$, but a 2-monad on the 2-category $\mathbf{Cat}$. Similarly for clubs over $\mathbf{S}$ and $\mathbf{S}^{op}$; while even for clubs over $\mathbf{G}$, as we said in §2.3, $\mathcal{K}\circ-$ is a 2-monad, not now on the 2-category $\mathbf{Cat}$, but on the 2-category $\mathbf{Cat}_g$. In more precise terms, everything in the $\mathbf{P}$, $\mathbf{S}$ and $\mathbf{S}^{op}$ cases is enriched over the cartesian closed category $\mathbf{Cat}$, while in the $\mathbf{G}$ case we have enrichment over the cartesian closed category $\mathbf{Gpd}$ of groupoids. Accordingly it behoves us to examine what becomes of the abstract notion of club in the case of categories enriched over some symmetric monoidal closed $\mathcal{V}$. We indicate briefly how this extension goes.

4.2. The $\mathcal{B}$ of §3.1 is now a finitely-complete $\mathcal{V}$-category; in order that *finite* completeness should make sense, we suppose that $\mathcal{V}$ (like $\mathbf{Set}$, $\mathbf{Cat}$, $\mathbf{Gpd}$, $\mathbf{Ab}$, $\mathbf{R\text{-}Mod}$, and so on) is *locally finitely presentable as a closed category* in the sense of [19]. By a *factorization system* $(\mathcal{E}, \mathcal{M})$ *on* $\mathcal{B}$ we mean a factorization system on the underlying ordinary category $\mathcal{B}_0$ which, in place of the usual unique-diagonal-fill-in property, satisfies the stronger property that

$$
\begin{array}{ccc}
\mathcal{B}(C, X) & \xrightarrow{\;\mathcal{B}(e, X)\;} & \mathcal{B}(B, X) \\
{\scriptstyle \mathcal{B}(C, m)}\big\downarrow & & \big\downarrow{\scriptstyle \mathcal{B}(B, m)} \\
\mathcal{B}(C, Y) & \xrightarrow[\mathcal{B}(e, Y)]{} & \mathcal{B}(B, Y)
\end{array}
\qquad (4.1)
$$

is a pullback in $\mathcal{V}$ whenever $e \in \mathcal{E}$ and $m \in \mathcal{M}$. It comes to the same thing, writing $\langle c, X \rangle$ for the cotensor product of $c \in \mathcal{V}$ with $X \in \mathcal{B}$, to say that $(\mathcal{E}, \mathcal{M})$ is a factorization system on $\mathcal{B}_0$ for which $\langle c, m \rangle \colon \langle c, X \rangle \to \langle c, Y \rangle$ lies in $\mathcal{M}$ whenever $m \in \mathcal{M}$ and c is finitely presentable — as is immediately clear, since the finitely-presentable c constitute a strong generator for $\mathcal{V}_0$. We call the factorization system $(\mathcal{E}, \mathcal{M})$ on $\mathcal{B}$ local if it is local in the sense of §3.1 as a factorization system on $\mathcal{B}_0$ and if moreover $\langle c, e \rangle \in \mathcal{E}$ whenever $e \in \mathcal{E}$ and c is finitely presentable. By a localization $\mathcal{A}$ of $\mathcal{B}$ we mean of course a full replete reflective sub-$\mathcal{V}$-category, with reflexion $\rho A \colon A \to rA$ say, for which the $\mathcal{V}$-functor r is left exact — that is, it preserves not only conical finite limits but also the cotensor products $\langle c, X \rangle$ with c finitely presentable.

The Cassidy-Hébert-Kelly result of §3.1 now generalizes to give a bijection between localizations of $\mathcal{B}$ and local factorization systems on $\mathcal{B}$. For on the one hand, given a localization $\mathcal{A}$ of $\mathcal{B}$, the $\rho A \colon A \to rA$ exhibit the underlying $\mathcal{A}_0$ as a localization of $\mathcal{B}_0$, providing a local factorization system on $\mathcal{B}_0$. To see that this is indeed a factorization system on $\mathcal{B}$, or equivalently that $\langle c, m \rangle \in \mathcal{M}$ when $m \in \mathcal{M}$ (it being understood that c is finitely presentable), first observe that $\langle c, - \rangle$ preserves pullbacks, so that $\langle c, m \rangle$ is by (3.1) the pullback of $\langle c, r(m) \rangle$; the latter, however, is a map in $\mathcal{A}_0$, since $\mathcal{A}$ is closed under limits in $\mathcal{B}$; so it lies in $\mathcal{M}$, whence so too does its pullback $\langle c, m \rangle$. It is a local factorization system on $\mathcal{B}$ because, for $e \colon B \to C$ in $\mathcal{E}$, the map $r\langle c, e \rangle$ is conjugate by the isomorphisms $r\langle c, A \rangle \to \langle c, rA \rangle$ and $r\langle c, B \rangle \to \langle c, rB \rangle$ to the invertible $\langle c, r(e) \rangle$, and hence is itself invertible, so that $\langle c, e \rangle \in \mathcal{E}$. On the other hand, a local factorization system $(\mathcal{E}, \mathcal{M})$ on $\mathcal{B}$, being one on $\mathcal{B}_0$, gives a localization $\mathcal{A}_0$ of $\mathcal{B}_0$, and hence a full replete sub-$\mathcal{V}$-category $\mathcal{A}$ of $\mathcal{B}$ whose objects are those of $\mathcal{A}_0$. For $A \in \mathcal{A}$ we have $A \to 1$ in $\mathcal{M}$ and hence $\langle c, A \rangle \to \langle c, 1 \rangle \cong 1$ in $\mathcal{M}$, so that $\langle c, A \rangle \in \mathcal{A}$. Thus $\mathcal{B}_0(\rho B, 1) \colon \mathcal{B}_0(rB, \langle c, A \rangle) \to \mathcal{B}_0(B, \langle c, A \rangle)$ is invertible, so that $\mathcal{V}_0(c, \mathcal{B}(\rho B, A)) \colon \mathcal{V}_0(c, \mathcal{B}(rB, A)) \to \mathcal{V}_0(c, \mathcal{B}(B, A))$ is invertible, whence (since the c form a strong generator of $\mathcal{V}_0$) $\mathcal{B}(\rho B, A)$ is invertible: which is to say that the ρB constitute a reflexion of $\mathcal{B}$ onto $\mathcal{A}$. This reflexion is a localization; for the comparison map $k \colon r\langle c, B \rangle \to \langle c, rB \rangle$ lies in $\mathcal{M}$ because it is a map in $\mathcal{A}_0$, and lies in $\mathcal{E}$ because its composite with the map $\rho\langle c, B \rangle$ of $\mathcal{E}$ is the map $\langle c, \rho B \rangle$ of $\mathcal{E}$, and thus is invertible.

4.3. Specialize now to the case where $\mathcal{V}$ is cartesian closed, so that the unit object for the tensor product in $\mathcal{V}$ is the terminal object 1. We have $\mathcal{B}/C$ as a $\mathcal{V}$-category, defined by its universal property as a comma-object in $\mathcal{V}$-**Cat**; its objects are again maps $f \colon B \to C$, and its hom-object is easily seen to be given by the pullback

$$
\begin{array}{ccc}
(\mathcal{B}/C)(f,f') & \longrightarrow & \mathcal{B}(B,B') \\
\downarrow & & \downarrow{\scriptstyle\mathcal{B}(B,f')} \\
1 & \xrightarrow{\ f\ } & \mathcal{B}(B,C)
\end{array} \qquad (4.2)
$$

Suppose that $f' \in \mathcal{M}$, and that $f = me$ where $m \in \mathcal{M}$ and $e \in \mathcal{E}$; we can write (4.2) as the exterior of

$$
\begin{array}{ccccc}
(\mathcal{B}/C)(f,f') & \longrightarrow & \mathcal{B}(X,B') & \xrightarrow{\ \mathcal{B}(e,B')\ } & \mathcal{B}(B,B') \\
\downarrow & & \downarrow{\scriptstyle\mathcal{B}(X,f')} & & \downarrow{\scriptstyle\mathcal{B}(B,f')} \\
1 & \xrightarrow{\ m\ } & \mathcal{B}(X,C) & \xrightarrow{\ \mathcal{B}(e,C)\ } & \mathcal{B}(B,C),
\end{array}
$$

whose right square is a pullback by (4.2); thus the left square too is a pull-back, comparison of which with (4.2) gives an isomorphism $(\mathcal{B}/C)(f,f') \cong (\mathcal{B}/C)(m,f')$. That is to say, $f \mapsto m$ gives a reflexion of the $\mathcal{V}$-category $\mathcal{B}/C$ onto its full subcategory $\mathcal{M}/C$.

It is again the case that the functor $\mathcal{B}/C \to \mathcal{A}/rC$ induced by $r\colon \mathcal{B} \to \mathcal{A}$ restricts to an equivalence $\mathcal{M}/C \to \mathcal{A}/rC$. We already know that the latter is essentially surjective on objects; it remains to show that r induces an isomorphism $(\mathcal{B}/C)(f,f') \cong (\mathcal{B}/rC)(r(f),r(f'))$ if f, $f' \in \mathcal{M}$. In the diagram

$$
\begin{array}{ccccc}
(\mathcal{B}/C)(f,f') & \longrightarrow & \mathcal{B}(B,B') & \xrightarrow{\ \mathcal{B}(B,\rho B')\ } & \mathcal{B}(B,rB') \\
\downarrow & & \downarrow{\scriptstyle\mathcal{B}(B,f')} & & \downarrow{\scriptstyle\mathcal{B}(B,r(f'))} \\
1 & \xrightarrow{\ f\ } & \mathcal{B}(B,C) & \xrightarrow{\ \mathcal{B}(B,\rho C)\ } & \mathcal{B}(B,rC),
\end{array}
$$

the left square is a pullback by (4.2) and the right square is one by (3.1); it follows that $(\mathcal{B}/C)(f,f') \cong (\mathcal{B}/rC)(\rho C.f, r(f'))$; and since $\rho C.f = r(f).\rho B$ with $r(f) \in \mathcal{M}$ and $\rho B \in \mathcal{E}$, it follows from the last paragraph that

$$
(\mathcal{B}/rC)(\rho C.f, r(f')) \cong (\mathcal{B}/rC)(r(f), r(f')).
$$

Since the equivalence $(\mathcal{M}/C)_0 \to (\mathcal{A}/rC)_0$ has as equivalence-inverse the functor $(\mathcal{A}/rC)_0 \to (\mathcal{M}/C)_0$ sending g to the m of (3.2), by general principles this functor has a unique enrichment to a $\mathcal{V}$-functor $\mathcal{A}/rC \to \mathcal{M}/C$ for which the isomorphism $g \to r(m)$ is $\mathcal{V}$-natural, and this $\mathcal{V}$-functor is an equivalence-inverse to $\mathcal{M}/C \to \mathcal{A}/rC$.

4.4. Still supposing $\mathcal{V}$ to be cartesian closed, consider now the situation of §3.2. The diagonal $\Delta\colon \mathcal{A} \cong [1,\mathcal{A}] \to [\mathcal{C},\mathcal{A}]$ is the $\mathcal{V}$-functor induced by $\mathcal{C} \to 1$;

note that the existence of such a diagonal requires, if not cartesian closedness of $\mathcal{V}$, at least that 1 should be the unit for the tensor product: otherwise $[1, \mathcal{A}]$ is not $\mathcal{A}$. The $\mathcal{V}$-functor E given by evaluation at $1 \in \mathcal{C}$ again provides a left adjoint for Δ, and again is left exact as a $\mathcal{V}$-functor; while Δ is again fully faithful, since the counit $E\Delta \to I$ of the adjunction is invertible. Thus we can apply the results of the last two paragraphs, and repeat all that was said in §3.2. The factorization system $(\mathcal{E}, \mathcal{M})$ on $[\mathcal{C}, \mathcal{A}]$ is given as there, as is the equivalence $\mathcal{M}/S \simeq \mathcal{A}/S1$. Note that, when we start from $g \in \mathcal{A}/S1$ and construct the αC by the pullbacks (3.5), we are really forming the pullback in $[\mathcal{C}, \mathcal{A}]$ of $\Delta g: \Delta K \to \Delta S1$, which we know exists pointwise; so there is a unique way of making T into a $\mathcal{V}$-functor $\mathcal{C} \to \mathcal{A}$ for which $\alpha: T \to S$ and $h: T \to \Delta K$ are $\mathcal{V}$-natural. Then again we have $T1 = K$ and $T! = hC$, provided that we choose the trivial pullback (3.5) when $C = 1$.

When $\mathcal{C}$ here is not small, $[\mathcal{C}, \mathcal{A}]$ does not exist as a $\mathcal{V}$-category; but this causes no real difficulty. As a locally-finitely-presentable cartesian closed category, $\mathcal{V}_0$ has the form $Lex[\mathcal{T}, \mathbf{Set}]$ where $\mathcal{T}^{op}$ is the full subcategory of $\mathcal{V}_0$ given by the finitely presentables. We have only to take $\mathbf{Set}'$ to be the category of sets in a universe large enough to contain $ob\mathcal{C}$, and to write $\mathcal{V}'$ for the cartesian closed category given by $\mathcal{V}_0' = Lex[\mathcal{T}, \mathbf{Set}']$, which has the same finitely-presentable objects as $\mathcal{V}$ (so that "finite completeness" doesn't change its meaning when a $\mathcal{V}$-category is seen as a $\mathcal{V}'$-category). Now $[\mathcal{C}, \mathcal{A}]$ exists as a $\mathcal{V}'$-category, as does $[\mathcal{C}, \mathcal{A}]/S$; while the reflective subcategory $\mathcal{M}/S$ of the latter, being equivalent to $\mathcal{A}/S1$, is an honest $\mathcal{V}$-category.

4.5. We now compare the local factorization system $(\mathcal{E}, \mathcal{M})$ on $[\mathcal{C}, \mathcal{A}]$, seen rather as a local factorization system on the underlying ordinary category $[\mathcal{C}, \mathcal{A}]_0$, with the local factorization system — let us call it here $(\mathcal{E}_0, \mathcal{M}_0)$ — arising as in §3.2 on the ordinary functor category $[\mathcal{C}_0, \mathcal{A}_0]$. We have the functor $\phi: [\mathcal{C}, \mathcal{A}]_0 \to [\mathcal{C}_0, \mathcal{A}_0]$ sending the $\mathcal{V}$-functor $T: \mathcal{C} \to \mathcal{A}$ to its underlying ordinary $T_0: \mathcal{C}_0 \to \mathcal{A}_0$, and sending the $\mathcal{V}$-natural $\alpha: T \to S$ to its underlying natural transformation $\alpha_0: T_0 \to S_0$ (which, having the same components as α, is often written simply as $\alpha: T_0 \to S_0$). The composite of ϕ with the $\Delta_0: \mathcal{A}_0 \to [\mathcal{C}, \mathcal{A}]_0$ underlying $\Delta: \mathcal{A} \to [\mathcal{C}, \mathcal{A}]$ is the ordinary diagonal $\Delta: \mathcal{A}_0 \to [\mathcal{C}_0, \mathcal{A}_0]$, while the composite with ϕ of the reflexion $E: [\mathcal{C}_0, \mathcal{A}_0] \to \mathcal{A}_0$ given by evaluation at 1 is the $E_0: [\mathcal{C}, \mathcal{A}]_0 \to \mathcal{A}_0$ underlying the reflexion $E: [\mathcal{C}, \mathcal{A}] \to \mathcal{A}$ given by evaluation at 1. It follows that, for $\alpha: T \to S$ in $[\mathcal{C}, \mathcal{A}]$, we have $\alpha \in \mathcal{E}$ if and only if $\alpha_0 \in \mathcal{E}_0$ and $\alpha \in \mathcal{M}$ if and only if $\alpha_0 \in \mathcal{M}_0$; so that the $(\mathcal{E}, \mathcal{M})$-factorization of α is equally the $(\mathcal{E}_0, \mathcal{M}_0)$-factorization of α_0.

Since evaluation at 1 induces equivalences $\mathcal{M}/S \to \mathcal{A}/S1$ and $\mathcal{M}_0/S_0 \to \mathcal{A}_0/S_01 = (\mathcal{A}/S1)_0$, the canonical $(\mathcal{M}/S)_0 \to \mathcal{M}_0/S_0$ induced by ϕ is an equivalence. In fact:

Proposition 4.1. *The equivalence $(M/S)_0 \simeq M_0/S_0$ is an isomorphism*

$$(M/S)_0 \qquad \cong \qquad M_0/S_0 \qquad\qquad (4.3)$$

of categories, since it is bijective on objects. That is to say, given $S:C \to A$ and $\alpha: P \to S_0$ in M_0, there is a unique $T:C \to A$ with $T_0 = P$ such that $\alpha: T \to S$ is V-natural and lies in M.

Proof. Recall that, for a V-functor T with $T_0 = P$, we have $TC = PC$ on objects, while Tf denotes Pf for a morphism f in C_0. Given now $\alpha: P \to S_0$ in M_0, we have the pullbacks (3.3), in which TC stands for PC and $T!$ for $P!$. By the remarks at the end of §4.4, there is a unique V-functor T given on objects by the TC for which $\alpha: T \to S$ and $h: T \to \Delta K$ are V-natural. So *a fortiori* $\alpha: T_0 \to S_0$ and $h: T_0 \to (\Delta K)_0 = \Delta K$ are natural; and now, by the remarks towards the end of §3.2, we have $T_0 = P$; this gives existence. As for uniqueness, if T is a V-functor with $\alpha: T \to S$ in M, we have the pullbacks (3.3), and we use again the remarks at the end of §4.4, noting now however that, for *any* V-functor $T:C \to A$, the $T!:TC \to T1$ are *automatically* the components of a V-natural $T \to \Delta T1$.

In consequence, there is no point in a separate consideration of M_0/S_0, which is now revealed as just the ordinary category underlying M/S, its equivalence with $A_0/S1 = (A/S1)_0$ merely underlying that of M/S with $A/S1$. A special case of Proposition 4.1 has been observed by Cockett: see §4.8 below.

Coming now to the generalization of §3.3, we observe that here composition is a V'-functor $[A,A] \times [A,A] \to [A,A]$, so that now $[A,A]$ is monoidal as a V'-category, as is $[A,A]/S$ for a monad S on A (by which of course we mean a V-monad.) Again we call such an S a *club* if the full subcategory M/S is closed under this monoidal structure; whereupon M/S too is monoidal as a V-category. Clearly we still have Proposition 3.1, and in fact S is a club if and only if its underlying ordinary monad (S_0, j, n) on A_0 is one. Now (4.3) is an isomorphism of monoidal categories, and of course we still have Proposition 3.4; moreover it is clear that

$$((ClubA)/S)_0 \qquad \cong \qquad (ClubA_0)/S_0. \qquad\qquad (4.4)$$

Of course $A/G = A/S1$ inherits from the equivalent M/S a monoidal structure as a V-category, the induced monoidal structure on the underlying $(A/G)_0 = A_0/G$ being that given in §3.4. It follows that a monoid (K,g) in A/G — that is, in A_0/G — called a *club over G* in §3.4, gives in fact a V-monad $(K,g)*-$ on A; which accords with §4.1 above.

4.6. The category of monoids in a monoidal V-category has in general no V-category structure: the category **Ab** of abelian groups is monoidal as an

Ab-category, but the category of monoids therein — to wit, the category of rings — is not an additive category.   It is,   however,   otherwise when $\mathcal{V}$ is cartesian closed;   for then we have a diagonal $\delta : X \to X \times X$ in $\mathcal{V}$. Here the category Mon$\mathcal{A}$ of monoids $(A, \alpha: 1 \to A, a: A \otimes A \to A)$ in $\mathcal{A}$ has a canonical $\mathcal{V}$-category structure, the hom-object (Mon$\mathcal{A}$)(A, B) of $\mathcal{V}$ being the joint equalizer of the maps

$$
\begin{array}{ccc}
\mathcal{A}(A, B) & \xrightarrow{\delta} & \mathcal{A}(A, B) \times \mathcal{A}(A, B) \\
{\scriptstyle \mathcal{A}(a,1)}\downarrow & & \downarrow{\scriptstyle \otimes} \\
\mathcal{A}(A \otimes A, B) & \xrightarrow[\mathcal{A}(1,b)]{} & \mathcal{A}(A \otimes A, B \otimes B)
\end{array}
$$

and of the maps

$$
\begin{array}{ccc}
\mathcal{A}(A, B) & \longrightarrow & 1 \\
& {\scriptstyle \mathcal{A}(\alpha,1)}\searrow & \downarrow{\scriptstyle \beta} \\
& & \mathcal{A}(1, B) \quad ;
\end{array}
$$

the verification that this is indeed a $\mathcal{V}$-category structure involves only the writing of a couple of large but transparent diagrams.

In particular, then , for a club S on $\mathcal{A}$ in the present enriched context, we have in fact a $\mathcal{V}$-category (**Club**$\mathcal{A}$)$/S$ of monoids in the monoidal $\mathcal{V}$-category $\mathcal{M}/S$ (or in the equivalent $\mathcal{A}/S1$).

4.7. The essential results above on enriched clubs survive without the hypothesis that $\mathcal{A}$ admit the cotensor products $\langle c, A \rangle$ for finitely presentable c, so long as the $\mathcal{V}$-category $\mathcal{A}$ admits conical finite limits; we lose only the precise generalization in §4.2 of the Cassidy-Hébert-Kelly result, which was not essential to our later observations. We then need, however, a direct proof that (4.1) is a pullback whenever $\mathcal{A}$ is reflective in a $\mathcal{V}$-category $\mathcal{B}$ with finite conical limits preserved by $r: \mathcal{B} \to \mathcal{A}$; for we used this in §4.3. But (4.1) is one face of a cube, whose opposite face is

$$
\begin{array}{ccc}
\mathcal{B}(C, rX) & \xrightarrow{\mathcal{B}(e, rX)} & \mathcal{B}(B, rX) \\
{\scriptstyle \mathcal{B}(C, r(m))}\downarrow & & \downarrow{\scriptstyle \mathcal{B}(B, r(m))} \\
\mathcal{B}(C, rY) & \xrightarrow[\mathcal{B}(e, rY)]{} & \mathcal{B}(B, rY),
\end{array} \tag{4.5}
$$

these faces being joined by four edges of the type $\mathcal{B}(C, \rho X)$; because $\mathcal{B}(C, -)$ and $\mathcal{B}(B, -)$ of (3.1) are pullbacks, it suffices to show that (4.5) is a pullback. This, however, is immediate when we observe that $\mathcal{B}(C, rX) \cong \mathcal{B}(rC, rX)$ and so on, with $r(e)$ invertible .

4.8. Suppose now that $\mathcal{A}$ is again a finitely-complete *ordinary* category. Various writers on computer science — see for instance Moggi [24] — have observed that many of the monads S on $\mathcal{A}$ arising in connexion with data types are what some call *strong*: not a very happy term (although it goes back to Kock's thesis), since it suggests rather a property of S than — what it really is — an extra structure borne by S. What is meant by a *strong endofunctor* $S: \mathcal{A} \to \mathcal{A}$ is an endofunctor S together with a natural transformation (the so-called *strength* of S) $\sigma_{AC}: SA \times C \to S(A \times C)$ satisfying evident associativity and unit axioms; while a *strong natural transformation* $(S, \sigma) \to (T, \tau)$ is a natural transformation $S \to T$ compatible in the obvious sense with σ and τ.

In fact these strong endofunctors (and in particular strong monads) can be treated perfectly well in our present enriched context. Re-name the ordinary category $\mathcal{A}$ as $\mathcal{A}_0$, and choose a category **Set** of sets large enough to contain as objects both $ob\mathcal{A}_0$ and all the hom-sets $\mathcal{A}_0(A, B)$; now take for $\mathcal{V}_0$ the locally-finitely-presentable category $[\mathcal{A}_0^{op}, \mathbf{Set}]$, to be called $\mathcal{V}$ when taken with its cartesian closed structure, and recall that the internal-hom of $\mathcal{V}$ is given by $[F, G]C = \mathcal{V}_0(F \times \mathcal{A}_0(-, C), G)$. Embed $\mathcal{A}_0$ in $\mathcal{V}_0$ by the Yoneda embedding, and write $\mathcal{A}$ for the full sub-$\mathcal{V}$-category of $\mathcal{V}$ determined by the objects of $\mathcal{A}_0$; then $\mathcal{A}(A, B) \in \mathcal{V}$ is given by

$$\mathcal{A}(A, B)C = \mathcal{V}_0(\mathcal{A}_0(-, A) \times \mathcal{A}_0(-, C), \mathcal{A}_0(-, B)) \cong \mathcal{A}_0(A \times C, B).$$

Now to give a $\mathcal{V}$-functor $S: \mathcal{A} \to \mathcal{A}$ is equally (see [9, p.464, Prop. 9.6]) to give a mere functor $S: \mathcal{A}_0 \to \mathcal{A}_0$ together with a natural transformation $\overline{\sigma}_{AB}: \mathcal{A}(A, B) \to \mathcal{A}(SA, SB)$ satisfying an associativity and a unit axiom; but to give such a natural transformation, with components $\overline{\sigma}_{AB}C: \mathcal{A}(A, B)C \to \mathcal{A}(SA, SB)C$ of the form $\mathcal{A}_0(A \times C, B) \to \mathcal{A}_0(SA \times C, SB)$, is by Yoneda equally to give a natural transformation $\sigma_{AC}: SA \times C \to S(A \times C)$ satisfying the two appropriate axioms — that is, a strength for S. Thus the strong endofunctors of $\mathcal{A}_0$ are revealed as nothing but the endo-$\mathcal{V}$-functors of $\mathcal{A}$; and similarly the strong natural transformations between these are just the $\mathcal{V}$-natural ones. So, in the light of §4.7 above, we have at our disposal the results of §4.5. These strong monads are further studied by Cockett in the unpublished draft manuscript [5], which contains Proposition 4.1 in this special case.

5.   The original clubs are clubs in the abstract sense; other examples and counter-examples

5.1. We continue to suppose that $\mathcal{A}$ is a finitely-complete category, possibly enriched over a locally-finitely-presentable cartesian closed $\mathcal{V}$, and in that case admitting at least finite *conical* limits, as in §4.7. Our first comment is that, although the category of monads on $\mathcal{A}$ has a terminal object $\Delta 1: \mathcal{A} \to \mathcal{A}$, this is never a club unless $\mathcal{A}$ is trivial; for the unit $I \to \Delta 1$ of this monad (recall that we use I for the identity functor $\mathcal{A} \to \mathcal{A}$) clearly does not lie in $\mathcal{M}$ unless $\mathcal{A}$ is equivalent to 1. It seems probable, although we haven't investigated this, that it is very rare for $\mathbf{Club}(\mathcal{A})$ to have a terminal object.

5.2. The identity monad $I : \mathcal{A} \to \mathcal{A}$ is always a club, the conditions of Proposition 3.1 being trivially satisfied. Here $\mathcal{A}/I1 \simeq \mathcal{A}$, and the monoidal structure o of §3.4 is just the cartesian one . So the clubs over I correspond to the monoids (M, i, m) in $\mathcal{A}$ (meaning "monoids for the cartesian structure"), and are the monads of the form $M \times - : \mathcal{A} \to \mathcal{A}$. Since $\mathcal{A}$ need not be cartesian closed, we see that the monoidal structure on $\mathcal{A}/S1$ for a club S need not be right-closed, in spite of the examples in §2.1 and §2.2. When , however, $\mathcal{A}$ *is* cartesian closed, the monoidal structure on $\mathcal{A}/S1$ is right-closed for each S of the form $M \times -$, as an easy calculation shows.

5.3. These monoids in $\mathcal{A}$ (for the cartesian monoidal structure) form a category $\mathrm{Mon}\mathcal{A}$, isomorphic to $(\mathbf{Club}\mathcal{A})/I$, which in the enriched case is itself a $\mathcal{V}$-category, as in §4.6. The forgetful $W : \mathrm{Mon}\mathcal{A} \to \mathcal{A}$ may have a left adjoint G, with $GA = (LA, iA, mA)$ say; and surely does so under various "reasonable" conditions on $\mathcal{A}$, since W preserves limits. When it is so, W is monadic, since it clearly creates W-absolute coequalizers — that this implies monadicity even in the enriched case, when W has a left adjoint as a $\mathcal{V}$-functor, is clear from the proof of Beck's theorem given by Barr and Wells in [1,§3.3]. So then we have the *free-monoid monad* $L = (L, k, r)$ on $\mathcal{A}$, which is a $\mathcal{V}$-monad in the enriched case.

It is usual in computer science to call this rather the *list* monad : but only if it has the further property that GA above is, for each A, the *algebraically-free* monoid on A. This notion was discussed in §23 of the author's article [18]. For a monoid $M = (M, i, m)$, a map $k : A \to M$ exhibits M as the algebraically-free monoid on A if it induces an isomorphism $M\text{-Alg} \to A\text{-Alg}$ of categories; here an A-algebra is an object B with a map $A \times B \to B$, while an M-algebra is a B with a map $M \times B \to B$ satisfying the associativity and unit conditions.

It follows from [18, Theorem 23.1] that the algebraically-free monoid M on A exists if and only if there are an object M and maps $i : 1 \to M$ and $q : A \times M \to M$ such that, for any $f : B \to C$ and any $g : A \times C \to C$ there is a unique h making commutative

$$
\begin{array}{ccc}
A \xrightarrow{i \times B} M \times B & \xleftarrow{q \times B} & A \times M \times B \\
{}^{f}\searrow \quad \downarrow h & & \downarrow A \times h \\
C & \xleftarrow[\quad g \quad]{} & A \times C \quad ;
\end{array}
$$

then $k = q(A \times i)$, and $m: M \times M \to M$ is the h obtained here when f is 1_M and g is q.

If k exhibits M as the algebraically-free monoid on A, it *a fortiori* exhibits it as the free monoid on A in the usual sense; the converse is false in general, but true if $\mathcal{A}$ is cartesian closed. If the cartesian closed $\mathcal{A}$ admits coproducts, we do indeed have the free and algebraically-free monoid, given by

$$LA = \sum_{n=0}^{\infty} A^n , \tag{5.1}$$

with the evident $k: 1 \to L$ and $r: L^2 \to L$; for all this, see [18, §23].

Cockett and Jay will show, in a forthcoming paper using results from [4], [11], and [12], that if a *distributive* category $\mathcal{A}$ (see §5.6 below) admits a list monad L, the latter is a club. We merely mention here that the techniques of [18] give very simply the following result, weaker in that it requires cocompleteness and an exactness property:

Proposition 5.1. *Let $\mathcal{A}$ be a cocomplete distributive cartesian closed category in which filtered colimits commute with finite limits. Then the list monad on $\mathcal{A}$ is a club.*

For the moment, however, we consider only the case $\mathcal{A} = \mathbf{Set}$; here we can speak in terms of elements, an element of LA being of the form $n[a_1, \cdots, a_n]$ with $n \in \mathbf{N}$ and $a_i \in A$; and now it is immediately obvious that k and r lie in $\mathcal{M}$, while L preserves all pullbacks. Here, then, the list monad L is certainly a club. Let us examine what the clubs on $\mathbf{Set}$ over L are, by considering the monoidal structure of §3.4 on $\mathbf{Set}/L1 = \mathbf{Set}/\mathbf{N}$.

An object $\Gamma: K \to \mathbf{N}$ of $\mathbf{Set}/\mathbf{N}$ may be identified with a graded set $K = (K_n)$; it is useful to think of $\Gamma\alpha$ for $\alpha \in K$ as the *arity* of α. We have $K * A = \sum K_n \times A^n$, and if H is a second object of $\mathbf{Set}/\mathbf{N}$ we have

$$(K \circ H)_n = \sum_{r_1 + \cdots + r_m = n} K_m \times H_{r_1} \times \cdots \times H_{r_m}; \tag{5.2}$$

but it is probably more convenient to use a notation like that of [14], writing an element of $K * A$ as $\alpha[a_1 \cdots, a_n]$ where $\alpha \in K$ with $\Gamma\alpha = n$ and the $a_i \in A$, and an element of $K \circ H$ as $\alpha[\beta_1 \cdots, \beta_n]$ where $\alpha \in K$ with $\Gamma\alpha = n$, and the $\beta_i \in H$; then $\Gamma(\alpha[\beta_1, \cdots, \beta_n]) = \sum \Gamma\beta_i$. Clearly $- * A$ has the right adjoint $\{A, -\}$, where $\{A, B\}_n = [A^n, B]$; and $- \circ H$ has a right adjoint $\{H, -\}$, of which $\{A, -\}$ above is the special case where each Γ is constant at 0.

It is convenient to denote a multiplication $K \circ K \to K$ making K into a $\circ$-monoid by $\alpha[\beta_1, \cdots, \beta_n] \mapsto \alpha(\beta_1, \cdots, \beta_n)$, while to give a unit for the $\circ$-monoid K is to give an element $\mathbf{1}$ of K_1; similarly it is convenient to denote an action $K * A \to A$ by $\alpha[a_1, \cdots, a_n] \mapsto \alpha(a_1, \cdots, a_n)$; in this notation the associativity and unit axioms take a simple form, along the lines of (2.3)–(2.5) of [14]. The clubs over L are the $K * -$ for such K; while in the alternative language of §3.4, such K themselves are the clubs over $\mathbf{N}$.

Such a monad $K * -$ is clearly finitary, so that its algebras are those for a Lawvere theory $\mathcal{K}$. We have an evident presentation of $\mathcal{K}$; the basic operations of arity n are the elements of K_n, and the basic equations are given by the associativity and unit laws

$$(\alpha(\beta_1, \cdots, \beta_n))(a_1, \cdots, a_m) = \alpha(\beta_1(a_1, \cdots, a_{m_1}), \cdots, \beta_n(\cdots, a_m)), \quad (5.3)$$

$$\mathbf{1}(a) = a; \quad (5.4)$$

here in (5.3) we have $\Gamma\alpha = n$, $\Gamma\beta_i = m_i$, and $m = \sum m_i$. Observe that in the equations (5.3) and (5.4), the same variables occur on each side, in the same order, without repetition.

The existence of a presentation with this latter property characterizes the Lawvere theories $\mathcal{K}$ which so arise. To see this we have only to go back to §3 of [14] and repeat the argument there, replacing clubs over $\mathbf{P}$ on $\mathbf{Cat}$ by the much simpler clubs over $\mathbf{N}$ on $\mathbf{Set}$.

In particular *free* Lawvere theores give clubs over L. One such is the theory of magmas — recall that for Bourbaki a magma is a set with a binary operation. If $\mathcal{K}$ is the corresponding club, computer scientists tend to think of $K * A$ less as the free magma on A than as the set **btree** A of binary trees with their leaves labelled by elements of A. Here one sees very clearly the "curious property" first mentioned above at the end of §2.1: to give an element of **btree** A is to give an element of **btree** 1 — that is, an unlabelled binary tree — and then to assign to the leaves of the tree elements of A, forming in their totality an element $n[a_1, \cdots, a_n]$ of LA. In precise terms, this is the assertion that

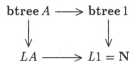

is a pullback, or that **btree** $\to L$ lies in $\mathcal{M}$.

Another free Lawvere theory is that given by one n-ary operation for each $n \geq 1$. An element of the corresponding $K * A$ may be seen as a general tree with leaves labelled by elements of A; again one observes the "curious property".

Among the non-free theories corresponding to clubs over L is that of semi-
-groups, where the only axiom is $mmxyz = mxmyz$; the theory of monoids
gives of course the club L itself. Again, the theory of M-sets for a monoid
M corresponds to a club over L; in fact $M \times -$ is, by §5.2, a club over I,
which is in turn a club over L, corresponding to the free Lawvere theory on
no operations. Another example is provided by the theory of a semigroup A
on which the monoid M acts; and so on.

5.4. Consider now the case where $\mathcal{A}$ is the cartesian closed category **Cat**.
Once again we have the list monad L given by (5.1), and once again we can
describe it in elementary terms: an object of LA is $n[a_1, \cdots, a_n]$ with $n \in \mathbf{N}$
and the a_i objects of A, while a morphism $n[a_1, \cdots, a_n] \to n[b_1, \cdots, b_n]$ is
$n[g_1, \cdots, g_n]$ where $g_i : a_i \to b_i$ in A. It is again clear that L preserves all
pullbacks, and that $k : I \to L$ and $r : L^2 \to L$ lie in $\mathcal{M}$, so that L is a club.
Here $L1$ is the set $\mathbf{N}$, seen as a discrete category; and the monoidal structure
of §3.4 on $\mathbf{Cat/N}$ is just like that of §2.1 on $\mathbf{Cat/P}$, with the permutations
ξ restricted to be identities. In fact $\mathbf{N}$ is a club over $\mathbf{P}$ in the sense of §2.1,
and *a fortiori* a club over $\mathbf{S}$ in the sense of §2.2.

To see that these clubs over $\mathbf{S}$, and hence those over $\mathbf{P}$ in the original sense,
are indeed clubs in the abstract sense of §3, we have only to observe that
$\mathbf{S} \circ - : \mathbf{Cat} \to \mathbf{Cat}$ is such a club S : for then the monoidal structure of
§3.4 on $\mathbf{Cat}/S1$ is clearly that of §2.2 on $\mathbf{Cat/S}$. It is, however, a simple
matter to check that $\mathbf{S} \circ -$ preserves all pullbacks, and that the unit and the
multiplication of this monad lie in $\mathcal{M}$.

One can check equally easily that the clubs over $\mathbf{S}^{op}$ of §2.2 are clubs in the
abstract sense: alternatively one may observe that, whenever S is a club on
Cat, so too is $\overline{S}$ given by $\overline{S}A = (SA^{op})^{op}$.

Since $\mathbf{S} \circ -$ is clearly a *2-monad* on **Cat**, it follows from §4.5 that each club
over $\mathbf{S}$ gives such a 2-monad; as do the clubs over $\mathbf{S}^{op}$.

When it comes to the clubs over $\mathbf{G}$ of §2.3, there is no "base" club like $\mathbf{P}$
or $\mathbf{S}$ to which we can refer them, since $\mathbf{G}$ is not a club. To see that such
a club $\mathcal{K}$ corresponds to a club in the abstract sense, we have to check this
directly for $\mathcal{K} \circ -$. However it is no harder to see that here, too, $\mathcal{K} \circ -$ preserves
all pullbacks, while the unit and the multiplication lie in $\mathcal{M}$. Moreover, for
$\alpha : F \to G : \mathcal{A} \to \mathcal{B}$, we can clearly define $\mathcal{K} \circ \alpha : \mathcal{K} \circ F \to \mathcal{K} \circ G : \mathcal{K} \circ \mathcal{A} \to \mathcal{K} \circ \mathcal{B}$
for *invertible* α, thus exhibiting $\mathcal{K} \circ -$ as a 2-monad on what in §2.3 we called
$\mathbf{Cat}_g$.

5.5. Consider the monad S on **Cat** whose algebras are those compact closed
categories whose tensor product $\otimes$ is strictly associative and whose unit J for

the tensor product is a strict unit; this monad was determined explicitly by Laplaza and the author in [20], of which see §3 and §9.

Every such compact closed category $\mathcal{A}$ is a $\mathcal{K}$-algebra for any club $\mathcal{K}$ over **G** in the sense of §2.3. The action $\mathcal{K} \circ \mathcal{A} \to \mathcal{A}$ takes, say, $P[A, B, C]$ where $\Gamma P = (- + +)$ to $A^* \otimes B \otimes C$, takes $P[f, g, h]: P[A, B, C] \to P[D, E, F]$ to $f^* \otimes g \otimes h$, and, for $k: P \to Q$ with Γk say the graph in the first paragraph of §2.3, takes $k_{AC}: P[A, A, C] \to QC$ to the evident $A^* \otimes A \otimes C \to C$; in general takes such a morphism to the canonical map made from the $d_A: J \to A \otimes A^*$, the $e_A: A^* \otimes A \to J$, the commutativities, and the identities — which is well-determined by the results of [20]. It follows that we have a map $\alpha: \mathcal{K}\circ- \to S$ of monads. If S were a club, we should after all have a common "base" club for these clubs over **G**, modulo verification that $\alpha \in \mathcal{M}$.

However, S is *not* a club: this fact, asserted in §10 of [20], was verified by John Power, to whom is due the following proof that $n: S^2 \to S$ is not in $\mathcal{M}$.

Recall the map $\tau: E(\mathcal{A}) \to [\mathcal{A}]$ of [20, p.195] and the map $\varepsilon: [\mathcal{A}] \to (S\mathcal{A})(J, J)$ of [20, p.207], and write $\sigma: E(\mathcal{A}) \to (S\mathcal{A})(J, J)$ for their composite; write also $\tau': E(S\mathcal{A}) \to [S\mathcal{A}]$ for the analogue of τ with $S\mathcal{A}$ in place of $\mathcal{A}$. For an object X of $S\mathcal{A}$, write X' for its image in $S^2\mathcal{A}$ under $jS\mathcal{A}: S\mathcal{A} \to S^2\mathcal{A}$. In particular, the identity J for $\otimes$ in $S\mathcal{A}$ gives an object J' of $S^2\mathcal{A}$, distinct from the identity for $\otimes$ in $S^2\mathcal{A}$. Now take for $\mathcal{A}$ the discrete category with two objects A and B. The maps $J \to J$ in $S\mathcal{A}$ are the elements of the free abelian monoid on $\sigma 1_A$ and $\sigma 1_B$. A map $J' \to J'$ in $S^2\mathcal{A}$ can be identified with a pair (f, ϕ) where $f: J \to J$ in $S\mathcal{A}$ and ϕ is an element of the free abelian monoid on $[S\mathcal{A}]$; consider the two such maps given by the pairs $(\sigma 1_A, \tau' \sigma 1_B)$ and $(\sigma 1_B, \tau' \sigma 1_A)$.

These are certainly different, since $\sigma 1_A \neq \sigma 1_B$. Yet $S!: S\mathcal{A} \to S1$ sends $\sigma 1_A$ and $\sigma 1_B$ each to $\sigma 1_*$, where $*$ is the unique object of 1; so that these maps become equal under $S^2!: S^2\mathcal{A} \to S^21$. They also become equal under $n\mathcal{A}: S^2\mathcal{A} \to S\mathcal{A}$, since this sends $(\sigma 1_A, \tau' \sigma 1_B)$ to $\sigma 1_A \otimes \sigma 1_B: J \otimes J \to J \otimes J$, which coincides with $\sigma 1_B \otimes \sigma 1_A$. Thus

$$
\begin{array}{ccc}
S^2\mathcal{A} & \xrightarrow{S^2!} & S^21 \\
{\scriptstyle n\mathcal{A}}\downarrow & & \downarrow{\scriptstyle n1} \\
S\mathcal{A} & \xrightarrow{S!} & S1
\end{array}
$$

is not a pullback, and n does not lie in $\mathcal{M}$.

5.6. We use the term *distributive category* in the strong sense originally introduced by Lawvere and Schanuel: $\mathcal{A}$ is distributive if it has finite limits and finite coproducts, and if the functor $\mathcal{A}/A \times \mathcal{A}/B \to \mathcal{A}/(A + B)$ sending (f, g) to $f + g$ is an equivalence. **Set** and **Cat** are obvious examples, as is any Grothendieck topos — but there are of course many others, not necessarily cocomplete.

Distributive categories have been recognized — see Cockett [4] and (with a weaker notion of distributivity) Walters [25] — as an apt setting for data structures in computer science. An object of such a category may be seen as a data type, and an endofunctor or a monad (like list) as a data-type constructor. Without further conditions, however, a distributive category lacks enough completeness and cocompleteness to ensure the existence of solutions to domain equations. Cockett's way around this is to consider only those distributive categories (called *locoses* in [4]) which admit the list monad; it seems that then $\mathcal{A}$ is "internally complete" enough to support the desired constructions, the necessary techniques to prove this being extensions of those developed by Cockett in [4] and pursued by Jay in [11] and [12]. The author is informed that they have a proof (as we mentioned in §5.3) that the list monad in any locos is a club. Cockett, in a May 1991 draft manuscript [6] made available to the author, has a proof that, if the strong monad S is a club in a locos $\mathcal{A}$, with a certain "boundedness" property (possessed by the list monad), and if F is an endofunctor with a map $\beta \colon F \to S$ in $\mathcal{M}$, then the algebraically-free monad T on F in the sense of [18], if it exists, is a club over S. The author further understands that Cockett has a proof of the existence of T, under conditions on S.

While awaiting the publication of these results, we merely mention here that the following, weaker in that it requires cocompleteness and an exactness property, is easily proved using the techniques of [18]:

Proposition 5.2. *Let $\mathcal{A}$ be a cocomplete distributive category in which filtered colimits commute with finite limits, and let F be a pullback-preserving finitary endo-functor of $\mathcal{A}$. Then the algebraically-free monad on F exists and is a club.*

References

[1] M.Barr and C.Wells, *Toposes, Triples and Theories,* Springer-Verlag, New York Berlin Heidelberg Tokyo, 1985.

[2] R.Blackwell, G.M.Kelly and A.J.Power, Two-dimensional monad theory, *J.Pure Appl. Algebra* 59(1989), 1-41.

[3] C.Cassidy, M.Hébert, and G.M.Kelly, Reflective subcategories, localizations, and factorization systems, *J.Austral. Math. Soc.* 38(Series A)(1985),

287-329.

[4] J.R.B.Cockett, List-arithmetic distributive categories; locoi, *J.Pure Appl. Algebra* 66(1990), 1-29.

[5] J.R.B.Cockett, Notes on strength, datatypes, and shape (draft manuscript), Macquarie Univ., May 1991.

[6] J.R.B.Cockett, The fundamental theorem of data structures for a locos (draft manuscript), Macquarie Univ., May 1991.

[7] B.J.Day, A reflection theorem for closed categories, *J.Pure Appl.Algebra* 2(1972), 1-11.

[8] S.Eilenberg and G.M.Kelly, A generalization of the functorial calculus, *J. Algebra* 3(1966), 366-375.

[9] S.Eilenberg and G.M.Kelly, Closed categories, in *Proc. Conf. on Categorical Algebra*(La Jolla, 1965), Springer-Verlag, Berlin Heidelberg New York, 1966; 421-562.

[10] P.J.Freyd and G.M.Kelly, Categories of continuous functors I, *J.Pure Appl. Algebra* 21(1972), 169-191.

[11] C.B.Jay, Tail recursion from universal invariants, in: D.H.Pitt et al.(eds), *Category Theory and Computer Science*, Paris, France, Sept.3-6, 1991, Proceedings; Springer Lecture Notes in Computer Science 530; 151-163.

[12] C.B.Jay, Matrices, monads, and the fast fourier transform, to appear.

[13] G.M.Kelly, Many-variable functorial calculus I, *Springer Lecture Notes in Mathematics* 281(1972), 66-105.

[14] G.M.Kelly, An abstract approach to coherence, *Springer Lecture Notes in Mathematics* 281(1972), 106-147.

[15] G.M.Kelly, A cut-elimination theorem, *Springer Lecture Notes in Mathematics* 281(1972), 196-213.

[16] G.M.Kelly, Coherence theorems for lax algebras and for distributive laws, *Springer Lecture Notes in Mathematics* 420(1974), 281-375.

[17] G.M.Kelly, Some remarks on categories with structure, Sydney Category Seminar Reports, July 1978.

[18] G.M.Kelly, A unified treatment of transfinite constructions for free algebras, free monoids, colimits, associated sheaves, and so on, *Bull. Austral. Math. Soc.* 22(1980), 1-83.

[19] G.M.Kelly, Structures defined by finite limits in the enriched context I, *Cahiers de Top. et Géom. Différentielle* 23(1982), 3-42.

[20] G.M.Kelly and M.L.Laplaza, Coherence for compact closed categories, *J. Pure Appl. Algebra* 19(1980), 193-213.

[21] G.M.Kelly and S. Mac Lane, Coherence in closed categories, *J. Pure Appl. Algebra* 1(1971), 97-140.

[22] J.Lambek, Deductive systems and categories II: Standard constructions and closed categories, *Springer Lecture Notes in Mathematics* 86(1969), 76-122.

[23] S.Mac Lane, Natural associativity and commutativity, *Rice Univ. Studies* 49 (1963), 28-46.

[24] E.Moggi, Computational lambda-calculus and monads, *Proc. IEEE Conf. on Logic in Computer Science*(1989), 14-23.

[25] R.F.C.Walters, *Categories and Computer Science*, Carslaw Publications, Sydney, 1991.

Penrose diagrams and 2-dimensional rewriting

Yves Lafont *

Abstract

A very natural generalization of word problems to higher dimensions has been proposed by Albert Burroni. In this framework, the monoidal category of finite sets with disjoint union is a *finitely presentable 2-monoid*. Here we exhibit a complete rewrite system for this presentation with twelve rules and fifty-six critical pairs. This can be used to eliminate variables in symbolic computation.

1. A finite presentation

Let $\mathbf{N}$ be the category of finite sets: objects are natural numbers and morphisms are maps. If n is an object of $\mathbf{N}$, we shall also write n for the identity on n. Addition defines an associative bifunctor $+ : \mathbf{N} \times \mathbf{N} \to \mathbf{N}$ with unit 0.

As a strict monoidal category, $\mathbf{N}$ is generated by the object 1 and the three following morphisms:

$$\mu : 2 \to 1, \qquad \eta : 0 \to 1, \qquad \tau : 2 \to 2 \text{ (transposition)}.$$

Indeed, any map $\varphi : n \to m$ can be decomposed into $\iota \circ \sigma \circ \pi$ where π is a permutation, σ is a monotone surjection and ι is a monotone injection. Note that, in this decomposition, π is not uniquely determined unless φ is injective, in which case σ is trivial. Now π is a product of transpositions which are of the form $p + \tau + q$. Similarly, σ is a product of maps of the form $p + \mu + q$, and ι is a product of maps of the form $p + \eta + q$.

---

*Laboratoire d'Informatique de l'Ecole Normale Supérieure (URA 1327 du CNRS), 45 rue d'Ulm, 75230 Paris Cedex 05, France. Email: lafont@dmi.ens.fr. This work was partially supported by the ESPRIT II BRA Programme of the EC under contract # 3003 (CLICS).

Our three generators satisfy the seven following equations:

$$\mu \circ (\mu + 1) = \mu \circ (1 + \mu) \qquad \mu \circ (\eta + 1) = 1 \qquad \mu \circ \tau = \mu$$

$$\tau \circ \tau = 2 \qquad (\tau + 1) \circ (1 + \tau) \circ (\tau + 1) = (1 + \tau) \circ (\tau + 1) \circ (1 + \tau)$$

$$\tau \circ (\mu + 1) = (1 + \mu) \circ (\tau + 1) \circ (1 + \tau) \qquad \tau \circ (\eta + 1) = 1 + \eta$$

All this is much better expressed with diagrams in Penrose style [PeR86]. Here, a map $\varphi : n \to m$ is represented by a diagram u with n inputs and m outputs. In particular, the generators correspond to atomic diagrams:

Penrose diagrams are obtained by horizontal and vertical composition of atoms, modulo the axioms of 2-categories, which means that, for example, the three following diagrams are identified:

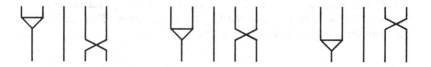

Now the equations (figure 1) are very intuitive. For example, the three first ones are laws of associativity, unit and commutativity. The two following ones are implicitly used in the well-known presentation of symmetric groups. Note also that three equations are asymmetrical, but the corresponding mirror images are derivable (figure 2). In fact, any equation which holds in **N** is derivable from those seven ones [Bur91], which means that **N** is finitely presented as a strict monoidal category.

This result is significant because **N** (in fact its dual) models the structural management of variables in equational reasoning. Typically, equations like $x \times (y + z) = x \times y + x \times z$ and $x \times 0 = 0$ can be represented as follows:

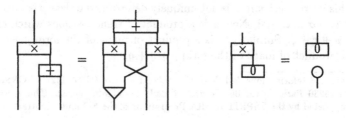

As a consequence, any finite first-order equational theory corresponds to a finitely presented strict monoidal category. More precisely, if the theory consists of n function symbols and m equations, the corresponding presentation has $n + 3$ generators and $m + 3n + 7$ equations. The $3n$ extra equations express commutation between each function symbol and the extra generators. For example, in the case of a binary function symbol, they are:

Our aim is to extend this kind of result to rewriting.

2.   A complete rewrite system

The Knuth-Bendix theorem gives a solution of the word problem in certain finite equational theories [KnB70, DeJ90]. The idea consists in orienting the equations in such a way that the corresponding rewrite relation is *noetherian* (there is no infinite reduction sequence) and *confluent* (if a term u reduces to v and to w, then v and w reduce to a common term). Then, two terms are equivalent in the theory if and only if they have the same *normal form*. Noetherianness ensures the existence of this normal form, and confluence ensures its uniqueness. Once noetherianness holds, confluence can be checked automatically by inspecting all conflicts between rewrite rules: there is only a finite number of those *critical pairs*. In order to make a critical pair confluent, one can add a new rule which corresponds to a derivable equation. This process of *completion* must be iterated until all conflicts are solved.

We adapted this methodology to Penrose diagrams, and rather than presenting the general theory (which is still not completely fixed), we discuss the example informally. The rewrite system for **N** (figure 3) has been obtained by Knuth-Bendix completion. The orientation of equations is not obvious: naïve choices would lead to an infinity of rules!

Theorem 1 *This system is noetherian and confluent.*

Proof :

To prove noetherianness, we define, for each diagram u representing a map $n \to m$, a strictly monotone map $\tilde{u} : W^n \to W^m$ where W is the set of positive integers, and W^n, W^m are equipped with the product order. This interpretation is defined on generators as follows:

and extended to composed diagrams in a straightforward way. For each rule $u \rightarrow v$, one checks that $\tilde{u}(x_1, \ldots, x_n) > \tilde{v}(x_1, \ldots, x_n)$ for the product order (figure 4). Therefore, if $u_0 \rightarrow u_1 \rightarrow u_2 \rightarrow \ldots$ is an infinite reduction sequence, one gets a strictly decreasing sequence

$$\tilde{u_0}(1, \ldots, 1) > \tilde{u_1}(1, \ldots, 1) > \tilde{u_2}(1, \ldots, 1) > \ldots$$

which is absurd.

Now, for confluence, it is enough to show that if a diagram u reduces in *one step* to v and to w, then v and w reduce, possibly in many steps, to a common diagram (*local confluence*). If the two given reductions are *disjoint*, this is obvious because rewriting is purely local (in particular, there is no built-in copying). When they overlap, it is enough to focus on the overlapping configuration, and there are only fifty-six such configurations, which are exhaustively handled in the appendix. For example, the first diagram handles a conflict between the rule 1 (associativity) and itself, the second one handles a conflict between the rules 1 and 2, etc.   □

3. Carrying on

In a later publication, we shall present a general theory, but we can already sketch the state of the art. First, 2-dimensional rewriting is at least as powerful as left-linear term rewriting, in the following sense:

Theorem 2 *Any complete left-linear rewrite system on terms corresponds to a complete rewrite system on Penrose diagrams.*

The point is that a linear term corresponds to a *tree*, *i.e.* a diagram using none of the three extra generators. Therefore, there will be no conflict between our twelve rules and the ones coming from the term rewriting system. There will be conflicts involving the extra commutation rules, but these are easily solved.

We have not yet solved the case of rules that are not left-linear, such as $x - x \rightarrow 0$, and we would also like to investigate commutative theories, since commutativity is a priori orientable (see our rule 11). But of course, terms are only a special case. We are mainly interested in theories such as *braids, knots* or *quantum groups*, for which Penrose diagrams are definitely needed [JoS91].

For such experiments in 2-dimensional symbolic computation, we are developing a software in the functional programming language CAML. It was used to check the confluence of our twelve rules (only few seconds for checking, a bit more for printing).

4. Acknowledgements

A non trivial part of this paper was the automatic drawing of Penrose diagrams in the appendix. I wish to thank Emmanuel Chailloux for his help in using the graphic library of CAML.

References

[Bur91] Burroni A., *Higher Dimensional Word Problem. Category Theory and Computer Science, LNCS 530, Springer-Verlag, 1991, pp. 94-105.*

[DeJ90] Dershowitz N., Jouannaud J.P., *Rewrite systems. Handbook of Theoretical Computer Science, Vol. B, Elsevier, 1990.*

[JoS91] Joyal A., Street R., *The Geometry of Tensor Calculus. Advances in Mathematics 88, 1991, pp. 55-112.*

[KnB70] Knuth D.E., Bendix P.B., *Simple word problems in universal algebras. Abstract Algebra (ed. J. Leech), Pergamon Press, 1970, pp. 263-297.*

[PeR86] R. Penrose & W. Rindler, *Spinors and space-time, Vol. 1: Two-spinor calculus and relativistic fields. Cambridge University Press, 1986.*

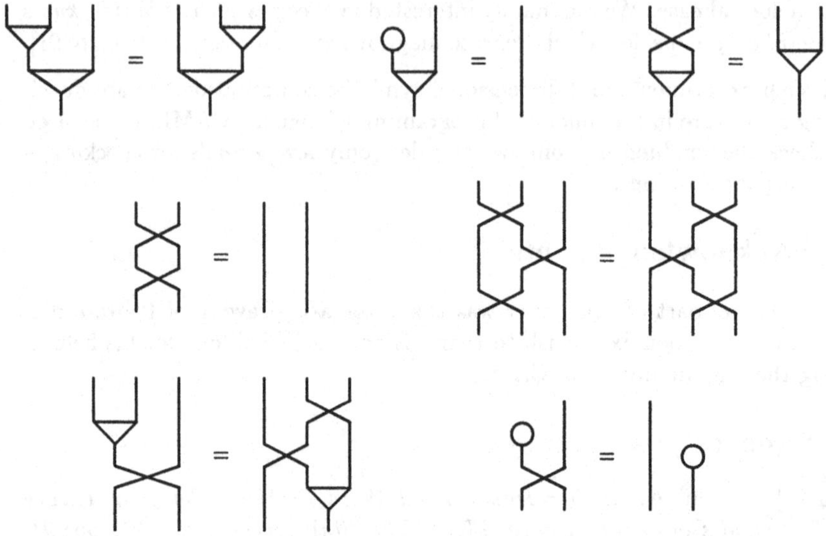

Figure 1. equations

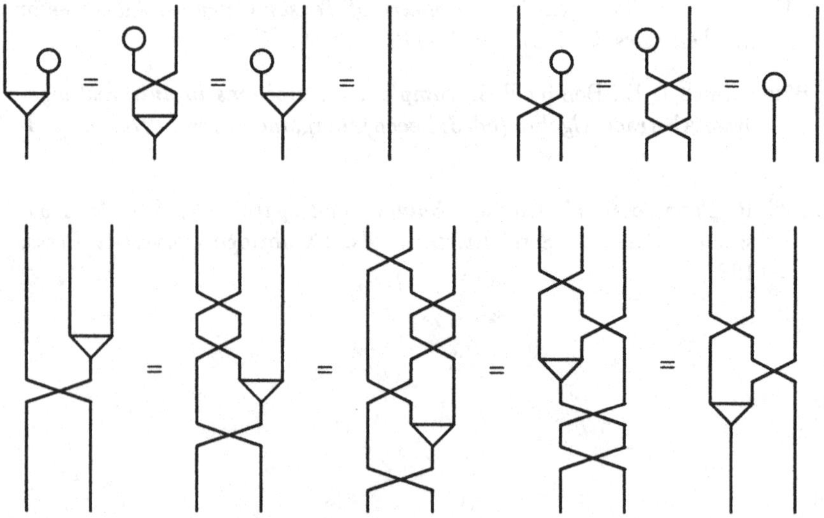

Figure 2. derived equations

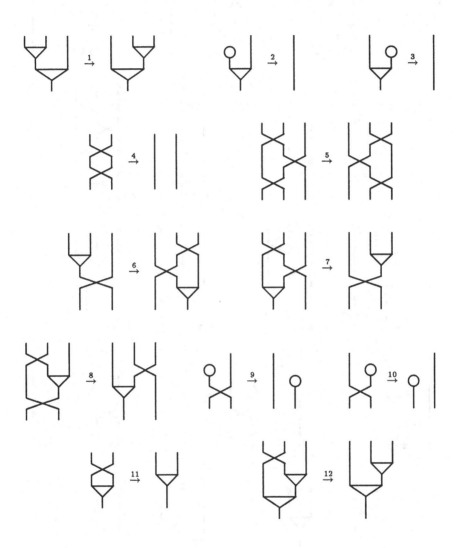

Figure 3. rewrite rules

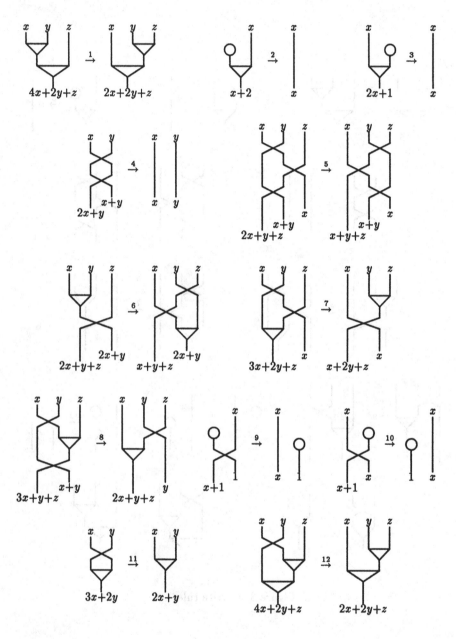

Figure 4. termination

Appendix: Confluence of the fifty-six critical pairs

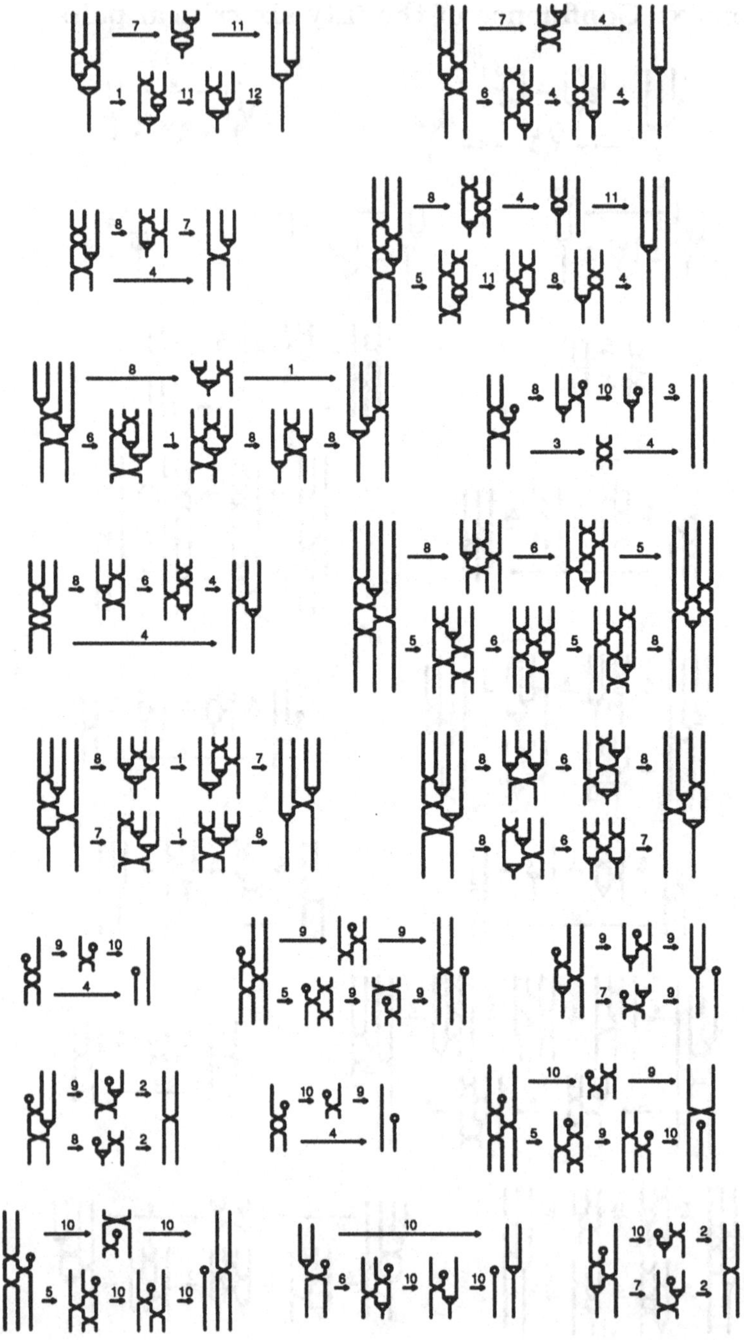

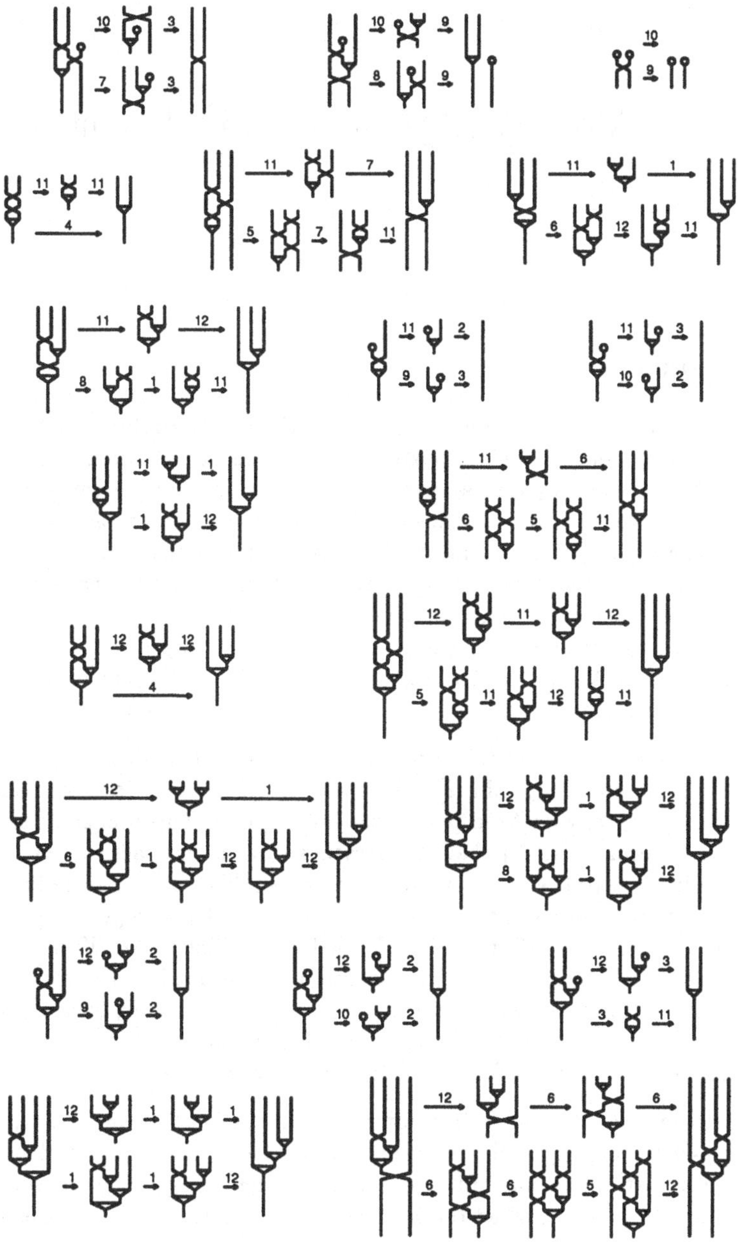

Strong Monads, Algebras and Fixed Points

Philip S. Mulry [*]

1. Introduction

There has been considerable recent interest in the use of algebraic methodologies to define and elucidate constructions in fixed point semantics [B], [FMRS], [Mu2]. In this paper we present recent results utilizing categorical methods, particularly strong monads, algebras and dinatural transformations to build general fixed point operators. The approach throughout is to evolve from the specific to the general case by eventually discarding the particulars of domains and continuous functions so often used in this setting. Instead we rely upon the structure of strong monads and algebras to provide a general algebraic framework for this discussion. This framework should provide a springboard for further investigations into other issues in semantics.

By way of background, the issues raised in this paper find their origins in several different sources. In [Mu2] the formal role of iteration in a cartesian closed category (*ccc*) with fixed points was investigated. This was motivated by the observation in [H-P] that the presence of a natural number object (*nno*) was inconsistent with *ccc's* and fixed points. This author introduced the notion of *onno* (ordered *nno*) which in semantic categories played a role as iterator and was precisely the initial T-algebra for T the strong lift monad. Using the *onno* a factorization of *fix* was produced and further it was shown *fix* was in fact a dinatural transformation. This was accomplished by avoiding the traditional projection/embedding approach to semantics. Similar results were extended to order-enriched and effective settings as well.

Turning to monads, their role in computation is not new. In particular, it was emphasized early in the development of topos theory that the partial map classifier was a strong monad. Little developed beyond this point because the partial map classifier classified too many maps. In [Mu1] many refinements

*Department of Computer Science, Colgate University, Hamilton, NY 13346. Present Address:Dept of Computer Science, University of Edinburgh, The King's Buildings, Edinburgh EH9 3JZ, e-mail psm@dcs.ed.ac.uk. This research was partially supported by NSF Grant CCR-9002251 and a Colgate University Picker Fellowship.

of this partial map classifier were described which in turn generated different strong monads T and distinct families of subobjects and partial maps. A partial computation was then described as a map $A \to TB$ for these various such T's, i.e. maps in the associated Kleisli category. This approach was later built upon in [Mo], with the subsequent introduction of other kinds of strong monads. A monadic lambda calculus was also introduced. In [Mu3] the lift monad was shown to be a special case of strong monads which arise naturally as refinements of partial map classifiers. The semantics of partial data types was developed in this context by considering both the Kleisli category and the category of Eilenberg-Moore algebras for the strong monad. In [C-P] a monadic calculus was utilized for general fixpoint computations in a higher-order constructive predicate logic setting. An important tool was the use of a fixed point object which was the initial algebra for the strong monad satisfying additional data.

In this paper we consolidate and generalize some of these developments. Getting to details, we begin in section two with the particular case of domains by introducing the notion of strong dinatural transformation and showing that fix is the unique dinatural transformation between appropriate bifunctors which is strong when variation is restricted to algebra maps. Following the approach in [Mu2] we produce a factorization of fix that utilizes and emphasizes the algebraic structure of the $onno$. The result is a straightforward extension of the results in [Mu2] and can be viewed as an algebraic analogue of Plotkin's characterization of fix. Motivated by this case, we proceed in section three to introduce the notion of algebraically strong dinatural transformation. Moving to a general algebraic setting we work in a cartesian closed category with strong monad and fixed point object. The central question considered is whether analogous results hold in this setting as well. We answer affirmatively by showing there exists a fixed point combinator FIX for any Eilenberg-Moore algebra D which acts on arbitrary endomaps of D. Utilizing the fixed point object, a factorization of FIX is constructed and we show that FIX is an algebraically strong dinatural transformation. If it is also a dinatural transformation it is the unique such one. When we restrict to the case of domains (continuous or effective), FIX agrees with fix. We finish with examples illustrating some of the obstructions to characterizing FIX, and also show the recent work in [F] is applicable here.

2.   Strong dinaturality and fix

We begin our discussion with a brief review of some relevant notions. Let $\mathbf{C}$ be an arbitrary category with endofunctor T.

Definition 2.1 A monad $T = <T, \eta, \mu>$ on $\mathbf{C}$ consists of an endofunctor T

and two natural transformations

$\eta : id \to T$ and $\mu : T^2 \to T$ so that

1) $\mu \circ \eta_T = id_T = \mu \circ T\eta$

2) $\mu \circ \mu_T = \mu \circ T\mu$

T is a strong monad for C if these exists a natural transformation $t_{A,B}$:
$TA \times B \to T(A \times B)$ so that

3) $t_{A,B} \circ (\eta_A \times id) = \eta_{A \times B}$

4) $t_{A,B} \circ (\mu_A \times id) = \mu_{A \times B} \circ T(t_{A,B}) \circ t_{TA,B}$

An Eilenberg-Moore algebra for monad T is a pair $< X, h >$ where $X \in C$
and $TX \xrightarrow{h} X$ is a map in C so that

5) $h \circ \eta = id$

6) $h \circ Th = h \circ \mu$.

The category of Eilenberg-Moore T-algebras for C, denoted $\mathcal{A}C$, has arrows
$< X, h_X > \to < Y, h_Y >$ so that the diagram commutes

$$
\begin{array}{ccc}
TX & \xrightarrow{Tf} & TY \\
\downarrow h_X & & \downarrow h_Y \\
X & \xrightarrow{f} & Y.
\end{array}
$$

Example 2.2. Let G be a group and T the endofunctor $G \times \_$ of **SET** (i.e.
$TX = G \times X$). T is a monad where $\eta(X) = (1, X)$ and $\mu(g_1, g_2, X) =$
$(g_1 g_2, X)$. Set X is a T algebra when it is a G-set, i.e. a set with action h.

There is a potential source of confusion in the use of the term algebra. An
algebra as defined in Definition 2.1 is an Eilenberg-Moore algebra and requires
that T be a monad. If T is simply an arbitrary endofunctor of C one also
calls a pair (X, q) an algebra where $q : TX \to X$ is an arbitrary map in C
and conditions (5) and (6) are not required to hold. Both notions are useful
and are utilized in this paper. When confusion seems possible we will call the
former notion a $\mathcal{E}$-$\mathcal{M}$ algebra and the latter simply an algebra. Obviously,
any $\mathcal{E}$-$\mathcal{M}$ algebra is an algebra but not conversely. To assist the reader the
map h will only be used in the case of an $\mathcal{E}$-$\mathcal{M}$ algebra and q in the case of
an algebra. Note in particular that an algebra (X, q) can also be an $\mathcal{E}$-$\mathcal{M}$
algebra (X, h), where $h \neq q$ and more than one $\mathcal{E}$-$\mathcal{M}$ algebra structure can
exist for object X in C.

Example 2.3. Let **DOM** denote the category of domains and continuous maps and let $T = ()_\perp$ denote the lift monad on **DOM**. Every retract of a domain is again a domain and every domain is a retract of its lifting. Thus $\mathcal{ADOM}$, the category of $\mathcal{E}$-$\mathcal{M}$ T-algebras, has the same objects as **DOM**. The maps, however, are different. If $\uparrow_D$ denotes the new bottom in $D_\perp$ then a map $D \xrightarrow{f} E$ in $\mathcal{ADOM}$ satisfies $fh_D = h_E Tf$. Since h_D is monotone, $h_D(\uparrow) \leq h_D \eta_D(\perp_D) = \perp_D$. This in turn implies $f(\perp_D) = fh_D(\uparrow) = h_E Tf(\uparrow) = h_E(\uparrow_E) = \perp_E$, i.e. f is strict. Thus $\mathcal{ADOM}$ is precisely **SDOM** the category of domains and strict maps.

The *onno*, N^∞, for **DOM** is the domain $0 \to 1 \to 2 \to \cdots \to \infty$, also known as the ordered natural numbers. In the case of **DOM** the following property holds. There exists a pair of maps $1 \xrightarrow{0} N^\infty \xrightarrow{s} N^\infty$ which is initial for data $1 \xrightarrow{d} D \xrightarrow{f} D$ where $d \leq fd$. In particular, note that $s(n) = n + 1$ if $n \neq \infty$, ∞ if $n = \infty$. The following is implicit in [Mu2].

Theorem 2.4. N^∞ is the initial $()_\perp$-algebra in **DOM**.

Proof : N^∞ is the $()_\perp$-algebra $s : N^\infty_\perp \to N^\infty$ where $s(\uparrow) = 0$, $s(\infty) = \infty$

and $s(n) = n + 1$ otherwise. Let (D, q_D) be any algebra then there exists a unique algebra map $N^\infty \to D$ where $f(n) = q_D^{n+1}(\uparrow_D)$. □

We note some additional details. The lift functor $()_\perp$ generates both a reflection **DOM** $\to \mathcal{ADOM}$ and a coreflection $\mathcal{KDOM} \to$ **DOM** (where $\mathcal{KC}$ denotes the Kleisli category associated to **C**) which play important roles in understanding the semantics of partial data types. Much of the above carries over directly to an effective setting, including the existence of N^∞ as the *onno*, thus providing a context for the discussion of effective semantics. For example N^∞ in the effective case now must have an effective presentation and can be characterized as the initial algebra for the effective lift monad (see examples 3.3 and 3.6). The interested reader can find further discussion of the functorial connections between $\mathcal{KC}$, **C**, and $\mathcal{AC}$ as well as the relationship of $()_\perp$ to the ambient topos in [Mu3].

There are other possible variations on the lift monad, each generating in turn a new domain. The initial algebras for these monads exist and describe other iterative models such as the flat and lazy natural numbers. Details can be found in [FMRS].

The notion of *onno* was put to good use in the following result in [Mu2]. We first recall a definition.

Definition 2.5. Given categories **C**, **A** and bifunctors $F, G : \mathbf{C}^{op} \times \mathbf{C} \to \mathbf{A}$, a dinatural transformation $T : F \xrightarrow{\cdot} G$ is a family of maps $T_d : F(d, d) \to$

$G(d, d)$ for all $d \in \mathbf{C}$ so that for every arrow $d \xrightarrow{f} e$ in $\mathbf{C}$, the following diagram commutes.

$$
\begin{array}{ccccc}
 & & F(d,d) \xrightarrow{T_d} G(d,d) & & \\
 & F(f,1) & & & G(1,f) \\
F(e,d) & \nearrow & & \searrow & G(d,e) \\
 & \searrow & & \nearrow & \\
 & F(1,f) & & & G(f,1) \\
 & & F(e,e) \xrightarrow{T_e} G(e,e) & &
\end{array}
$$

Theorem 2.6. In order enriched cartesian closed categories $\mathbf{C}$ with *onno*, *fix* can be factored as $ev \circ T$ where for any $D \in \mathbf{C}$, $T_D : D^D \to D^{N^\infty}$ and $ev_D : D^{N^\infty} \to D$. In addition, *fix* is a dinatural transformation.

Proof :   See [Mu2].

In the theorem ev_D is the left adjoint of the diagonal map and T is defined by utilizing the universal property of N^∞ as an initial algebra. T is neither a natural nor dinatural transformation in general except in the special case where the variation is strict. As pointed out in [Mu2], it is an open question whether *fix* is the unique dinatural $()^0 \xrightarrow{..} ()$ in **DOM** or other similar semantic categories. (An easy counterexample exists for the category of complete lattices.) We now introduce a notion that allows us to fine-tune the previous result and arrive at a well known characterization of *fix*.

Definition 2.7. Given the categories and bifunctors of Definition 2.5, a strong dinatural transformation $T : F \xrightarrow{..} G$ is a family of maps T_d for all $d \in \mathbf{C}$, so that for every arrow $d \xrightarrow{f} e$ the diagram commutes

$$
\begin{array}{ccc}
F(d,d) \xrightarrow{T_d} G(d,d) & & \\
 & G(1,f) & \\
\nearrow & \searrow & \\
P & & \nearrow \quad G(d,e) \\
\searrow & G(f,1) & \\
 & \nearrow & \\
F(e,e) \xrightarrow{T_e} G(e,e) & &
\end{array}
$$

where P is the pullback of $F(d, f)$ and $F(f, e)$.

Note: One could provide a definition of strong dinaturality which doesn't require the existence of such pullbacks, however for all the examples we consider such pullbacks exist.

Proposition 2.8. Every strong dinatural transformation T is a dinatural transformation.

Proof :   Since $F(d,f) \circ F(f,d) = F(f,e) \circ F(e,f)$, the pair $F(f,d)$, $F(e,f)$ factors through P and thus the hexagon in definition 2.5 commutes.   □

Theorem 2.9. In **DOM** *fix* is the unique dinatural transformation $()^{()} \overset{..}{\to} id$ which is strong when variation is restricted to maps in $\mathcal{ADOM}$.

Proof :   By Theorem 2.6 *fix* is a dinatural transformation. The pullback

$$
\begin{array}{ccc}
P & \to & D^D \\
\downarrow & & \downarrow f^D \\
E^E & \overset{\to}{E^f} & E^D
\end{array}
$$

exists in **DOM** and consists of pairs $D \overset{g}{\to} D$, $E \overset{h}{\to} E$ satisfying $f \circ g = h \circ f$.

Suppose F is both a dinatural and a strong dinatural transformation $()^{()} \overset{..}{\to} id$ for maps f in $\mathcal{ADOM}$. Then F produces fixed points and the diagram

$$
\begin{array}{ccc}
& \nearrow D^D \overset{F_D}{\to} D & \\
P & & \downarrow f \\
& \searrow E^E \overset{F_E}{\to} E &
\end{array}
$$

commutes which in turn is equivalent to the following rule: F produces fixed points and if $f \circ g = h \circ f$ and f is an algebra map, then $f \circ F_D (g) = F_E (h)$. Since f is an algebra map is equivalent to f being strict (Example 2.3) this rule is exactly equivalent to Plotkin's characterization of *fix* (see [Mu2]). Thus *fix* is also a restricted strong dinatural transformation and uniquely so.   □

Perhaps the requirement of restricting to variation along algebra(strict) maps is unnecessary and a characterization of *fix* as the unique strong dinatural transformation $()^{()} \overset{..}{\to} id$ is possible. The following example shows this is not so(see also example 3.14).

Example 2.10. In the case of **DOM**, *fix* is not a strong dinatural transformation $()^{()} \overset{..}{\to} ()$. For example let $\mathbf{D} = \mathbf{E} = 0 \to 1$, let d, f equal $\lambda x.1$ and $e = id_E$. The equation $fd = ef$ holds however $fix(d) = 1, fix(e) = 0$ and $f(fix(d)) = f(1) = 1 \neq 0 = fix(e)$.

As the last example illustrates, restricting the variation to strict maps (algebra maps) plays a special role. In light of this, we consider in the next section similar questions raised in the context of categories of algebras and introduce the notion of algebraically strong dinatural transformation.

3.   Algebraically strong dinaturality and *FIX*

In this section we generalize the results of the previous section to the case of
an arbitrary strong monad for a cartesian closed category with a fixed point
object. In so doing we provide a fixed point semantics for various models of
monadic computation. Throughout we deal with a ccc **C** with strong monad
T. The following definition in [C-P] generalizes the notion of *onno* in [Mu2].

Definition 3.1. A fixed point object in **C** is

1. an initial algebra $TZ \overset{\sigma_Z}{\to} Z$;

2. a global element $1 \overset{\omega}{\to} Z$ which is the unique fixed point of $\sigma_Z \circ \eta_Z$:
   $Z \to Z$.

The definition above differs from that found in [C-P] since here ω was the
unique fixed point of $\eta_Z \circ \sigma_Z$ (i.e. $\omega \epsilon TZ$). The following result indicates this
does not create a problem.

Lemma 3.2. Given definition 3.1, $\eta_Z \omega$ is the unique fixed point of $\eta_Z \circ \sigma_Z$.
Conversely, if ω' is the unique fixed point of $\eta_Z \circ \sigma_Z$, then $\sigma_Z \omega'$ is the unique
fixed point of $\sigma_Z \circ \eta_Z$.

Proof :   The proof is routine. First, given definition 3.1, $\eta_Z \sigma_Z(\eta_Z \omega) =$

$\eta_Z(\sigma_Z \eta_Z \omega) = \eta_Z \omega$. To show uniqueness suppose $\eta_Z \sigma_Z \alpha = \alpha$ then $\sigma_Z \eta_Z \sigma_Z \alpha =$
$\sigma_Z \alpha$ so by uniqueness $\sigma_Z \alpha = \omega$. Thus $\alpha = \eta_Z \sigma_Z \alpha = \eta_Z \omega$. The other direction
is completely analogous.   □

Example 3.3. The *onno* N^∞ is the fixed point object for **DOM**, **SFP**, etc.
where s in Example 2.4 is the algebra map. Note in particular that $\eta \circ s = s$
so ∞ is the unique fixed point of $\eta \circ s$. A more interesting example found
in [Mu3] involves placing an effective presentation on N^∞. In that paper we
pointed out that N^∞ is the initial Σ - algebra for Σ the effective lift monad.
Thus it becomes the fixed point object for effective domains, **PER**, etc. The
following description of N^∞ for the case of **PER** appeared in [Mu4] (see also
[FMRS]). Let $T(n)$ denote $T_n n$ where T_n is the nth Turing machine. Let
$L = L^0 = \{n | T(n) \text{ diverges }\}$, $L^i = \{n | T^i(n) \text{ converges but } T^{i+1}(n) \text{ diverges}$
$\}$. Note that the complement of L is recursively enumerable but L is not. N^∞
is now the total equivalence realation on **N** defined by $[0] = L$, $[i] = L^i$ and
$[\infty] = \{n | T^i(n) \text{ converges for all i }\}$.

Recently Freyd has introduced the notion of a T-invariant object in an algebraically compact category.

Definition 3.4. A category **C** is algebraically compact if every covariant endo-functor has both an initial algebra and a final coalgebra and they are canoni-cally isomorphic. If A is the (necessarily) T-invariant object which is both an initial algebra and a final coalgebra for a covariant T and $q : TA \to A$ denotes the initial algebra structure map then we can up to isomorphism denote the final coalgebra by $q^{-1} : A \to TA$.

Example 3.5. We note that in our context any such T-invariant object, A, for T a monad is a fixed point object. In fact the argument already appears in [F] for the case of $T = \Sigma$ but for completeness we include the general argument here. The object A is already an initial algebra so we only need produce the unique global object ω. Since T is a monad and A is T-invariant, the unit η_1 exists and is a coalgebra and thus there exists a unique map $u : 1 \to A$ making the diagram commute.

$$
\begin{array}{ccc}
1 & \xrightarrow{\;u\;} & A \\
\eta_1 \downarrow & & \downarrow q^{-1} \\
T1 & \xrightarrow{\;Tu\;} & TA
\end{array}
$$

The diagram also commutes with q^{-1} replaced with η_A. We have $q\eta_A u = qTu\eta_1 = qq^{-1}u = u$. So u is a fixed point of $q\eta_A$. Suppose $q\eta_A v = v$. Then $q^{-1}q\eta_A v = q^{-1}v$ and $Tv\eta_1 = \eta_A v = q^{-1}v$, so the above diagram commutes with u replaced by v. Since u was the unique such map, v = u and thus u is the required ω.

Example 3.6. We now use these ideas to give a different description of the initial Σ-algebra N^∞ of Example 2.4 as a **PER**. In [FMRS] it is pointed out that N^∞ is also a final Σ-coalgebra and from its description is easily seen to be a T-invariant object. The total partial equivalence relation for N^∞ can be described as a map $N \to N^\infty$. In [Mul] it was pointed out that in the recursive topos (and therefore **PER**), partial recursive functions corresponded to maps $N \to \Sigma(N)$. Let $T : N \to \Sigma(N)$ denote the partial recursive function of Example 2.4. Since N^∞ is final there exists a unique map u making the diagram a map of coalgebras.

$$
\begin{array}{ccc}
N & \xrightarrow{\;u\;} & N^\infty \\
T \downarrow & & \downarrow s^{-1} \\
\Sigma N & \xrightarrow{\;\Sigma u\;} & \Sigma N^\infty
\end{array}
$$

The unique map u gives the desired description.

The question now arises as to whether a fixed point operator exists in this general setting, whether that operator has a factorization by utilizing the fixed point object and whether dinaturality plays a role. In fact, all of the above do occur.

Throughout this section we utilize the following convention:

Convention. **C** is a ccc with fixed point object Z and strong monad T.

We begin with a lemma about $\mathcal{E}$-$\mathcal{M}$ T-algebras.

Lemma 3.7. B an $\mathcal{E}$-$\mathcal{M}$ T-algebra implies B^A is an $\mathcal{E}$-$\mathcal{M}$ T-algebra.

Proof :   If $TB \xrightarrow{h_B} B$ is the $\mathcal{E}$-$\mathcal{M}$ algebra, then we have the composite

$h_B \circ T(eval) \circ t_{B^A,A}$: $T(B^A) \times A \rightarrow B$. Eval is the map $B^A \times A \rightarrow B$ and $t_{B^A,A}$ is the occurrence of t guaranteed for T strong. Taking the lambda abstraction produces the required structure map $T(B^A) \rightarrow B^A$.   □

Example 3.8. In the case of **DOM** the above construction produces the canonical $()_\perp$ algebra on B^A, namely the retract $(B^A)_\perp \rightarrow B^A$ of its lifting.

Corollary 3.9. For any $A, B \epsilon C, TB^A$ is a $\mathcal{E}$-$\mathcal{M}$ T-algebra.

Proof :   It is well known that the monad creates a reflection from **C** into

$\mathcal{A}C$ the category of $\mathcal{E}$-$\mathcal{M}$ algebras. In particular, the reflection of B into $\mathcal{A}C$ is $T^2B \xrightarrow{\mu_T} TB$.   □

We also note that Lemma 3.7 does not imply that $\mathcal{A}C$ is also cartesian closed. The category $\mathcal{A}\mathcal{D}\mathcal{O}\mathcal{M}$ is an easy counterexample.

Care must be taken in producing a fixed point combinator in this setting. Starting with an $\mathcal{E}$-$\mathcal{M}$ algebra (X, h_X) and fixed point object Z we wish to describe a factorization of a map $FIX_X : X^X \rightarrow X$ that utilizes Z. First note that we wish to interpret X^X in **C** rather than $\mathcal{A}C$, otherwise the map FIX_X may become too restrictive. For example, if T is the lift monad in **DOM**, then algebra maps are strict, $\perp$ is always the trivial fixed point, and FIX_X reduces to the constant $\perp$ map. With arbitrary maps $X \xrightarrow{f} X$, however, we need to find a corresponding algebra in order to find a role for Z in the factorization of FIX. For a general X in **C** we have no mechanism for finding such an algebra.

If we restrict our attention to $\mathcal{A}C$ then we fare better because an associated algebra is readily available for any endomap in **C**. Namely, suppose (A, h) is an $\mathcal{E}$-$\mathcal{M}$ algebra and $A \xrightarrow{f} A$ is any endomap in **C** then $(A, f \circ h)$ is a new algebra.

Example 3.10. Let $D\epsilon$ **DOM**, $T = ()_{\perp}$ and $D \xrightarrow{f} D$ be an endomap in **C**. Then assuming the canonical algebra on D we generate the map $g : TD \to D$ where $g \uparrow = f \perp$ and $gd = fd$ otherwise. Here we let $\uparrow \neq \perp_D$ denote the new bottom generated by T. The map g is just the algebra associated to f.

Before we state the main result of this section the above considerations and those at the end of section 2 require a more refined notion of strong dinatural transformation.

Definition 3.11. Given the categories and bifunctors of Definition 2.7 as well as a strong monad on C, T is an algebraically strong dinatural transformation if the diagram of Definition 2.7 commutes when the variation map f must come from category $\mathcal{A}C$. More precisely F,G are bifunctors $F, G : \mathcal{A}C^{op} \times \mathcal{A}C \to \mathbf{A}$ and f is a map $< d, h_d > \to < e, h_e >$ in $\mathcal{A}C$.

We stress that the diagram referred to in definition 3.11 is interpreted in category **A**. In particular when **A** is **C**, a *ccc*, the diagram is interpreted in **C** rather than $\mathcal{A}C$ which in general is not a *ccc*.

Theorem 3.12. For any *ccc* **C** with strong monad and fixed point object Z there exists a fixed point combinator FIX_D for any $\mathcal{E}$-$\mathcal{M}$ T-algebra (D, h_D). The combinator is an algebraically strong dinatural transformation $()^{()} \xrightarrow{\cdot} ()$. In addition, FIX has a factorization utilizing the fixed point object. If FIX is also a dinatural transformation with $(Z, h_Z) \in \mathcal{A}C$, and (Z, σ^{-1}) is also a final coalgebra with σ^{-1} a map in $\mathcal{A}C$, then FIX is the unique such one which is also algebraically strong.

Proof :   In an analogous manner to section one we produce a factorization

of FIX_D utilizing the fixed point object Z. The map $H_D : D^D \to D^Z$ is defined as follows: $H_D(g)$ is the unique algebra map $u : Z \to D$ where D is the algebra generated by g, i.e. the composite $g \circ h_D : TD \to D$. Composing H with the map $D^\omega : D^Z \to D$ where ω is the fixed point of Z produces the desired combinator and factorization. For example $FIX_D(g) = u\omega$ so $gFIX_D(g) = gu\omega = u\sigma\eta_Z\omega = u\omega = FIX_D(g)$.

We now address the issue of strong dinaturality. Both factorization and strong dinaturality can be illustrated in the following diagram where P is the usual pullback.

$$
\begin{array}{ccccc}
 & D^D & \xrightarrow{H_D} & D^Z & \xrightarrow{D^\omega} & D \\
P \nearrow & & & \downarrow f^Z & & \downarrow f \\
 \searrow & E^E & \xrightarrow{H_E} & E^Z & \xrightarrow{E^\omega} & E
\end{array}
$$

We begin by noting the right square commutes since $()^\omega : ()^Z \xrightarrow{\cdot} ()$ is a natural transformation. In general we should not expect the left (truncated) hexagon to commute but for our special case it does.

Lemma 3.13. If f is a map in $\mathcal{AC}$ then the left hexagon commutes.

Proof :   Suppose (d, e) is an element in P, i.e. $f \circ d = e \circ f$. Clockwise we

have $f \circ H_D(d)$ and the other direction yields $H_E(e)$. We cannot appeal yet to the initiality of Z because f is only known to be an algebra map between the original $\mathcal{E}$-$\mathcal{M}$ algebras (D, h_D) and (E, h_E). The following diagram, however, shows that f is an algebra map for the generated algebras.

$$
\begin{array}{ccc}
TD & \xrightarrow{\;Tf\;} & TE \\
h_D \downarrow & & \downarrow h_E \\
D & \xrightarrow{\;f\;} & E \\
d \downarrow & & \downarrow e \\
E & \xrightarrow{\;f\;} & E
\end{array}
$$

The top square commutes because f is in $\mathcal{AC}$ and the lower square because $(d, e) \epsilon P$. Thus f is an algebra map from $(D, d \circ h_D)$, to $(E, e \circ h_E)$. By the initiality of Z we have $f \circ H_D(d) = H_E(e)$. The lemma is proven.   □

By the lemma H is an algebraically strong dinatural transformation. We emphasize in particular that the diagram (and thus the transformation) exists in $\mathbf{C}$ (not $\mathcal{AC}$) so FIX is acting on arbitrary endomaps $D \to D$ of $\mathcal{E}$-$\mathcal{M}$ algebra (D, h_D). The composition of an algebraically strong dinatural transformation with a natural transformation is still an algebraically strong dinatural so FIX is algebraically strong.

We now prove the uniqueness condition. Suppose $W : ()^{()} \overset{..}{\to} ()$ is both dinatural and algebraically strong dinatural and let $d : D \to D$ be an arbitrary endomap in $\mathbf{C}$ of $\mathcal{E}$-$\mathcal{M}$ algebra D. We show that W must in fact be FIX. There exists a unique algebra map $g : Z \to (D, d \circ h_D)$ which in turn generates the commutative diagram

$$
\begin{array}{ccc}
Z & \xrightarrow{\;g\;} & D \\
\downarrow \eta_Z & & \downarrow \eta_D \\
TZ & \xrightarrow{\;Tg\;} & TD \\
\downarrow \sigma & & \downarrow d \circ h \\
Z & \xrightarrow{\;g\;} & D
\end{array}
$$

The top square commutes by the naturality of η and the bottom since g is an algebra map. Consider the diagram

$$
\begin{array}{ccc}
 & Z^Z \xrightarrow{\ W_Z\ } Z & \\
P \Big\langle \ \ \ \ \ \ \ \ \ \ & & \downarrow g \\
 & D^D \xrightarrow{\ W_D\ } D &
\end{array}
$$

By the above diagram $(d \circ h \circ \eta_D, \sigma \circ \eta_Z) = (d, \sigma \circ \eta_Z)\epsilon P$. Unfortunately although Z is in $\mathcal{AC}$, g is generally not a map in $\mathcal{AC}$. Consider instead the map $g' = h \circ Tg \circ \sigma^{-1}$. Since Tg and h are always maps in $\mathcal{AC}$ and σ^{-1} is by assumption, g' is also a map in $\mathcal{AC}$. We have $g'\sigma\eta_Z = hTg\eta_Z = h\eta_D g = g = dg'$. Therefore the pair $(d, \sigma \circ \eta_Z)\epsilon P$ where g is replaced by g' in the above diagram. First we consider $W_Z(\sigma \circ \eta_Z)$. Since W is a dinatural transformation it is a fixed point operator [Mu2]. Thus $W_Z(\sigma \circ \eta_Z)$ is a fixed point of $\sigma \circ \eta_Z$. There is only one such fixed point, however, namely ω and thus going clockwise we arrive at $g'(\omega)$. Since W is an algebraically strong dinatural and g' is a map in $\mathcal{AC}$, we have $W_D(d) = g'(\omega)$. $W_D(d)$ is a fixed point of d and thus $W_D(d) = dW_D(d) = dg'(\omega) = g(\omega) = FIX_D(d)$ and we are done.   □

As pointed out in [CP], for Moggi's computational semantics, fixpoint combinators are only needed for certain types, namely those of the form $\alpha \to TB$. The following result is now a corollary of Theorem 3.12.

Corollary 3.14. In the presence of a fixed point object there exists a combinator $Y_{\alpha,\beta} : ((\alpha \to T\beta) \to (\alpha \to T\beta)) \to \alpha \to T\beta$.

Proof :   By corollary 3.9, $\alpha \to T\beta$ is an $\mathcal{E}$-$\mathcal{M}$ T-algebra and so by theorem

3.12 the result follows.   □

It should be pointed out that it would also make sense to define the notion of algebraic dinatural transformation by simply replacing definition 2.7 by definition 2.5 in the wording of definition 3.11. The proof of theorem 3.12 would then also show that FIX is an algebraic dinatural transformation (the equation $fd = ef$ is replaced by the associative law $f(gf) = (fg)f$). However since this weakens rather than strengthens the notion of dinatural transformation, it does not appear to help refine our characterization of FIX. The following examples illustrate some of the difficulties that arise in using the notions of dinaturality and strong dinaturality to characterize *fix*. Let f, d, e be as above.

Example 3.15. Not every algebraically strong dinatural transformation produces a fixed point combinator. In **DOM** let $W(d) = d(\bot)$. Then $fW(d) = fd(\bot) = ef(\bot) = e(\bot) = W(e)$. Thus W is an algebraically strong dinatural transformation yet for an arbitrary non-strict d clearly fails to yield a fixed point if $d(\bot) \neq dd(\bot)$. This example does not contradict the uniqueness condition of theorem 3.12 because W is not a dinatural transformation.

Example 3.16. Some strong monads that have initial algebras may not produce fixed point objects. For example in **SET** the strong monad $TX = 1 + X$ has an initial algebra, namely the natural numbers, yet clearly T has no fixed point object. This need not imply there is no fixed point combinator as Example 3.17 shows but trivially this is the case here.

Example 3.17. Fixed point combinators need not require the existence of a fixed point object. The following example is due to R. Crole. Let **C** be ω cpo's with monotone maps and T be the lift monad. Then it is well known that **C** has least fixed points, however **C** has no fixed point object for T. The usual candidate, N^∞, fails because it is only weakly initial. For a given lift algebra A, there may be many maps from N^∞ to A.

Example 3.18. The uniqueness result in theorem 3.12 is dependent on the uniqueness condition on ω in definition 3.1. If **C** is the category of complete lattices with maps preserving sups and infs(including empty ones) and T = $()_\bot$ then N^∞ is no longer a fixed point object since N^∞ has two fixed points, 0 and ∞. These in turn produce two fixed point combinators which are both dinatural and algebraically strong dinatural. These correspond to the least(fix) and greatest (G) fixed point combinators respectively which in this case trivialize to bottom and top combinators respectively. Changing **C** where now maps preserve only nonempty sups/infs, the usual fixed point object, N^∞, exists and both fix and G are dinatural transformations. Theorem 3.12 remains valid however as now fix is algebraically strong while G is not. For example if L is the lattice $0 \to 1$, $e : L \to L$ is a map in $\mathcal{AC}$ where $e(0) = 0 = e(1)$. In fact e is idempotent so we have $e \circ e = id \circ e$ and so if G is algebraically strong $eG(e)$ must equal $G(id)$. We have however $eG(e) = e(0) = 0 \neq 1 = G(id)$. Actually there is another strong monad at work here namely T = $()_\top$ which adds a new top to the lattice. For this monad there exists a fixed point object as well, namely $N^{-\infty} = 0 \leftarrow -1 \leftarrow -2 \leftarrow \cdots \leftarrow -\infty$. Now FIX agrees with G and is both dinatural and algebraically strong dinatural relative to the new strong monad.

Example 3.19. Consider the monad T defined on **DOM** as follows, $TX = \Sigma \vee X$ where we use the notation of [FMRS]. A T algebra is a pair (t,a) consisting of a, an element of A, and a (not necessarily strict) endomap, t, of A. The flat natural numbers $N_\bot$ is not the initial algebra for T defined on **DOM** but rather for T defined on $\mathcal{ADOM}$ (where t is now strict) and this is where $N_\bot$ is a fixed point object. In this case ω is just $\bot$ and consequently FIX_D as constructed in theorem 3.12 takes an arbitrary endomap d to $\bot_D$.

While FIX is not a dinatural transformation in **DOM**, it is in $\mathcal{ADOM}$ and is algebraically strong.

Example 3.20. Let **C** denote **BDOM**, the *ccc* of bottomless domains(ie those domains not necessarily having a bottom element). Clearly there is no fix combinator for **C**. Simply take D to be the bottomless domain with 2 unrelated points then the twist map has no fixed point. The category of T algebras for T the lift monad is simply $\mathcal{ADOM}$ and so by theorem 3.12 if we restrict to domains D, the objects of $\mathcal{ADOM}$, and arbitrary endomaps of D in **BDOM** we get a fixed point combinator FIX. These endomaps agree exactly with those in **DOM** and FIX is simply fix.

As the last few examples illustrate, the notion of strong dinatural transformation is too restricted to characterize FIX while the notion of algebraically strong dinatural transformation alone is too encompassing. What we have provided in this paper is a general means of generating fixed point combinators on algebras in the presence of strong monads with fixed point objects along with a partial algebraic characterization of these combinators. In particular the algebraically compact categories of Freyd will generate such fixed point objects. In a future paper we will investigate more closely the connections of the above work to that found in [F] and also examine other monads of interest in categorical semantics.

REFERENCES

[B] M. Barr, *Fixed Points in Cartesian Closed Categories*, **Theoretical Computer Science** 70 (1990) 65-72.

[BFSS] E. S. Bainbridge, P. J. Freyd, A. Scedrov, P. J. Scott, *Functorial Polymorphism*, **Theoretical Computer Science** 70 (1990).

[C-P] R. L. Crole, A. M. Pitts, *New Foundations for Fix Point Computations*, **Proceedings of LICS**, University of Pennsylvania, 1990.

[F] P. Freyd, *Algebraically Complete Categories*, preprint, 1991.

[FMRS] P. Freyd, P. Mulry, G. Rosolini, D. Scott, *Extensional PERs*, 1990, to appear in **Information and Computation**.

[HP] H. Huwig, A. Poigne, *A note on inconsistencies caused by fixpoints in a cartesian closed category*, preprint, 1988.

[Mo] E. Moggi, *Computational Lambda-Calculus and Monads*, **Proceedings of LICS**, Asilomar, 1989.

[Mu1] P. S. Mulry, *Generalized Banach-Mazur Functionals in the Topos of Recursive Sets*, **Journal of Pure and Applied Algebra** 26 (1982), 71-83.

[Mu2] P. S. Mulry, *Categorial Fixed Point Semantics*, **Theoretical Computer Science** 70 (1990), 85-97.

[Mu3] P. S. Mulry, *Monads and Algebras in the Semantics of Partial Data Types*, 1989, to appear in **Theoretical Computer Science**.

[Mu4] P. S. Mulry, *Remarks on Modest Sets and Computation*, preprint, 1989.

Semantics of Local Variables

P. W. O'Hearn*   R. D. Tennent[†]

Abstract

This expository article discusses recent progress on the problem of giving sufficiently abstract semantics to local-variable declarations in Algol-like languages, especially work using categorical methods.

1 Introduction

One of the first things a beginning programmer learns is how to "declare" a new variable, as in the following Algol-like block:

$$\textbf{new } x\!:\!\textsf{int}. \ \ x := 0 \ ; \cdots ; x := x + 1 \ ; \cdots$$

It might be thought that a satisfactory semantic interpretation for such declarations would be well established by now. But when free identifiers, possibly of higher-order type, can appear within the bodies of such declaration blocks, there are serious difficulties. The aim of this expository paper is to survey recent progress on this problem, especially work using categorical methods.

The traditional denotational-semantic approach [9, 19, 18] is to assume a denumerable set L of "locations" (abstract storage addresses) and construct a set of states (also known as "stores") along the following lines:

$$S \ = \ L \rightarrow (V + \{unused\}) \, ,$$

where V is the set of *storable values*. (For simplicity, we will assume there is a *single* type of storable value, say, integers.) The idea is that each state $s \in S$ records, for every location $\ell \in L$, *either* the $v \in V$ currently stored at that location, *or* the fact that location ℓ is not currently "in use." Then the

*School of Computer and Information Science, Syracuse University, Syracuse, New York, USA 13244. This author attended the Durham Symposium with the aid of a travel stipend from the S.E.R.C.

†Department of Computing and Information Science, Queen's University, Kingston, Ontario, Canada K7L 3N6. This author was supported by an operating grant from the Natural Sciences and Engineering Research Council of Canada.

effect of a variable declaration is to bind the declared variable identifier to any currently unused location for the execution of the block body.

This kind of interpretation of local variables is adequate to show the correctness of the usual style of implementation of block structure [9]; however, many authors [3, 16, 13, 5, 14, 2, 8] have criticized it as being insufficiently abstract. For example, consider the simple equivalence

$$\begin{array}{c} \textbf{new } \iota. \ \iota := 0; \\ C \end{array} \quad \equiv \quad C \,,$$

when identifier ι is not free in command C. The equivalence is a consequence of the inaccessibility of the "new" location to non-local entities in Algol-like languages. Unfortunately, the equivalence fails in general in the traditional semantics. To see the problem, assume that the initial state marks a location ℓ as being unused and ℓ is chosen as the meaning of ι in the block; but in general the meaning of C need not respect the convention that ℓ is not accessible to non-local entities and might, for example, branch on whether it is "unused."

Some researchers [9, 5] have addressed this kind of difficulty by attempting to define the set of "accessible" locations or *storage support* of semantic entities, in order to ensure that the "new" location allocated for a variable declaration is not already accessible in the block body. However, this has proved to be technically complex and delicate. Phrases of basic type, such as expressions and commands, can be treated fairly straightforwardly; but the storage support of a procedure is much more difficult to define, particularly if an extensional treatment of procedures is desired.

An alternative is to adopt a semantical approach in which storage supports are explicit and logically precede any assignment of meanings to phrases; this is the possible-world method of [16] and [13].

2   Possible Worlds

The basic idea underlying possible-world semantics is to parameterize interpretations by semantic entities termed "possible worlds" that determine certain "local" aspects of meanings. To treat storage support, we assume that each world is a *natural number*, interpreted as the number of locations currently in use; in implementation terms, this is the number of variables on the run-time stack. Then, for each world w, we can interpret basic phrase types **var** (variables), **exp** (expressions), and **comm** (commands) as the following computational domains:

$$[\![\textbf{var}]\!]w \ = \ \{0, 1, 2, \ldots, w - 1\}, \text{ discretely ordered};$$

$$[\![\mathbf{exp}]\!]w \;=\; S(w) \rightarrow V_\perp;$$

$$[\![\mathbf{comm}]\!]w \;=\; S(w) \rightarrow S(w)_\perp,$$

where $S(w) = [\![\mathbf{var}]\!]w \rightarrow V$ is the set of w-states, $A_\perp$ for any set A is the "lifted" domain obtained by adding a new least element $\perp$, and, for any sets or domains D and E, $D \rightarrow E$ is the domain of (continuous) functions from D to E, ordered pointwise.

Intuitively:

- a declared variable in world w denotes a location (natural number) less than w;

- an expression meaning in world w is a function from w-states to values (or to $\perp$, to allow for non-termination of evaluation), where a w-state is a vector of storable values with exactly w components; and

- a command meaning in world w is a function from w-states to w-states (or to $\perp$).

The worlds w determine the storage support of the basic semantic entities in a simple and natural way. Procedures will be discussed in Section 3.

To cope with variable declarations, we must define how to map phrase meanings from non-local to local worlds. This is most coherently done by defining a *category* $\mathbf{W}$ of possible worlds: the objects are natural numbers and the morphisms from w to x are the injective functions from $\{0, 1, \ldots, w-1\}$ to $\{0, 1, \ldots, x-1\}$, with functional composition as the composition in $\mathbf{W}$. We will explain later why *all* injections (and not merely the inclusions) are needed.

We can now extend the interpretations of basic phrase types θ to *functors* $[\![\theta]\!]$ from $\mathbf{W}$ to the category $\mathbf{D}$ of directed-complete posets and continuous functions by defining morphism parts $[\![\theta]\!]f$ for every $\mathbf{W}$-morphism $f\colon w \rightarrow x$. We define

$$[\![\mathbf{var}]\!]f\colon [\![\mathbf{var}]\!]w \rightarrow [\![\mathbf{var}]\!]x$$

to be the function f. To define

$$[\![\mathbf{exp}]\!]f\colon [\![\mathbf{exp}]\!]w \rightarrow [\![\mathbf{exp}]\!]x,$$

we note that, for any state $s \in S(x)$, $(f\,;s) \in S(w)$ is the corresponding "non-local part" of s, where ; denotes composition in diagrammatic order, and then

$$[\![\mathbf{exp}]\!]f\,e\,s \;=\; e(f\,;s)$$

for any $e \in [\![\mathbf{exp}]\!]w$ and $s \in S(x)$.

Similarly, $[\![\mathbf{comm}]\!]f\,c\,s$ is $\perp$ if $c(f;s) = \perp$, and otherwise is the state $s' \in S(x)$ such that, for any $\ell \in [\![\mathbf{var}]\!]x$, $s'(\ell) = c(f\,;s)(\ell')$ if $\ell = f(\ell')$ and $s'(\ell) = s(\ell)$ otherwise; that is, the effect of executing a non-local command meaning c is to change the non-local part of the state in the manner determined by c, but to preserve the local part of the state invariant.

To deal with free identifiers and environments, we note that finite products of functors from $\mathbf{W}$ to $\mathbf{D}$ are definable pointwise: for any $F, G \colon \to \mathbf{D}$ and any world w, $(F \times G)(w) = F(w) \times G(w)$, and similarly for the morphism part. Then, for any assignment π of phrase types to a finite set of identifiers, we can define $[\![\pi]\!]$ to be the product of the functors for the types assigned by π to these identifiers. Then $[\![\pi]\!]w$ is the domain of π-compatible environments appropriate to world w, and $[\![\pi]\!]f$ for $\mathbf{W}$-morphisms $f \colon w \to x$ maps such w-environments to the appropriate x-environments.

We use the notation $(\pi \mid \iota \mapsto \theta)$ to denote the type assignment that is like π except that identifier ι is assigned type θ, and similarly for the environment $(u \mid \iota \mapsto m)$, and the state $(s \mid \ell \mapsto v)$.

Finally, any phrase X that is well-formed and of type θ in the context of phrase-type assignment π (notation: $\pi \vdash X \colon \theta$) is interpreted (using a valuation function $[\![\cdot]\!]_{\pi\theta}$) as a *natural transformation* $[\![X]\!]_{\pi\theta} \colon [\![\pi]\!] \overset{\cdot}{\to} [\![\theta]\!]$; i.e., the right-hand side of the following diagram commutes for all $\mathbf{W}$-morphisms $f \colon w \to x$:

$$
\begin{array}{ccc}
w & [\![\pi]\!]w \xrightarrow{\;[\![X]\!]_{\pi\theta}(w)\;} [\![\theta]\!]w \\[2mm]
\Big\downarrow f & \quad [\![\pi]\!]f \Big\downarrow \qquad\qquad\qquad \Big\downarrow [\![\theta]\!]f \\[2mm]
x & [\![\pi]\!]x \xrightarrow{\;[\![X]\!]_{\pi\theta}(x)\;} [\![\theta]\!]x
\end{array}
$$

The commutativity requirement ensures that the parameterized interpretation of X is *uniform* with respect to all changes of world f. The phrase-type assignment (π) and phrase-type (θ) subscripts on phrase-valuation brackets will often be omitted when they are obvious from context.

Representative syntax rules and semantic valuations are presented in Tables 1 and 2, respectively. The semantic arguments in the latter are as follows: w is any $\mathbf{W}$-object (natural number), u is an environment in $[\![\pi]\!]w$, and $s \in S(w)$ is a state appropriate to world w.

$$\frac{\pi \vdash X:\mathbf{var} \quad \pi \vdash E:\mathbf{exp}}{\pi \vdash X := E:\mathbf{comm}} \;\; Assignment \qquad \frac{(\pi \mid \iota \mapsto \mathbf{var}) \vdash C:\mathbf{comm}}{\pi \vdash \mathbf{new}\,\iota.\,C:\mathbf{comm}} \;\; New$$

$$\frac{\pi \vdash C_1:\mathbf{comm} \quad \pi \vdash C_2:\mathbf{comm}}{\pi \vdash C_1\,;C_2:\mathbf{comm}} \;\; Sequencing \qquad \frac{\pi \vdash X:\mathbf{var}}{\pi \vdash X:\mathbf{exp}} \;\; Dereferencing$$

Table 1. Selected Syntax Rules.

$$[\![X := E]\!]w\,u\,s \;=\; \begin{cases} (s \mid \ell \mapsto v), & \text{if } [\![E]\!]w\,u\,s = v \in V \\ & \text{and } [\![X]\!]w\,u = \ell \\ \bot, & \text{if } [\![E]\!]x\,u\,s = \bot \end{cases}$$

$$[\![C_1\,;C_2]\!]w\,u\,s \;=\; \begin{cases} [\![C_2]\!]w\,u\,s', & \text{if } [\![C_1]\!]w\,u\,s = s' \in S(w) \\ \bot, & \text{if } [\![C_1]\!]w\,u\,s = \bot \end{cases}$$

$$[\![X]\!]_{\pi\,\mathbf{exp}}w\,u\,s \;=\; s(\ell), \text{ where } \ell = [\![X]\!]_{\pi\,\mathbf{var}}w\,u$$

Table 2. Sample Valuations.

We can now discuss the interpretation of the variable-declaration block. Let $v_0 \in V$ be a standard initial value for declared variables, and $f: w \to w+1$ be the inclusion of $\{0,1,\ldots,w-1\}$ in $\{0,1,\ldots,w\}$; then $[\![\mathbf{new}\,\iota.\,C]\!]_{\pi\,\mathbf{comm}}\,w\,u\,s \;=\; f\,;s'$ when

$$[\![C]\!]_{(\pi|\iota\mapsto\mathbf{var})\mathbf{comm}}(w+1)(u')(s \mid w \mapsto v_0) \;=\; s',$$

where $u' = \big([\![\pi]\!]\,f\,u \mid \iota \mapsto w\big)$, and is $\bot$ otherwise.

Notice that the meaning of the block in world w is defined in terms of the meaning of the body of the block in world $w+1$. In the body of the block, the declared variable denotes the "new" location, and all other free identifiers denote meanings obtained by applying the morphism part of the relevant functor to the non-local meanings in the environment. After execution of the body, the new location is discarded. We leave it as an exercise for the reader to prove the validity of the problematical equivalence discussed in Section 1 using this semantics.

We conclude this section by discussing the naturality of $[\![\mathbf{new}\,\iota.\,C]\!]$; i.e., the commutativity of

$$
\begin{array}{ccc}
[\![\pi]\!]w & \xrightarrow{\;[\![\mathbf{new}\,\iota.\,C]\!]w\;} & [\![\mathbf{comm}]\!]w \\[2pt]
{\scriptstyle[\![\pi]\!]f}\big\downarrow & & \big\downarrow{\scriptstyle[\![\mathbf{comm}]\!]f} \\[2pt]
[\![\pi]\!]x & \xrightarrow{\;[\![\mathbf{new}\,\iota.\,C]\!]x\;} & [\![\mathbf{comm}]\!]x
\end{array}
$$

for every $f\colon w \to x$. Note that $[\![\mathbf{new}\,\iota.\,C]\!]w$ and $[\![\mathbf{new}\,\iota.\,C]\!]x$ bind ι to possibly *different* locations; the former uses $w \in [\![\mathbf{var}]\!](w + 1)$, and the latter uses $x \in [\![\mathbf{var}]\!](x + 1)$. The key to showing naturality is to relate w and x in a suitable fashion. This is done by extending f with the identity id_1 on 1; i.e., using the **W**-morphism $(f + \mathrm{id}_1)\colon w + 1 \longrightarrow x + 1$ that renames $w \in [\![\mathbf{var}]\!](w+1)$ to $x \in [\![\mathbf{var}]\!](x+1)$, where, for any $f\colon w \to x$ and $g\colon y \to z$, the **W**-morphism $(f + g)\colon w + y \longrightarrow x + z$ is given by

$$
(f + g)(i) \;=\; \begin{cases} f(i), & \text{if } i < w \\ x + g(j), & \text{if } i = w + j \end{cases}
$$

for $i \in [\![\mathbf{var}]\!](w + y)$. Naturality of $[\![C]\!]$ with respect to $(f + \mathrm{id}_1)$ gives us the commutativity of

$$
\begin{array}{ccc}
[\![\pi]\!](w + 1) & \xrightarrow{\;[\![C]\!](w + 1)\;} & [\![\mathbf{comm}]\!](w + 1) \\[2pt]
{\scriptstyle[\![\pi]\!](f + \mathrm{id}_1)}\big\downarrow & & \big\downarrow{\scriptstyle[\![\mathbf{comm}]\!](f + \mathrm{id}_1)} \\[2pt]
[\![\pi]\!](x + 1) & \xrightarrow{\;[\![C]\!](x + 1)\;} & [\![\mathbf{comm}]\!](x + 1)
\end{array}
$$

which shows that the specific choice of fresh location does not affect the meaning of $\mathbf{new}\,\iota.\,C$, and then naturality of $[\![\mathbf{new}\,\iota.\,C]\!]$ follows easily.

The important role of "location renaming" in this discussion explains why we used all injective maps, and not just inclusions, as morphisms in the category of worlds: it is forced on us by naturality. This even shows up on the level of terms, via equivalences valid in the model; e.g.,

$$
\mathbf{new}\,x.\,\mathbf{new}\,y.\,C \;\equiv\; \mathbf{new}\,y.\,\mathbf{new}\,x.\,C.
$$

The principle that the specific "names" of declared locations are irrelevant turns out to be technically quite important; it plays a central role in [5, 2, 8] as well. In their treatment of support, Meyer and Sieber consider a semantic framework very similar to possible worlds, but using inclusion maps, rather than all injections, as the morphisms between worlds. They must then cut down the model by restricting to semantic entities invariant under certain location-renaming permutations. Here, the additional maps in our category of worlds provide us with stronger uniformity conditions.

It should be mentioned that, for languages *without* procedures, the kind of equivalences we have been discussing can be validated using "traditional" models. The main advantage of the possible-world approach to variable allocation is that it extends smoothly to procedural and other types. By building storage support into the "semantical spaces" (i.e., functors), one can find a notion of procedure meaning, based on exponentiation in functor categories, that itself incorporates support. This is the topic of the next section.

3   Procedures

In this section, we consider how *procedural* types of the form $\theta \rightarrow \theta'$ are dealt with in the possible-world framework. A procedure defined in world w might be called in any world x accessible from w using a **W**-morphism $f: w \rightarrow x$; for example, there might be variable declarations occurring between the point of definition and the point of call. But it is the domain structure determined by f and x that should be in effect when the procedure body is executed. This suggests that we cannot in general define $[\![\theta \rightarrow \theta']\!]w$ to be just $[\![\theta]\!]w \rightarrow [\![\theta']\!]w$, where the arrow on the right is the exponentiation in **D**. In general, the meaning of a procedure defined in world w must be a *family* of functions, indexed by changes of world $f: w \rightarrow x$, and suitably constrained by a uniformity condition.

The key fact is that, for *any* small category **W**, the functor category $\mathbf{D}^{\mathbf{W}}$ is Cartesian closed. This is proved for any complete Cartesian closed category **D** in [10]; the special case of **D** being the category of directed-complete posets and continuous functions is treated in [13].

Finite products in $\mathbf{D}^{\mathbf{W}}$ are definable pointwise. We now give a general definition of exponentiation and corresponding interpretations of abstraction and application satisfying the usual laws of the typed lambda calculus. For any $F, G: \rightarrow \mathbf{D}$ and world w, $p \in (F \rightarrow G)(w)$ is a *family* $p(\cdot)$ of functions, indexed by **W**-morphisms with domain w, such that $p(f: w \rightarrow x)$ is a continuous function from $F(x)$ to $G(x)$ and the family satisfies the following uniformity condition: for all **W**-morphisms $f: w \rightarrow x$ and $g: x \rightarrow y$,

$$p(f)\,;G(g) \;=\; F(g)\,;p(f\,;g)$$

i.e., the following diagram commutes:

$$
\begin{array}{ccc}
F(x) & \xrightarrow{\;p(f)\;} & G(x) \\
{\scriptstyle F(g)}\big\downarrow & & \big\downarrow{\scriptstyle G(g)} \\
F(y) & \xrightarrow{\;p(f\,;g)\;} & G(y)
\end{array}
$$

The domain $(F \to G)(w)$ is ordered pointwise, and the morphism part of $F \to G$ is defined as follows: for any $f: w \to x$, $p \in (F \to G)(w)$, and $g: x \to y$,

$$(F \to G)\,f\,p\,g \;=\; p(f\,;g).$$

We can now interpret procedural types by induction on the structure of types; for any phrase types θ and θ', $[\![\theta \to \theta']\!] = [\![\theta]\!] \to [\![\theta']\!]$, where the arrow on the right is the just-described exponentiation in the functor category $\mathbf{D}^{\mathbf{W}}$. The valuations for abstraction and application are then as follows; for worlds w and environments $u \in [\![\pi]\!]w$,

$$[\![\lambda\iota\!:\!\theta.\,P]\!]_{\pi(\theta\to\theta')}\,w\,u\,f\,a \;=\; [\![P]\!]_{(\pi|\iota\mapsto\theta)\theta'}\,x\left([\![\pi]\!]\,f\,u \,\middle|\, \iota \mapsto a\right)$$

for all $f: w \to x$ and $a \in [\![\theta]\!]x$, and

$$[\![P\,Q]\!]_{\pi\theta'}\,w\,u \;=\; [\![P]\!]_{\pi(\theta\to\theta')}\,w\,u\,(\mathrm{id}_w)\bigl([\![Q]\!]_{\pi\theta}\,w\,u\bigr),$$

where id_w is the identity $\mathbf{W}$-morphism on world w. Note that these definitions are applicable with *any* category $\mathbf{W}$ of possible worlds.

The above treatment of procedures is *not* in general extensional in the usual sense; however, for *any* category $\mathbf{W}$, the functor category $\mathbf{S}^{\mathbf{W}}$ (where $\mathbf{S}$ is the category of sets and functions) is a topos [4, 7, 1], and so there is an interpretation of logical connectives such as implication ($\Rightarrow$), universal quantification ($\forall$), and identity ($\equiv$) that validates the laws of intuitionistic logic, and also the following *formal* law of extensionality:

$$\bigl(\forall\iota\!:\!\theta.\,P(\iota) \equiv Q(\iota)\bigr) \;\Rightarrow\; P \equiv Q$$

where ι is not free in $P, Q: \theta \to \theta'$. This allows us to reason about procedures in (intuitionistic) programming logics as if they were extensional. One such system will be discussed in Section 6.

4 States and Contravariance

At this point the reader may be wondering why functors for commands and expressions were defined pointwise in Section 2, whereas procedures were treated using exponentiation of functors in Section 3. Would it not be possible to obtain a more systematic treatment by using exponentiation in the definition of base types as well? That is, one might try to define a "states" functor S and proceed as follows:

$$[\![exp]\!] = S \rightarrow V_\perp$$

$$[\![comm]\!] = S \rightarrow S_\perp,$$

where $(\cdot)_\perp$ is a pointwise "lifting" operation on functors from $\mathbf{W}$ to $\mathbf{D}$. However, this is not as straightforward as it might appear.

We can of course define a states functor as an exponentiation of the form $S = [\![var]\!] \rightarrow V$, where V here is the obvious constant functor (of values). If w is a world and $s \in S(w)$, the uniformity condition on s requires commutativity of

$$
\begin{array}{ccc}
[\![var]\!]w & \xrightarrow{\;s(\mathrm{id}_w)\;} & V \\
{\scriptstyle [\![var]\!]f}\Big\downarrow & & \Big\downarrow{\scriptstyle \mathrm{id}_V} \\
[\![var]\!]x & \xrightarrow{\;s(f)\;} & V
\end{array}
$$

for any $\mathbf{W}$-morphism $f: w \rightarrow x$. But this does not constrain $s(f)$ outside the image of $[\![var]\!]f$, and so there are far too many states; states $s, s' \in S(w)$ would be distinguished even if they agree on all locations in $[\![var]\!]w$. There would even be many "global" states in $S(0)$, whereas one would expect $S(0)$ to be a singleton, containing only the "empty" state in which there are no locations.

An alternative is to use $S(w) = [\![var]\!]w \rightarrow V$, as before, and, for any $f: w \rightarrow x$ in $\mathbf{W}$ and $s \in S(w)$, define

$$S(f)\,s\,i = \begin{cases} s(j), & \text{if } f(j) = i \\ v_0, & \text{otherwise,} \end{cases}$$

for some fixed storable value v_0. But it is difficult to justify the arbitrary choice of v_0; furthermore, this allows too many expression and command meanings. For example, if $e \in [\![exp]\!]w$ and $f: w \rightarrow w + 1$ is the inclusion, commutativity of

$$
\begin{array}{ccc}
S(w) & \xrightarrow{\;e(\mathrm{id}_w)\;} & V_\perp \\
{\scriptstyle S(f)}\big\downarrow & & \big\downarrow {\scriptstyle \mathrm{id}_{V_\perp}} \\
S(w+1) & \xrightarrow{\;e(f)\;} & V_\perp
\end{array}
$$

in no way constrains $e(f)$ on states $s \in S(w+1)$ for which $s(w) \neq v_0$, and so the value yielded by a non-local expression meaning can be affected by changing the contents of a local variable.

Similarly, if $c \in [\![\mathbf{comm}]\!]w$, commutativity of

$$
\begin{array}{ccc}
S(w) & \xrightarrow{\;c(\mathrm{id}_w)\;} & S(w)_\perp \\
{\scriptstyle S(f)}\big\downarrow & & \big\downarrow {\scriptstyle S(f)_\perp} \\
S(w+1) & \xrightarrow{\;c(f)\;} & S(w+1)_\perp
\end{array}
$$

does not preclude $c(f)$ from changing the contents of the local variable (whenever this differs from v_0). For example, c could be such that $c(f)(s)$ returns a state in which the contents of every variable is v_0, thus changing the contents of every variable whose contents in s is not v_0, including the local variable. This would invalidate expected equivalences such as the one mentioned in the Introduction.

In fact, it seems most natural to have a *contravariant* states functor: a state in a "big" world can be converted to one in a "small" world by simply dropping locations. We can make S from Section 2 into a functor from $\mathbf{W}^{\mathrm{op}}$ to $\mathbf{D}$ by defining $S(f)s = f\,;s$, when $f: w \to x$ in $\mathbf{W}$ and $s \in S(x)$.

But then how can *covariant* functors $[\![\mathbf{exp}]\!]$ and $[\![\mathbf{comm}]\!]$ be constructed from *contravariant* S? A somewhat mysterious answer to these questions is given in [20] by defining a variant of exponentiation in functor categories, which we will term *contra-exponentiation*, that constructs a covariant functor from contravariant functors.

For example, a contra-exponential of the form $S \to V_\perp$ is defined just like the usual exponential, but with a reversal of vertical arrows in uniformity diagrams to account for contravariance. That is, $e(\cdot) \in (S \to V_\perp)w$ is a family of functions, indexed by morphisms out of w, such that

$$
\begin{array}{ccc}
S(x) & \xrightarrow{\;\; e(f) \;\;} & V_\perp \\[2pt]
\Big\uparrow{\scriptstyle S(g)} & & \Big\uparrow{\scriptstyle \mathrm{id}_{V_\perp}} \\[2pt]
S(y) & \xrightarrow{\;\; e(f\,;g) \;\;} & V_\perp
\end{array}
$$

commutes, where $f\colon w \to x$ and $g\colon x \to y$ in $\mathbf{W}$. The morphism part is defined as before to yield a *covariant* functor:

$$(S \to V_\perp)\,f\,m\,g \;=\; m(f\,;g)$$

where $f\colon w \to x$, $m \in (S \to V_\perp)w$, and $g\colon x \to y$. It can be seen that all of the $e(f)$ for $f\colon w \to x$ are in fact determined by $e(\mathrm{id}_w)$, and so the "pointwise" definition of $[\![\mathbf{exp}]\!]$ given earlier is isomorphic to $S \to V_\perp$.

We can also use contra-exponentiation to define a covariant functor $S \to S_\perp$; the relevant diagram is

$$
\begin{array}{ccc}
S(x) & \xrightarrow{\;\; c(f) \;\;} & S(x)_\perp \\[2pt]
\Big\uparrow{\scriptstyle S(g)} & & \Big\uparrow{\scriptstyle S(g)_\perp} \\[2pt]
S(y) & \xrightarrow{\;\; c(f\,;g) \;\;} & S(y)_\perp
\end{array}
$$

However, it is evident that one of the problems with a covariant S arises here as well: this definition does not preclude non-local command meanings from changing the values of local variables. To get a definition of $[\![\mathbf{comm}]\!]$ comparable to that in Section 2, we need to impose the additional requirement on $c \in (S \to S_\perp)w$ that, for every $f\colon w \to x$, when $c(f)$ is defined it must preserve the values of locations not in the range of f.

Thus, contra-exponentiation allows us to build covariant functors from a contravariant states functor, but a further restriction is needed to get a satisfactory definition of $[\![\mathbf{comm}]\!]$. It is not known whether there is a more uniform way of dealing with these problems. Contra-exponentiation will be encountered again in Section 6, where it will be used to define $[\![\mathbf{comm}]\!]$ in a setting where a pointwise definition is *not* appropriate.

5   Generalized Variables

Reynolds [16] and Oles [13] argue that it is preferable to treat variables in Algol-like languages without assuming that states are structured by locations.

This can be achieved by defining a category $\mathbf{X}$ of possible worlds with *sets* (of allowed states) W, X, ... as objects. (We assume that the small collection of $\mathbf{X}$-objects contains the set V of values, and is closed under cartesian products.) To interpret variable declarations we use morphisms that "expand" a world W to a world $W \times V$; the extra V-valued component holds the value of a new variable. Roughly, a state $x \in X$ is related to $\langle x, v \rangle \in X \times V$ similarly to the way $s \in S(w)$ is related to $(s \mid w \mapsto v) \in S(w+1)$ in the semantics of Section 2.

In general, the $\mathbf{X}$-morphisms from W to X will be pairs f, Q where f is a function from X to W and Q is an equivalence relation on X such that the following is a product diagram in the category of sets and functions:

$$ W \xleftarrow{\quad f \quad} X \xrightarrow{\quad q \quad} X/Q $$

where q maps every element of X to its Q-equivalence class. Intuitively, f extracts the small stack embedded in a larger one, and Q relates large stacks with identical "expansion components." This is just the category-theoretic way of saying that larger stacks are formed from smaller ones by adding independent components for local variables. It can be verified that $X \cong W \times X/Q$, and that the restriction of f to each Q-equivalence class is bijective.

The identity morphism id_W on an object W has as its two components: the identity function on W, and the universally-true binary relation on W. The composition $(f, Q) \, ; \, (g, R) : W \rightarrow Y$ of $\mathbf{X}$-morphisms $f, Q : W \rightarrow X$ and $g, R : X \rightarrow Y$ has as its two components: the functional composition of f and g, and the equivalence relation on Y that relates $s_0, s_1 \in Y$ just if they are R-related and Q relates $g(s_0)$ and $g(s_1)$. To see that these components have the required product property, let $h, S = (f, Q) \, ; \, (g, R)$ and note that

$$ \begin{aligned} Y & \cong & X \times Y/R \\ & \cong & (W \times X/Q) \times Y/R \\ & \cong & W \times (X/Q \times Y/R) \\ & \cong & W \times Y/S, \end{aligned} $$

by the isomorphism between Y/S and $X/Q \times Y/R$ established by the function mapping $[y]_S$ to $\big([g(y)]_Q, [y]_R \big)$, whose inverse is the function mapping $\big([x]_Q, [y]_R \big)$ to $\big\{ y' \in [y]_R \mid g(y') Q x \big\}$.

We can now define a states functor $S : \mathbf{X}^{\mathrm{op}} \rightarrow \mathbf{D}$ as follows: $S(W) = W$, discretely-ordered, and $S(f, Q) = f$. Then, we can use contra-exponentiation to define $[\![\mathbf{exp}]\!]$ as $S \to V_\perp$ and $[\![\mathbf{comm}]\!]$ as a restriction of $S \to S_\perp$ (or use pointwise definitions).

All of this seems very similar to the location-based approach: components of products, instead of locations, are used to hold the values of variables. Furthermore, an n-fold product V^n can be viewed as a world that results from n variable declarations, and a state

$$\langle v_0, ..., v_{n-1} \rangle \in V^n$$

would correspond to a state $s \in S(n)$ with $s(i) = v_i$ for $i \in \{0, ..., n-1\}$, where $S(n)$ is as in Section 2. There are even morphisms that are similar to the "location renaming" maps needed for the naturality of variable declarations; e.g., isomorphisms on worlds of the form $W \times V \times V$ that exchange the two V-valued components.

However, this analogy between locations and components of products cannot be pushed too far. The reason is that the component structure is only used when defining local variables and there is no *a priori* structure on states that allows us to identify "location-like" components independently of declarations. Consider, for example, a world $V \times V$. If locations are components then this should correspond to a world with two locations. But in the state-set category of worlds, $V \times V$ is isomorphic to V if V is infinite. So how many locations does the world have, one or two?

The answer is neither. In fact, the state-set category of worlds seems to preclude identification of entities that behave like locations. But what then should $[\![\mathbf{var}]\!]$ be? The approach taken in [16, 13] is to treat variables in the following more general way:

$$[\![\mathbf{var}]\!] = (V \to [\![\mathbf{comm}]\!]) \times [\![\mathbf{exp}]\!] .$$

Intuitively, a variable consists of two components, the first of which allows for changing the state when a storable value is supplied to it, and the second allows for accessing the state and yielding a stored value. Thus, the main difference between the natural-number category of worlds and the state-set category is that the former is oriented to a view of variables as locations, while the latter is oriented to states as abstract "objects" whose internal structure is deliberately left unspecified. These different views have important ramifications on how the type of variables is interpreted.

Even though the type of generalized variables does not consist of locations, the interpretation of declarations requires us to manufacture an appropriate new variable that has the location-like property that assigning a value to it results in a state in which the value in its expression component is the value just assigned. To do this, we make use of the product structure of morphisms.

A new variable $(a, e) \in [\![\text{var}]\!](W \times V)$ is defined as follows, assuming pointwise definitions of $[\![\text{comm}]\!]$ and $[\![\text{exp}]\!]$:

$$a\,(f, Q)\,v\,s \;=\; s'$$

$$e\,\langle x, v \rangle \;=\; v$$

where $f : W \times V \to Y$, $v \in V$, $s \in S(Y)$, and s' is the unique state in $S(Y)$ such that $s\,Q\,s'$ and if $f(s) = \langle x, v' \rangle$ then $f(s) = \langle x, v \rangle$. (Uniqueness of s' follows from the bijectivity requirement on the f part of f, Q.) The idea behind the definition of a is that the V-component of s' is changed to v, while the X-component and the "expanded" part coming from f, Q remain unchanged. With these definitions, the semantics of **new** can be treated essentially as in Section 2, using an expansion $W \to W \times V$ in place of the inclusion $w \to w + 1$.

6   Non-Interference

The support-oriented approach to local variables represents a significant improvement on the models based on marked stores. However, there remain simple equivalences involving procedures that are not treated properly. For example, the following block should be equivalent to **diverge** (i.e., a never-terminating command):

> **new** x. $x := 0$;
> $P(x := x + 2)$;
> **if even** x **then diverge**

when x is not free in P : **comm** $\to$ **comm**. Intuitively, the block diverges because the *only* access to x that can result from the call to P must come from using the argument $x := x + 2$. This argument preserves the evenness of x and x is even before the call, so it should be even on termination. Therefore, the test in the conditional will succeed (if the call to P terminates) and the block will diverge; if the call does not terminate, neither will the block, and so in every case the block diverges.

This example (attributed in [8] to Allen Stoughton) fails in support-oriented interpretations because it is possible for P to denote a procedure that preserves non-local variables but sets to 1 any local variables whose values are changed by its argument. It also fails for models based on the state-set category of worlds from the preceding section. As with the location-based category, this can be shown by exhibiting a $p \in [\![\text{comm} \to \text{comm}]\!]X$ such that $p(f)(c)(\text{id})(s, v) = (s', 1)$ if $c(\text{id})(s, v) = (s', v')$ and $v' \neq v$, and $c(\text{id})(s, v)$ otherwise, where $f : X \to X \times V$ is the expansion morphism; with the state-set category it requires some work to show that such a p actually exists.

While this block is not itself a realistic program, it illustrates how the semantic problems with local variables involve principles that are important in programming practice. The procedure P is supplied with an argument that provides a *limited capability* for access to the local variable, and one reasons that the *only* way that the procedure call can access the variable is via this limited capability. This kind of informal reasoning is familiar from programming with "modules" or "objects." A module typically has a number of internal variables, which are accessed and modified by clients through the use of explicitly "exported" procedures only. Often, the exports provide only a limited capability for accessing the variables, in a way that presents to the client a view of the module that does not reveal details of its internal structure.

In this section we will consider a model that has arisen from work on finding a satisfactory semantic interpretation of the "specification logic" of [15, 17], which is a formal system for proving equivalences and partial-correctness properties of programs in Algol-like languages with higher-order procedures. For our purposes, the most important rule of the logic is one for reasoning about local-variable declarations; for clarity, we present it in natural-deduction format:

Local-Variable Declarations:

$$\cfrac{\begin{bmatrix} \iota \,\#\, E_1, \ldots, \iota \,\#\, E_m \\ C_1 \,\#\, \iota, \ldots, C_n \,\#\, \iota \end{bmatrix} \\ \vdots \\ \{P\}\, C\, \{Q\}}{\{P\}\, \mathbf{new}\ \iota.\, C\, \{Q\}}$$

where the bound variable identifier ι is not free in assertions P or Q, in phrases E_i or C_j, or in uncancelled assumptions. Here, $\{P\}\, C\, \{Q\}$ is the familiar *Hoare triple* [6], and $C \,\#\, E$ ("C doesn't interfere with E") essentially asserts that every way of using C preserves any value obtained by using E. The rule says that, when reasoning about a locally-declared variable, it is valid to assume that a non-local entity and the declared variable don't interfere with one another.

The divergence of the troublesome block above can in fact be proved in specification logic by deriving $\{\mathbf{true}\}\, B\, \{\mathbf{false}\}$, where B is the block. This is done using the rule for variable declarations, and instantiating R to $\mathbf{even}\ x$ and C to $x := x + 2$ in the following axiom [11, 12]:

Non-Interference Abstraction:

$$P \,\#\, R \ \& \ \{R\}\, C\, \{R\} \ \Rightarrow \ \{R\}\, P(C)\, \{R\}$$

where P: **comm** $\rightarrow$ **comm**.

Meyer and Sieber [8] have shown how to modify support-oriented interpretations in a way that handles the "evenness" example as well. They use logical relations to cut down the model, restricting to procedures that preserve invariant properties of locations outside their support. For example, in the above equivalence the evenness of x is a property of a variable (x) outside the support of P, and the requirement is that any call to P must preserve the evenness of x if its argument does. The Non-Interference Abstraction axiom relates non-interference to these ideas of Meyer and Sieber about invariants.

The model for specification logic is based on a modification of the state-set category of worlds: the morphisms from X to Y are again pairs f, Q, where f is a function from Y to X, and Q is an equivalence relation on Y, but the restriction of f to a Q-equivalence class is only required to be injective, instead of bijective. We call this modified category $\mathbf{X'}$. The definition of composition given in Section 5 works here as well.

We still have the expansion maps from W to $W \times V$. But now we also have morphisms that "restrict" the possible world; for any $X' \subset X$, the *restriction morphism* $f, Q: X \rightarrow X'$ has as its components: the inclusion function f from X' to X and the everywhere-true binary relation Q on Y. We shall see how these additional morphisms give us stronger uniformity conditions.

A contravariant states functor is defined as in Section 5. Expression meanings can then be treated using contra-exponentiation: $[\![\mathbf{exp}]\!] = S \rightarrow V_\bot$. But now, because of the possibility of restricting to a subset of states, it is possible to prove that expressions cannot have side effects, not even "temporary" ones. For any $w \in W$, there is a "restriction" morphism $f: W \rightarrow \{w\}$ such that $S(f)$ is the inclusion of $\{w\}$ in W, and commutativity of the following diagram shows that the value of $e \in [\![\mathbf{exp}]\!]W$ at $w \in W$ is determined in the singleton world $\{w\}$ in which w is the *only* state:

$$
\begin{array}{ccc}
W & \xrightarrow{\;e(\mathrm{id}_W)\;} & V_\bot \\
{\scriptstyle S(f)}\big\uparrow & & \big\uparrow{\scriptstyle \mathrm{id}_{V_\bot}} \\
\{w\} & \xrightarrow[\;e(f)\;]{} & V_\bot
\end{array}
$$

This property plays an important role in reasoning about assignments. It is also reflected in the following equivalence:

$$
\begin{array}{ccc}
\textbf{if } x = 0 & & \textbf{if } x = 0 \\
\quad\textbf{then } P(x) & \equiv & \quad\textbf{then } P(0) \\
\quad\textbf{else } 1 & & \quad\textbf{else } 1
\end{array}
$$

where $P: \textbf{exp} \rightarrow \textbf{exp}$. The idea is that execution of $P(x)$ never changes the value of x, so it should be equivalent to executing $P(0)$ in a state where the value of x is 0. This equivalence is not valid in the models discussed in previous sections because the semantic domains for expressions in these models allow for the possibility of "temporary" side effects, where the values of global variables are altered during expression evaluation and then restored on termination of the evaluation.

Command meanings must be treated somewhat differently. Because of restriction morphisms in $\mathbf{X}'$, $S \rightarrow S_\perp$ is too small, whereas previously it was too large; the commutativity of

$$
\begin{array}{ccc}
W & \xrightarrow{\; c(\mathrm{id}_W) \;} & W_\perp \\
{\scriptstyle S(f)}\Big\uparrow & & \Big\uparrow{\scriptstyle S(f)_\perp} \\
\{w\} & \xrightarrow[\; c(f) \;]{} & \{w\}_\perp
\end{array}
$$

for any $w \in W$ would preclude commands from changing states at all!

Instead, a *partial* contra-exponentiation $\rightharpoonup$ can be defined in $\mathbf{D}^{\mathbf{X}'}$, and then commands are interpreted using $S \rightharpoonup S$. The uniformity requirement on $c \in (S \rightharpoonup S)(W)$ is that, for all $\mathbf{X}'$-morphisms $f: W \rightarrow X$ and $g: X \rightarrow Y$,

$$
c(f \,;\, g) \,;\, S(g) \;\subseteq\; S(g) \,;\, c(f)
$$

where the $c(\cdot)$ are *partial* functions on states, and the ordering is graph inclusion. This is *semi*-commutativity of

$$
\begin{array}{ccc}
S(X) & \xrightarrow{\; c(f) \;} & S(X) \\
{\scriptstyle S(g)}\Big\uparrow & \supseteq & \Big\uparrow{\scriptstyle S(g)} \\
S(Y) & \xrightarrow[\; c(f \,;\, g) \;]{} & S(Y)
\end{array}
$$

which allows command meanings to be less-defined in more-restricted worlds. Notice that, in general, $F \to G_\perp$ is *not* isomorphic to $F \to G$ in functor categories. Furthermore, as before, an additional condition is needed to preclude non-local commands from changing local variables, or, more generally, from violating local constraints on state changes [20, 12].

An example will be helpful to illustrate the uniformity condition on $\to$. Consider a command meaning $c(\cdot) \in [\![\text{comm}]\!](X \times V)$ corresponding to an assignment statement $x := x + 1$, where x denotes the local component in $W \times V$. The partial function for $c(\text{id}_{W \times V})$ maps $\langle w, n \rangle \in W \times V$ to $\langle w, n+1 \rangle$. But we also need to define $c(f)$ for any $\mathbf{X}'$-morphism with domain $W \times V$, including any restriction morphism. Suppose that we restrict to the world

$$Y = \{\langle w, v \rangle \in W \times V \mid v \text{ is even}\}.$$

If $f: W \times V \to Y$ is the restriction morphism and $c(f)\langle w, n \rangle$ is defined, then it must be equal to $c(\text{id}_{W \times V})\langle w, n \rangle$ by the uniformity requirement on $S \to S$. But this is impossible, because $\langle w, n + 1 \rangle$ is not in Y (assuming that n is even). Instead, $c(f)\langle w, n \rangle$ is undefined. More generally, if morphisms g and h are such that $s = S(h)s'$ and $c(g)s$ is defined but not in the range of $S(h)$, then $c(g ; h)s'$ must be undefined.

Now consider the composite $x := x + 1 ; x := x + 1$, and its semantical counterpart $c ; c$, with c as above. Sequential composition is interpreted componentwise, so for command meanings c_1 and c_2, $(c_1 ; c_2)(f)$ is just the compositition $c_1(f) ; c_2(f)$ of the partial functions for the components. Thus, we get that $(c ; c)(\text{id}_{W \times V})\langle w, n \rangle$ is $\langle w, n + 2 \rangle$. However, $(c ; c)f\langle w, n \rangle$ is undefined, where f is the restriction to Y, because $c(f)\langle w, n \rangle$ is undefined; the attempt to "stray" from Y, even at an intermediate state, leads to divergence.

This example shows how state-set restrictions and $\to$ interact: a command meaning at a restricted world terminates only if all the states encountered during "execution" are within the world. This is the key idea behind the semantics of non-interference in [20]. Given $c \in [\![\text{comm}]\!]W$ and $e \in [\![\text{exp}]\!]W$, we define

$$c \,\#\, e \iff \text{for all } f: W \to X \text{ and } s \in X, \ c(f)s = c(f ; g)s$$

where g is the restriction to

$$\{s' \in X \mid e(f)s' = e(f)s\}.$$

Intuitively, $c \,\#\, e$ holds iff any (terminating) execution of c can be restricted to a world for which the value of e is invariant. For discussion of axioms that make use of this invariance at intermediate states, see [17, 20].

In this model [[var]] is again treated using the generalized form of variable discussed in the previous section. The rule for variable declarations can then be validated by showing that non-local entities don't interfere with the expression component of a local variable, and that no assignment to a local variable interferes with a non-local entity. For the generalization of # to higher types, and validation of Non-Interference Abstraction, see [12].

Our final example ties together these ideas about non-interference, restriction morphisms, and local variables:

$$\textbf{new } x. \ x := 0;$$
$$P(x)$$

is equivalent to $P(0)$, where x is not free in P: $\textbf{exp} \rightarrow \textbf{comm}$. This equivalence fails in other models. A sketch of its validity here is as follows. First, since P is non-local, none of its calls interfere with the local variable, unless the argument does. But the argument x is an expression, which doesn't cause any side effects, so $P(x)$ doesn't interfere with (the expression component of) x. Then, by the definition of non-interference given above, the result of evaluating $P(x)$ in a state $s \in W \times V$ is equivalent to evaluating the call in the restricted world

$$\{(w, v) \in W \times V \mid v = 0\},$$

using an appropriate restriction morphism. But in this restricted world identifier x and 0 have the same denotation: both are (families of) functions that return 0 in *any* state. This means that x can be replaced by 0 in the call, and the equivalence follows.

7 Discussion

We have concentrated here on models based explicitly on the possible-world approach of [16, 13]; but, it is interesting to note that, in a parallel line of development, Meyer and Sieber [8] have used methods that can be fairly said to have possible worlds at their core (combined with other principles for restricting models). If we count [8] as within the approach, then in fact the most satisfactory local-variable models are all based on possible worlds.

At present, there is no single "best" model for local variables. We have seen that the non-interference model validates a number of test equivalences that fail in other models. However, Arthur Lent has pointed out to us that there are equivalences that fail in this model because it is sensitive to intermediate states encountered during program execution, such as

$$\textbf{new } x. \ x := 0; \qquad \equiv \qquad \textbf{new } x. \ x := 0;$$
$$P(x := x + 2) \qquad\qquad P(x := x + 1 \ ; x := x + 1)$$

where x is not free in P: **comm** $\rightarrow$ **comm**. This fails in the non-interference model, essentially because the different arguments to P are not semantically equal, yet it *is* valid in the location-based model of Sections 2 and 3, the Reynolds/Oles state-set model of Section 5, and the models of [8].

Another significant point is state-sets *versus* locations. The major difference between the models of Sections 5 and 2 is that the former is predisposed against having a primitive type of locations in the language, while the latter is not. (Of course, whether one considers this a strength or weakness of the model depends on one's philosophy of what variables should be.)

An equivalence that fails in *all* known models is the following variation on Example 7 from [8]: execution of

$$\textbf{new } x. \;\; x := 0;$$
$$P(x := x + 1)$$

where x is not free in P: **comm** $\rightarrow$ **comm** might set x to any value; however, because P cannot *read* from x and x is discarded after execution of the body, the block should be equivalent to $P(\textbf{skip})$. In each of the models discussed in this paper, a counterexample to this equivalence may be obtained by considering a procedure meaning p that, given a command argument c and a state s, terminates with result s if $c(s) = s$, and diverges otherwise. Intuitively, such a p reads from local variables, such as x in this example, because the operation of testing states for equality depends on their values.

The possible-world approach has resulted in models that better explain the sense in which a declared variable is "new." Nevertheless, the best current models are still far from perfect. It seems that the semantic ideas on support, non-interference, and preservation of invariants developed in [16, 13, 8, 20, 12] might prove to be just special cases of some more general and fundamental principle that is still lacking from our understanding of the interactions between non-local procedures and local variables in Algol-like languages.

Acknowledgements

We have benefitted from discussions with Eugenio Moggi, Arthur Lent, and Kurt Sieber, and comments on a draft by David Andrews.

References

[1] J. L. Bell. *Toposes and Local Set Theories*. Oxford University Press, 1988.

[2] S. D. Brookes. A fully abstract semantics and a proof system for an ALGOL-like language with sharing. In A. Melton, editor, *Mathematical Foundations of Programming Semantics*, volume 239 of *Lecture Notes in Computer Science*, pages 59–100, Manhattan, Kansas, 1985. Springer-Verlag, Berlin.

[3] J. Donahue. Locations considered unnecessary. *Acta Informatica*, 8:221–242, 1977.

[4] R. Goldblatt. *Topoi, The Categorial Analysis of Logic*. North-Holland, Amsterdam, 1979.

[5] J. Y. Halpern, A. R. Meyer, and B. A. Trakhtenbrot. The semantics of local storage, or what makes the free-list free? In *Conf. Record 11th ACM Symp. on Principles of Programming Languages*, pages 245–257, Austin, Texas, 1984. ACM, New York.

[6] C. A. R. Hoare. An axiomatic basis for computer programming. *Comm. ACM*, 12(10):576–580 and 583, 1969.

[7] J. Lambek and P. J. Scott. *Introduction to Higher-Order Categorical Logic*. Cambridge University Press, Cambridge, 1986.

[8] A. R. Meyer and K. Sieber. Towards fully abstract semantics for local variables: preliminary report. In *Conf. Record 15th ACM Symp. on Principles of Programming Languages*, pages 191–203, San Diego, California, 1988. ACM, New York.

[9] R. E. Milne and C. Strachey. *A Theory of Programming Language Semantics*. Chapman and Hall, London, and Wiley, New York, 1976.

[10] E. Nelson. On exponentiating exponentiation. *J. of Pure and Applied Algebra*, 20:79–91, 1981.

[11] P. W. O'Hearn. *The Semantics of Non-Interference: a Natural Approach*. Ph.D. thesis, Queen's University, Kingston, Canada, 1990.

[12] P. W. O'Hearn and R. D. Tennent. Semantical analysis of specification logic, part 2. Technical Report 91-304, Department of Computing and Information Science, Queen's University, Kingston, Canada, 1991. To appear in revised form in *Information and Computation*.

[13] F. J. Oles. *A Category-Theoretic Approach to the Semantics of Program-
      ming Languages.* Ph.D. thesis, Syracuse University, Syracuse, 1982.

[14] F. J. Oles. Type algebras, functor categories and block structure. In
      M. Nivat and J. C. Reynolds, editors, *Algebraic Methods in Semantics,*
      pages 543–573. Cambridge University Press, Cambridge, 1985.

[15] J. C. Reynolds. *The Craft of Programming.* Prentice-Hall International,
      London, 1981.

[16] J. C. Reynolds. The essence of Algol. In J. W. de Bakker and J. C. van
      Vliet, editors, *Algorithmic Languages,* pages 345–372, Amsterdam, 1981.
      North-Holland, Amsterdam.

[17] J. C. Reynolds. Idealized Algol and its specification logic. In D. Néel, ed-
      itor, *Tools and Notions for Program Construction,* pages 121–161. Cam-
      bridge University Press, Cambridge, 1982.

[18] D. S. Scott. Mathematical concepts in programming language semantics.
      In *Proc. 1972 Spring Joint Computer Conference,* pages 225–34. AFIPS
      Press, Montvale, N.J., 1972.

[19] J. E. Stoy. *Denotational Semantics: The Scott-Strachey Approach to Pro-
      gramming Language Theory.* The MIT Press, Cambridge, Massachusetts,
      and London, England, 1977.

[20] R. D. Tennent. Semantical analysis of specification logic. *Information
      and Computation,* 85(2):135–162, 1990.

Using fibrations to understand subtypes

Wesley Phoa *

Abstract

Fibrations have been widely used to model polymorphic λ-calculi. In this paper we describe the additional structure on a fibration that suffices to make it a model of a polymorphic λ-calculus with subtypes and bounded quantification; the basic idea is to single out a class of maps, the *inclusions*, in each fibre. Bounded quantification is made possible by a imposing condition that resembles local smallness. Since the notion of inclusion is not stable under isomorphism, some care must be taken to make everything strict.

We then show that PER models for λ-calculi with subtypes fit into this framework. In fact, not only PERs, but also any full reflective subcategory of the category of modest sets ('PERs' in a realizability topos), provide a model; hence all the small complete categories of 'synthetic domains' found in various realizability toposes can be used to model subtypes.

1. Introduction

1.1. What this paper is about

At the core of object-oriented programming, and related approaches to programming, are the notions of subtyping and inheritance. These have proved to be very powerful tools for structuring programs, and they appear—in one form or another—in a wide variety of modern programming languages.

One way of studying these notions formally is to use the framework of typed λ-calculus. That is, we start with a formal system (say, a version of the polymorphic λ-calculus) and extend it by adding a notion of type inclusion, together with suitable rules; we obtain a larger system. We can then use the methods of mathematical logic to study the properties of the system.

---

*Laboratory for the Foundations of of Computer Science, University of Edinburgh, Edinburgh, EH9 3JZ, Scotland. This work was partially supported by an 1851 Research Fellowship.

It's also of interest to study the semantics of such systems. For example, PERs (partial equivalence relations on N, or on other partial combinatory algebras) can be used to model certain typed λ-calculi with subtyping. Models are useful: not just as a check that our systems are consistent, but as a way of getting some insight into the meaning and consequences of the rules we've chosen. Sometimes they also let us prove technical results.

Category theory provides a useful general framework for talking about the semantics of typed λ-calculi, and the notion of model. For example, models of the simply typed λ-calculus correspond to cartesian closed categories; and models of various polymorphic λ-calculi correspond to various sorts of fibrations (or indexed categories, if you prefer). That is, to construct a model we need only construct an appropriate sort of—possibly fibred—category.

In fact, the categorical structures that arise can be studied without reference to any formal systems. For example, we can study cartesian closed categories quite independently of the simply typed λ-calculus; the theory can be viewed as an alternative formulation of the same notions, less suitable as a programming language but more amenable to mathematical (rather than logical or syntactic) analysis. This can be viewed as a syntax-free way of studying the notion of higher-order function.

This paper is a tentative attempt to find a corresponding categorical framework for the polymorphic λ-calculus with subtyping and bounded quantification. That is, the emphasis is on defining appropriate categorical structure; our definitions are motivated by certain rules appearing in existing formal systems such as Quest and Ponder, but the formalism of typed λ-calculus plays a strictly secondary role; in fact, no complete formal systems are presented.

For a category theorist, as opposed to a logician, this is the more natural point of view. We'll also see that it's a useful one: for example, it leads to the discovery that various categories of 'synthetic domains' can be used to model subtyping. We also describe a couple of simple consequences of our axioms that have some type-theoretic meaning; they are obtained by categorical, rather than logical, means.

Throughout this paper we rely heavily on the language of fibrations (for which see [Ben85] and [HRR90]). The internal language of a topos is used nowhere; nonetheless, it was a useful tool in obtaining the results of the final section, where we are dealing with particular categories that live inside realizability toposes (see [Hy88], [HRR90] for an explanation of the connection). I am not sure how to talk about subtyping in the internal language of a topos.

A funny feature   One reason for this is an aspect of this paper which will seem very odd to a category theorist, but which seems to be unavoidable at the moment. Namely: our notion of 'inclusion' is not stable under isomorphism.

This is a basic feature of the PER model: in it, types $X, Y, \ldots$ are modelled by partial equivalence relations $R, S, \ldots$, and we say that $X \leq Y$ ('X is a subtype of Y') if $R \subseteq S$—a relation that is clearly not preserved when we compose with isomorphisms. And this does reflect a pragmatic 'fact': subsumption or coercion should involve a trivial amount of computation, preferably none at all, whereas isomorphisms can be arbitrarily complicated functions.

However, this does make the category theory look a bit unnatural. We are forced to talk about equality (rather than just isomorphism) of objects; we have to use split (rather than cloven) fibrations, so that reindexing functors commute on the nose. . . This is all possible, of course, but it requires some extra care.

As Lawvere has forcefully pointed out, this may reflect a basic conflict between the categorical perspective and the computer scientists' view of subtyping—in particular, the subsumption rule and the notion of 'canonical' inclusions.

Actually, the definitions that follow can be 'loosened' to give a notion of subtyping that *is* stable under composition with isomorphisms (though one then has to give up the subsumption rule in favour of explicit coercions). In fact, they were phrased with this possibility in mind. I won't do this, though, as I don't know of any examples that would justify it.

Unanswered questions   There are many; let me just mention a couple of the more obvious problems:

- Proving a soundness theorem for a particular system.

- Fitting ideal models and retraction models into the framework.

- Investigating (synthetic) domain-theoretic models in which, following Reynolds and Rosolini, type inclusions are taken to be strict maps.

- Generalizing the notion of type inclusion to encompass 'provable isomorphisms' between types (so that, for example, if $X \leq S \times T$ then $X \leq T \times S$).

- Integrating subtyping with recursive types (as in Quest): see [AC90].

- Extending subtyping to the full calculus of constructions.

These are the subject of current research and will not be discussed further here. Some are problematic to say the least.

A note on the literature There is quite a large body of work on typed λ-calculus, its semantics, PER models, records, subtyping and inheritance. I have not attempted to give a full bibliography at the end of the paper; however, more comprehensive lists of references appear in [CL90] and [deP90], both of which I found very useful.

Of particular interest, however, is the paper [M-OM90] of Martí-Oliet and Meseguer (I must thank the authors for drawing it to my attention). It mostly consists of a categorical study of 'first-order' inclusions, which behave like subset inclusions rather than subquotients; but in section 5, they do briefly consider the notion of subtype studied here, for the simply typed λ-calculus. Their conditions correspond to our axioms **uniqueness, identities, composition, contra-co** and **products**; they also mention that these axioms appear in Taylor's thesis [Tay86]. So the original features of the present paper are the generalisation to fibrations, the way we handle polymorphism and the account we give of powerkinds and bounded quantification.

Acknowledgements I began this work while trying to understand Valeria de Paiva's report on subtyping in Ponder; a further spur was Pino Rosolini's paper on the semantics of Quest. Both of them gave me advice and encouragement, as did the members of Bob Walters' Categories in Computer Science seminar. I have also benefited from correspondence with Martin Abadi, Giuseppe Longo and Narciso Marti-Oliet, and discussions with Mike Johnson, Bart Jacobs and Thomas Streicher. I should acknowledge the assistance of the Australian Research Council in preparing an earlier draft of this paper.

I was grateful for the chance to present this work at the LMS Symposium on Categories in Computer Science. Useful comments from the other participants helped me understand the subject a little better; so did several of the other talks. And finally, I must thank the referee for a number of remarks and corrections.

1.2.   Fibrations and polymorphic λ-calculus

We'll be concerned with systems like $F\omega$ (plus subtypes and bounded quantification) or a weakened version of Quest: cf. [CL90]. But I won't set down any formal systems here; instead, I'll just give an informal description. So some familiarity with typed λ-calculus is already assumed.

Here are the rough features of our system.

- We have *kinds* $\mathbb{I}$, $\mathbb{J}$... which are closed under the formation of product kinds and arrow kinds: $\mathbb{I} \times \mathbb{J}, \mathbb{I} \to \mathbb{J}$.

- We have *types* $S, T \ldots$ which are closed under the formation of product types and arrow types: $S \times T, S \to T$.

- We can have families of types indexed by kinds: $T(X)[X \in \mathbb{I}]$. In particular, there is a kind $\mathbb{T}$ of types with a 'generic type' $G(X)[X \in \mathbb{T}]$ indexed by $\mathbb{T}$.

- We can form products of families of types indexed by kinds; for example, given the $\mathbb{I}$-indexed family of types $T(X)[X \in \mathbb{I}]$ we can form the polymorphic type $\Pi X \in \mathbb{I}.T(X)$.

- We have appropriate *terms* $t : T$ of each type.

Note that we take products as primitive, rather than encoding the binary product $S \times T$ as $\Pi X \in \mathbb{T}.(S \to T \to X) \to X$; from a category-theoretic point of view, our approach seems more natural.[1]

One can use fibrations to model systems like this; for a proper account, see the thesis of Bart Jacobs: [Jac91]. There is no room to outline the theory of fibrations here, but see [Ben85] for a long discussion or [HRR90] for a fairly brief summary.

The sort of structure we need is as follows:[2]

1. A cartesian closed category $\mathcal{K}$ of *kinds*.

2. A fibration $\begin{smallmatrix} \mathcal{T} \\ \downarrow \\ \mathcal{K} \end{smallmatrix}$ over $\mathcal{K}$ of *types*, or rather families of types indexed by kinds. This fibration must have the following additional structure:

   - Each fibre $\mathcal{T}^{\mathbb{I}}$ over each kind $\mathbb{I} \in \mathcal{K}$ is cartesian closed.

   - The fibration is cloven, so that given any $f : \mathbb{I} \to \mathbb{J}$ in $\mathcal{K}$ we have a 'substitution' or 'reindexing' functor $f^* : \mathcal{T}^{\mathbb{J}} \to \mathcal{T}^{\mathbb{I}}$; furthermore, each f^* should preserve products and exponentials.

   - The fibration is complete: that is, we are given a right adjoint Π_f for each substitution functor f^*, satisfying the Beck-Chevalley condition.

   - There is a kind $\mathbb{T}$ and a 'generic type' $G \in \mathcal{T}^{\mathbb{T}}$ such that any type $X \in \mathcal{T}^{\mathbb{I}}$ is (isomorphic to) f^*G for a unique arrow $f : \mathbb{I} \to \mathbb{T}$.

---

[1]It also means we don't need to verify that the encoding actually has the property of being a categorical product; whether or not this is the case will depend on the model.

[2]Somewhat weaker conditions would do, cf. [Jac91, chapter 3]; however, the conditions that follow are the simplest to state.

For example, we could take $\mathcal{K} = \mathit{Sets}$, and let $\mathcal{T}^1$ be the category of $\mathbb{I}$-indexed families of PERs and uniform families of maps. This gives the usual PER model for $F\omega$: cf. [Hy88].

Later on, in order to characterise 'inclusions', we'll impose stronger conditions on our fibration as well as adding extra structure.

Σ-types To handle abstract data types, and for other reasons, it's useful to have types of the form $\Sigma X.T(X)$; these corresponds to *left* adjoints to substitution functors (and hence, like binary products, should be taken as primitive rather than defined in terms of Π and $\rightarrow$). I believe that they can be fitted into the framework proposed in this paper; but, for simplicity, we won't discuss them further.

2.   Modelling subtypes using fibrations

2.1.   Rules for subtypes and bounded quantification

To motivate the categorical definitions that will appear soon, we need to consider the sort of rules that we'd like a notion of subtyping to satisfy. The following rules (not spelled out precisely!) are based on the ones in Ponder and Quest, and many other systems.

A subtyping judgement looks like

$$S \leq T$$

where S, T are types, possibly varying over a fixed kind $\mathbb{I}$. What kinds of subtyping judgements ought to be derivable?

In the rules that follow we will leave out variable restrictions, and often any mention of types being explicitly indexed over kinds—for example, we will write $\Pi X.T(X)$ instead of $\Pi X \in \mathbb{I}.T(X)$ when the rule in question can be understood without knowing anything about the particular kind $\mathbb{I}$ over which we are quantifying.

The first rule is the *subsumption* rule:

$$\frac{t : S \quad S \leq T}{t : T}$$

Actually, we could replace this rule by some rules dealing with *explicit coercions* (see [CL90]); this approach would be easier to model, and somewhat more natural from a categorical point of view, but we do not consider it here.

Type inclusion should be reflexive and transitive; we would also like to have the *contra-co* subtyping rule

$$\frac{S \leq T \quad S' \leq T'}{T \to S' \leq S \to T'}$$

as well as several rules relating subtypes to polymorphic types: for example,

$$\overline{\Pi X.T(X) \leq T[X := S]}$$

$$\overline{\Pi X.T \to S(X) \leq T \to \Pi X.S(X)}$$

$$\frac{T \leq S(X)}{T \leq \Pi X.S(X)}$$

To encode records (see [CL90]) we need **Top** and rules for products. **Top** is a type such that

$$T \leq \textbf{Top}$$

for any type T. And for products, we simply need

$$\frac{S \leq S' \quad T \leq T'}{S \times T \leq S' \times T'}$$

as well as

$$\frac{U \leq S \times T \quad t : U}{\textbf{fst}(t) : S}$$

and similarly for **snd**.

Bounded quantification   Having bounded quantification means being able to form polymorphic types of the form $\Pi X \leq T.S(X)$. It clearly suffices to have, for each type T, a *power kind* $\downarrow T$, 'the kind of subtypes of T'; we could then just form Π-types indexed by $\downarrow T$; that is, types having the form $\Pi X \in \downarrow T.S(X)$.

In the presence of polymorphism, all this has to interact properly with the fact that there's a generic type G (over $\mathbb{T}$) which has an associated power kind $\downarrow G$. One might also like to form the 'kind of supertypes' $\uparrow T$, and hence polymorphic types of the form $\Pi T \leq X.S(X)$; these do not seem to be so useful in programming.

Note that the notion of powerkind is useful even if we do not need to form types of the form $\Pi X \leq T.S(X)$. For example, we might be interested in subtyping judgements of the more general form

$$\Gamma \vdash S \leq T$$

where S, T are types and Γ is a list of subtyping constraints of the form

$$X \leq U \qquad \text{or} \qquad U \leq X$$

where U is a type and X is a type variable. But a (sub)typing judgement depending on a subtyping constraint $X \leq U$ is essentially an assertion about types varying over the kind $\downarrow U$; similarly, a judgement with a hypothesis $U \leq X$ can be interpreted as a judgement about types varying over $\uparrow U$. So if powerkinds exist, we can make sense of such judgements.[3]

Rules for subkinds In Quest there is also a notion of subkind, which we might write $\mathbb{I} \subseteq \mathbb{J}$. We won't discuss it in any detail here, but simply indicate briefly how it fits into the framework. It raises some issues that haven't been resolved yet.

The subkind relation should be reflexive and transitive; it should probably behave well with respect to products of kinds, but it need not satisfy the contra-co rule. It is related to subtyping via the following rules:

$$\frac{S \leq T}{\downarrow S \subseteq \downarrow T}$$

$$\frac{\mathbb{I} \subseteq \mathbb{J}}{\Pi X \in \mathbb{J}.S(X) \leq \Pi X \in \mathbb{I}.S(X)}$$

There is also a kind subsumption rule in Quest which is not considered here. In fact, we will soon see that the notion of subkind may not be necessary in our framework.

2.2. Characterising inclusions

From the point of view of fibrations, what could a model of such a system look like? As I mentioned earlier, we need to be able to model notions of type inclusion that are not stable under isomorphism, and this requires some care.

So as before, we'd like a fibration $\begin{smallmatrix} \mathcal{T} \\ \downarrow \\ \mathcal{K} \end{smallmatrix}$ of the sort described earlier. But we'll also need to impose the following conditions:

---

[3]Interestingly, Ponder allows judgements of this form but has no notion of powerkind; its type discipline resembles System F rather than $F\omega$. Jacobs has raised the question of whether we can make sense of subtyping constraints without assuming the existence of powerkinds, perhaps by some sort of 'two-level' indexing.

split The fibration is split:[4] that is, if $f: \mathbb{I} \to \mathbb{J}$ and $g: \mathbb{J} \to \mathbb{K}$ then $(gf)^* = f^*g^*$ (for a cloven fibration, we'd only have $(gf)^* \cong f^*g^*$).

Π-functors compose Π-functors should also compose 'on the nose': that is, $\Pi_{gf} = \Pi_g\Pi_f$.

Subtypes are modelled as follows: For each fibre $T^{\mathbb{I}}$ over $\mathbb{I}$, we are given a class of maps

$$S \hookrightarrow T$$

in $T^{\mathbb{I}}$, called **inclusions**; such a map witnesses the fact that S is a subtype of T. These classes of maps should satisfy the following conditions:

uniqueness For any objects S, T, there is at most one inclusion $S \hookrightarrow T$ (this seems necessary to get a coherent interpretation of subsumption; it would not be needed if we were dealing with explicit coercions).

stability Substitution functors f^* preserve inclusions.

identities Every identity map is an inclusion (though an isomorphism need not be an inclusion).

composition A composite of inclusions is an inclusion.

The last three conditions say that the inclusions form a (usually non-full) subfibration of the fibration of types over kinds, with the same class of objects; the first, that they form a fibred preorder over $\mathcal{K}$.

Now we need to say how inclusions must behave with respect to arrow types, binary products and **Top**.

contra-co If $S \hookrightarrow T$ and $S' \hookrightarrow T'$ are inclusions, then so is the induced map $S'^T \to T'^S$.

products If $S \hookrightarrow S'$ and $T \hookrightarrow T'$ are inclusions, then so is the induced map $S \times T \to S' \times T'$.

top There is a specified terminal object **Top** in each fibre, such that $f^*\mathbf{Top} = \mathbf{Top}$ for any $f: \mathbb{I} \to \mathbb{J}$ in $\mathcal{K}$, and such that for any object T in any fibre, the unique map $T \to \mathbf{Top}$ is an inclusion. (It follows that **Top** is a terminal object in the subcategory of inclusions.)

---

[4]Bob Walters has pointed out that instead of using a split fibration, we might as well just talk about an ordinary functor $\mathcal{K}^{op} \to \mathcal{C}at$. We prefer to phrase everything in terms of fibrations so that we can generalise to possible notions of inclusion that may be stable under isomorphism, and because some people prefer talking about fibrations rather than indexed categories anyway.

And inclusions must behave properly with respect to products. Let $f: \mathbb{I} \to \mathbb{J}$ be any arrow in $\mathcal{K}$, and let $S \in T^{\mathbb{I}}$, $T \in T^{\mathbb{J}}$ be any two objects in the fibres over $\mathbb{I}$ and $\mathbb{J}$; then:

counits The component $\epsilon_S : f^* \Pi_f S \to S$ of the counit of $f^* \dashv \Pi_f$ at S is an inclusion.

mates If $f^* T \hookrightarrow S$ is an inclusion, then so is its mate $T \to \Pi_f S$.

strengths Recall that f^* preserves products, and consider the map

$$f^*(T \times \Pi_f(S^{f^*T})) \cong f^* T \times f^* \Pi_f(S^{f^*T}) \to f^* T \times S^{f^*T} \to S$$

where the middle map is $f^* T \times \epsilon_{S^{f^*T}}$ and the rightmost map is evaluation; let its mate be the map $T \times \Pi_f(S^{f^*T}) \to \Pi_f S$; then its exponential transpose $\Pi_f(S^{f^*T}) \to (\Pi_f S)^T$ is an inclusion.

Of course, all of these conditions are just categorical formulations of the subtyping rules sketched above.

There is a nice—if unsurprising—feature of our axioms. Namely, we never need to assume the existence of new maps between types that need not otherwise be there; our conditions only say that certain *existing* maps should be inclusions. These correspond to Mitchell's *retyping* functions (see [Mit90]), though the connection has not been fully investigated.

2.3.   Incorporating bounded quantification

To interpret bounded quantification we need, for each type T, a kind $\downarrow T$ that parametrises the subtypes of T. This should be characterised by a universal property resembling local smallness, or a 'definability' condition for type inclusions (see [Ben85]); namely, there should be a 'generic subtype' of T in the fibre over $\downarrow T$. Forgetting about the generic type for the moment, we want:

powerkinds For each type $T \in T^{\mathbb{I}}$ there is a kind $\downarrow T$, a map $\epsilon: \downarrow T \to \mathbb{I}$ and an inclusion $T' \hookrightarrow \epsilon^* T$ in the fibre $T^{\mathbb{I}T}$. This data has the following universal property: if $f: \mathbb{J} \to \mathbb{I}$ and $S \hookrightarrow f^* T$ is any inclusion in $T^{\mathbb{J}}$ then there is a unique map $\kappa: \mathbb{J} \to \downarrow T$ such that $\epsilon \kappa = f$, $S = \kappa^* T'$ and

$$S = \kappa^* T' \hookrightarrow \kappa^* \epsilon^* T = f^* T$$

is the same map as $S \hookrightarrow f^* T$.

One can define generic supertypes in a completely symmetrical way; this is left as an exercise for the reader.

Powerkinds in the presence of a generic type We now only want a generic subtype for the generic type $G \in \mathcal{T}^{\mathbf{T}}$. We should then be able to obtain generic subtypes for all other types, as long as we have pullbacks in the category of kinds. For all this to work, G must be a generic type in the 'strict' sense.

pullbacks $\mathcal{K}$ has pullbacks.

strict generic For any $\mathbb{I} \in \mathcal{K}$ and any $T \in \mathcal{T}^{\mathbb{I}}$ there is a unique map $\eta \colon \mathbb{I} \to \mathbb{T}$ with $T = \eta^* G$.

Now we ask for a 'generic powerkind'. That is, we must have a kind $\downarrow G \in \mathcal{K}$, a map $\alpha \colon \downarrow G \to \mathbb{T}$, an object $G' \in \mathcal{T}^{\downarrow G}$ and an inclusion $G' \hookrightarrow \alpha^* G$ satisfying:

powerkinds2 Let $T \in \mathcal{T}^{\mathbb{I}}$, and define the kind $\downarrow T$ by the pullback square

$$
\begin{array}{ccc}
\downarrow T & \xrightarrow{\xi} & \downarrow G \\
\epsilon \downarrow & & \downarrow \alpha \\
\mathbb{I} & \xrightarrow{\eta} & \mathbb{T}
\end{array}
$$

so that we have an inclusion $T' \hookrightarrow \xi^* \alpha^* G = \epsilon^* T$ where $T' = \xi^* G'$; then this data has the universal property described in the **powerkinds** condition.

2.4.   Incorporating subkinds

Building a notion of subkind into our structure is very similar. That is, we must single out a class of 'inclusion maps' in $\mathcal{K}$ satisfying certain conditions. Writing $\mathbb{I} \sqsubseteq \mathbb{J}$ for an inclusion of kinds, we can state these as follows:

subkind-uniqueness There is at most one kind inclusion between any two kinds.

subkind-identities Identity maps are kind inclusions.

subkind-composition The composite of two kind inclusions is a kind inclusion.

subkind-powerkinds If $S \hookrightarrow T$ is a type inclusion, there is a kind inclusion $\downarrow S \sqsubseteq \downarrow T$ in $\mathcal{K}$.

subkind-restrictions If $m \colon \mathbb{I} \sqsubseteq \mathbb{J}$ is a kind inclusion, $f \colon \mathbb{J} \to \mathbb{K}$ is any map and $T \in \mathcal{T}^{\mathbb{J}}$, there is a type inclusion $\Pi_f T \hookrightarrow \Pi_{fm} m^* T$.

Proposition 2.4.1 *The **subkind-restrictions** condition is redundant; it holds for any map* $m: \mathbb{I} \to \mathbb{J}$ *between kinds.*

Proof. We will define a natural map $\Pi_f T \to \Pi_{fm} m^* T$ and show that it's a type inclusion.

Firstly, by the **counits** condition we have a type inclusion $f^* \Pi_f T \hookrightarrow T$; by the **stability** condition we can apply the substitution functor m^* to obtain a type inclusion $(fm)^* \Pi_f T = m^* f^* \Pi_f T \hookrightarrow m^* T$; and by the **mates** condition, the corresponding map $\Pi_f T \to \Pi_{fm} m^* T$ must be a type inclusion.   ∎

The **subkind-powerkinds** condition is sort of redundant too; a candidate for the kind inclusion already exists:

Proposition 2.4.2 *If* $S \hookrightarrow T$ *is a type inclusion, there is a canonical map of kinds* $\downarrow S \to \downarrow T$.

Proof. Suppose $S \hookrightarrow T$ is a type inclusion in $T^{\mathbb{I}}$, and let $\phi : \downarrow S \to \mathbb{I}$ be the arrow guaranteed to exist by **powerkinds**; then by **stability** we have an inclusion $\phi^* S \hookrightarrow \phi^* T$ in $T^{\downarrow S}$. But **powerkinds** also tells us that there is an inclusion $S' \hookrightarrow \phi^* S$ in $T^{\downarrow S}$, namely the generic subtype of S. By **composition** there is then an inclusion $S' \hookrightarrow \phi^* T$; so by **powerkinds** again, there is a unique arrow $\kappa : \downarrow S \to \downarrow T$ such that

$$\kappa^* (T' \hookrightarrow \epsilon^* T) = S' \hookrightarrow \phi^* T$$

∎

So our categorical definitions can actually be used to prove simple theorems. In this case, we have shown that the notion of kind inclusion is not really necessary; if we are prepared to forgo the **kind-uniqueness** condition (and hence kind subsumption), we may as well consider arbitrary maps of kinds to be kind 'inclusions'—the rules we want will still be valid.

Similarly, we might like to regard all arrows between types as type inclusions. There are two problems with this idea:

- This idea does not seem computationally realistic (type inclusions ought to be computationally trivial; but then, what about kind inclusions?);

- If we allow all arrows between types to be type inclusions, powerkinds may not exist—indeed, I don't know of any models in which they do, at least in the case $\mathcal{K} = Sets$.

3.   Examples

3.1.   PER models

We will see how the classical PER models for subtyping fit into the framework described above. Let A be a partial combinatory algebra—for example, $\mathbb{N}$ (indices for partial recursive functions) or some model of the untyped λ-calculus. We will model types as partial equivalence relations (or **PERs**) on A. All this is essentially well-known; it's only the category theory that's new.

For PERs R, S on A there are, at first sight, at least two possible notions of inclusion:

- The usual one:
$$aRb \longrightarrow aSb;$$

  that is, $R \subseteq S$.

- A stronger one: $R \subseteq S$ and
$$aSb \wedge aRa \wedge bRb \longrightarrow aRb$$

  as well.

(There may be other notions, for special A; we won't consider any.)

Recall that PERs correspond, roughly speaking, to modest sets in the realizability topos on A. Then the second notion corresponds to the notion of $\neg\neg$-closed monomorphism between modest sets; that is, every $\neg\neg$-closed subobject of a modest set has a unique representative of that form. So it has a good categorical characterization. Unfortunately, it does not seem to be a very useful notion: for example, if we were to adopt it, the contra-co subtyping rule would fail.

So by **inclusion** of PERs we mean the first sort of inclusion. Speaking categorically, this is not such a natural notion; for example, it is well known that *any* map between two PERs factors as a composite of isomorphisms and inclusions. However, it turns out to be pretty useful—it gives a simple model of subtyping that's pretty easy to work with. In fact, it is in some sense the 'standard' model: see [CL90].

First we have to define a fibration $\begin{smallmatrix} P \\ \downarrow \\ \mathcal{K} \end{smallmatrix}$. We do this as follows:

- $\mathcal{K}$ is *Sets*, the ordinary category of sets.

- If $\mathbb{I}$ is a set, the fibre $\mathcal{P}^{\mathbb{I}}$ of $\mathcal{T}$ over $\mathbb{I}$ is the category of $\mathbb{I}$-indexed families of PERs on A and uniform families of maps.

- The kind $\mathbb{T}$ of all types is the set $\mathbb{P}$ of all PERs on A.

- The generic type $G \in \mathcal{P}^{\mathbb{P}}$ is the family $\{R\}_{R \in \mathbb{P}}$ of all PERs on A.

This has the 'strict' structure required. (Recall that Π-functors are calculated as ordinary intersections of PERs: see [LM91] and [Hy82]).

Let $T' = \{R_i\}_{i \in \mathbb{I}}, T = \{S_i\}_{i \in \mathbb{I}} \in \mathcal{P}^{\mathbb{I}}$ be two $\mathbb{I}$-indexed families of PERs; then there is an **inclusion** $T' \hookrightarrow T$ exactly when $R_i \subseteq S_i$ for all $i \in \mathbb{I}$; that is, an inclusion is a uniform family of maps $R_i \to S_i$ tracked by the identity.

To handle powerkinds we have:

- $\downarrow G = \{\langle S, R\rangle : S, R \text{ are PERs and } S \subseteq R\}$

- $\alpha : \downarrow G \to \mathbb{P}$ is the projection $\langle S, R\rangle \mapsto R$

$G' \in \mathcal{P}^{\downarrow G}$ and $G' \hookrightarrow \alpha^* G$ are obvious.

All the conditions can be verified by direct calculation, if products and exponentials are defined as PERs in the standard way and Π-functors are defined using intersections; **Top** is defined to be the maximal PER, namely $A \times A$. The details are omitted.

Observe that if $\{R_i\} \in \mathcal{P}^{\mathbb{I}}$ then the powerkind $\downarrow\{R_i\}$ is the set

$$\{\langle i, S, R_i\rangle : S \subseteq R_i\}.$$

This is what we ought to expect.

Note the problem in extending all this to the calculus of constructions: Now we must index, not just over *Sets*, but over A-valued assemblies (e.g. ω-sets); and Π-functors are no longer calculated simply as intersections.

Interpreting subkinds   Kind inclusion $\mathbb{I} \subseteq \mathbb{J}$ can be interpreted as ordinary subset inclusion. The only problem is the **subkind-powerkinds** condition; we may need to redefine powerkinds to get an actual subset inclusion. The details are omitted.

3.2. Models arising from PER models

Sometimes we're interested, not in all PERs on A, but only in certain special ones: for example, ones that can be regarded as '(pre)domains': see, e.g., [Ph90]. That is, if (by abuse of language) we let $\mathbb{P}$ denote the category of modest sets in the realizability topos on A, then we're interested in a full subcategory

$$\mathbb{C} \rightarrowtail \mathbb{P}$$

of $\mathbb{P}$. The object-of-objects need not be a 'set' ($\neg\neg$-sheaf or 'uniformizable' object: see [Hy82]); it could be any subobject of the object of modest sets. We'll consider the situation when this category is *reflective*: cf. [Ph?].

Then we have a map

$$r: \mathbb{P} \to \mathbb{P}$$

that is the object part of the internal functor $\mathbb{P} \to \mathbb{C} \to \mathbb{P}$, where the first arrow is the internal reflection functor and the second arrow is the inclusion.

Proposition 3.2.1 *Exponentials and indexed products in $\mathbb{C}$ agree with the ordinary ones for PERs (modest sets); we may compute the indexed products or exponentials of objects of $\mathbb{C}$ in the category of modest sets, so that the resulting objects again lie in $\mathbb{C}$ and are products/exponentials in that category.*

Proof. This is a well-known fact about full reflective subcategories. ∎

Now we can define a new model using a construction appearing in [Ph91b]: that is, we define a fibration $\begin{smallmatrix} \mathcal{C} \\ \downarrow \\ \mathcal{K} \end{smallmatrix}$ as follows.

- $\mathcal{K}$ is *Sets* again.

- $\mathcal{C}^{\mathbb{I}}$ is the category of $\mathbb{I}$-indexed families of PERs; however, now the maps $\{R_i\} \to \{S_i\}$ are uniform families of maps $rR_i \to rS_i$.

- $\mathbb{T} = \mathbb{P}$ as before.

Note that Π-functors, binary products and exponentials in each fibre, can be calculated very simply; for example,

$$\{S_i\}^{\{R_i\}} = \{rS_i^{rR_i}\}$$

and

$$\{R_i\} \times \{S_i\} = \{rR_i \times rS_i\}.$$

We'll illustrate indexed products by calculating the Π-functor along a map $\mathbb{I} \to 1$: Let $\{R_i\} \in \mathcal{C}^{\mathbb{I}}$ and let S be a PER. Then uniform families of maps of PERs $rS \to rR_i$ correspond to maps $rS \to \Pi_i rR_i$ of PERs, where Π_i is the $\mathbb{I}$-indexed product of ordinary PERs: we could define $\Pi_i\{rR_i\} = \bigcap_i rR_i$.

So maps $\{S\} \to \{R_i\}$ in $\mathcal{C}^{\mathbb{I}}$, i.e. uniform families of maps of PERs $rS \to rR_i$, correspond to uniform families of maps of PERs $rS \to \Pi_i rR_i \cong r\Pi_i rR_i$; that is, to maps $S \to \Pi_i rR_i = \bigcap_i rR_i$ in $\mathcal{C}^1$. Therefore, in the fibration $\underset{\mathcal{K}}{\downarrow}$, we can define the Π-functor for $\mathbb{I} \to 1$ as $\Pi_i\{R_i\} = \bigcap_i rR_i$; this works.[5]

The **inclusions** are the families of maps tracked by the identity; that is, we have an inclusion $\{R_i\} \hookrightarrow \{S_i\}$ if

$$rR_i \subseteq rS_i$$

for all $i \in \mathbb{I}$. And if we put

$$\downarrow G = \{\langle S, R \rangle : rS \subseteq rR\}$$

and so on, it all carries through. The main observations are

- '$rS \subseteq rR$' is a $\neg\neg$-closed predicate, so $\downarrow G$ can be regarded as an ordinary set and not just an assembly (i.e. ω-set); and

- products and exponentials may be calculated in basically the same way as for PERs, and so satisfy the required properties for the same reasons as before.

(The first remark was not used explicitly; it simply explains why the constructions work, and why things don't work if we allow arbitrary arrows to be type inclusions—because then the resulting 'powerkinds' are no longer ordinary sets, i.e. $\neg\neg$-sheaves.)

In [Ros91], Rosolini showed that if the object-of-objects of $\mathbb{C}$ was a $\neg\neg$-sheaf then $\mathbb{C}$ gives rise to a model of subtyping; so we have basically generalised Rosolini's result to arbitrary reflective $\mathbb{C}$.

3.3.   Subtypes and constructions on types

If $\mathbb{C}$ can be regarded as a category of predomains, it will probably support various other constructions: for example,

$$T \mapsto T_\perp \qquad \text{lifting (computable partial map classifier)}$$

---

[5]Note that we *cannot* put $\Pi_i\{R_i\} = \bigcap_i R_i$.

We might then like to have rules like

$$\frac{S \leq T}{S_\perp \leq T_\perp}$$

That is, we'd like the functor $(-)_\perp$ to preserve inclusions.

Now $T_\perp$ is characterised by universal properties, i.e. only (at first sight) up to isomorphism; so why should it preserve inclusions, which may not be stable under isomorphism? This is a general problem; luckily, for the models we've mentioned, we know how to handle it: we have an explicit description of $T_\perp$ as a PER, and we can verify the condition directly.

A different kind of solution is suggested by [Ros91]: Rosolini showed using the uniformity principle that if $\mathbb{C}$ is a full subcategory of $\mathbb{P}$ whose object-of-objects is a $\neg\neg$-sheaf, then certain internal functors $\mathbb{C} \to \mathbb{C}$ may be assumed to preserve inclusions. One can deduce from this that fibred functors $\begin{smallmatrix} \mathcal{C} \\ \downarrow \\ \mathcal{K} \end{smallmatrix} \to \begin{smallmatrix} \mathcal{C} \\ \downarrow \\ \mathcal{K} \end{smallmatrix}$ arising from these internal functors must preserve inclusions.

The fibration just described was essentially the externalisation of the category $\mathbb{C}$, equivalent to $\mathbb{C}$, described in [Ph91b]; its object-of-objects is simply $\mathbb{P}$, so Rosolini's result applies here too. That is, for any full reflective subcategory $\mathbb{C}$ of $\mathbb{P}$, 'internally definable' fibred endofunctors $\begin{smallmatrix} \mathcal{C} \\ \downarrow \\ \mathcal{K} \end{smallmatrix} \to \begin{smallmatrix} \mathcal{C} \\ \downarrow \\ \mathcal{K} \end{smallmatrix}$ preserve inclusions.

References

[AP90]   M. Abadi, G. Plotkin: A PER model of polymorphism and recursive types, in: Proc. of 5th Annual Symposium on Logic in Computer Science, 1990.

[AC90]   R. Amadio, L. Cardelli: *Subtyping recursive types*, DEC SRC Research Report 62, August 1990.

[Ben85] J. Benabou: Fibred categories and the foundations of naive category theory, *J. Symbolic Logic*, vol. 50, no. 1 (1985) 10–37.

[CL90]   L. Cardelli, G. Longo: *A semantic basis for Quest*, DEC SRC Research Report 55, February 1990.

[Fai89] J. Fairbairn: Some types with inclusion properties in $\forall, \to, \mu$, University of Cambridge Computer Laboratory Technical Report No. 171, 1989.

[FMRS90] P. Freyd, P. Mulry, G. Rosolini, D. S. Scott: Extensional PERs, in: Proc. of 5th Annual Symposium on Logic in Computer Science, 1990.

[Gir90]   J.-Y. Girard: *Intérpretation fonctionelle et élimination des coupoures dans l'arithmetique d'ordre supérieur*, Thèse de doctorat d'état, Université Paris VII, 1972.

[Hy82]   J. M. E. Hyland: The effective topos, in: *The L.E.J. Brouwer Centenary Symposium* (ed. A. S. Troelstra, D. van Dalen), North-Holland, 1982.

[Hy88]   J. M. E. Hyland: A small complete category, *Annals of Pure and Applied Logic* 40 (1988) 135-165.

[HRR90]   J. M. E. Hyland, E.P. Robinson, G. Rosolini: The discrete objects in the effective topos, *Proc. Lond. Math. Soc.* (3) 60 (1990) 1–36.

[Jac91]   B. Jacobs: *Categorical Type Theory*, dissertation, University of Nijmegen, 1991.

[LM91]   G. Longo, E. Moggi: Constructive natural deduction and its 'ω-set' interpretation, *Math. Struct. in Comp. Science* (1991), *vol.* 1.

[M-OM90] N. Martí-Oliet, J. Meseguer: *Inclusions and Subtypes*, CSL Technical Report SRI-CSL-90-16, SRI International, December 1990.

[Mit90]   J. Mitchell: Polymorphic type inference and containment, in *Logical Foundations of Functional Programming* (ed. G. Huet), Prentice-Hall, 1990.

[deP90]   V. C. V. de Paiva: *Subtyping in Ponder (preliminary report)*, University of Cambridge Computer Laboratory Technical Report No. 203, August 1990.

[Ph90]   W. K.-S. Phoa: Effective domains and intrinsic structure, in: Proc. of 5th Annual Symposium on Logic in Computer Science, 1990.

[Ph91a]   W. K.-S. Phoa: From term models to domains, in: *Proc. of Theoretical Aspects of Computer Software (Sendai, 1991)*, Springer LNCS 526.

[Ph91b]   W. K.-S. Phoa: Two results on set-theoretic polymorphism, in: *Proc. of Category Theory and Computer Science (Paris, 1991)*, Springer LNCS 526.

[Ph9?]   W. K.-S. Phoa: Building domains from graph models, to appear in *Math. Struct. in Comp. Science*.

[Ph?]   W. K.-S. Phoa: Reflectivity for categories of synthetic domains, draft.

[Rob89] E. Robinson: How complete is PER? in: Proc. of 4th Annual Symposium on Logic in Computer Science, 1989.

[Ros91] G. Rosolini: An Exper model for Quest, to appear in the proceedings of Mathematical Foundations of Programming Semantics, 1991.

[Ros?] G. Rosolini: About modest sets, to appear.

[Tay86] P. Taylor: *Recursive Domains, Indexed Categories and Polymorphism*, dissertation, University of Cambridge, 1986.

Reasoning about Sequential Functions via Logical Relations

Kurt Sieber *

1. Introduction

In his seminal paper [Plo77], Plotkin introduced the functional language PCF ('programming language for computable functions') together with the standard denotational model of cpo's and continuous functions. He proved that this model is computationally adequate for PCF, but *not* fully abstract.

In order to obtain full abstraction he extended PCF by a parallel conditional operator. The problem with this operator is, that it changes the nature of the language. All computations in the original language PCF can be executed sequentially, but the new operator requires expressions to be evaluated in parallel.

Here we address the problem of full abstraction for the original *sequential* language PCF, i.e. instead of extending the language we try to improve the model. Our approach is to cut down the standard model with the aid of certain logical relations, which we call *sequentiality relations*. We give a semantic characterization of these relations and illustrate how they can be used to reason about sequential (i.e. PCF-definable) functions. Finally we prove that this style of reasoning is 'complete' for proving observational congruences between closed PCF-expressions of order ≤ 3. Technically, this completeness can be expressed as a full abstraction result for the sublanguage which consists of these expressions.

2. Sequential PCF

In [Plo77], PCF is defined as a simply typed λ-calculus over the ground types ι (of integers) and o (of Booleans). For the sake of simplicity we throw out the

---

*Current affiliation: Institut für Angewandte Informatik und formale Beschreibungsverfahren, Universität Karlsruhe. Permanent address: FB 14 Informatik, Universität des Saarlandes, W-6600 Saarbrücken, Germany.

Booleans and simulate them by integers; 0 stands for 'true', all other integers for 'false'. Hence the set *Type* of all types τ is defined by

$$\tau ::= \iota \mid \tau_1 \to \tau_2$$

and the constants c with their types are

$$\begin{aligned}
\underline{n} : \quad & \iota & &\text{for each } n \in \mathbb{N}, \\
succ, pred : \quad & \iota \to \iota, \\
cond : \quad & \iota \to \iota \to \iota \to \iota, \\
Y_\tau : \quad & (\tau \to \tau) \to \tau & &\text{for each type } \tau.
\end{aligned}$$

PCF-expressions $M, N, \ldots$ are defined to be the well-typed λ-expressions over these constants and over a set of typed variables $x^\tau, y^\tau, \ldots$. A PCF-*program* is a closed PCF-expression of ground type ι. Instead of $cond\, M\, N\, P$ we also use the more familiar notation **if** M **then** N **else** P **fi**. Ω_τ is used as an abbreviation for the (diverging) expression $Y_\tau(\lambda\, x^\tau . x)$.[1]

The standard denotational semantics for PCF associates a cpo D^τ with each type τ, namely

$$\begin{aligned}
D^\iota &= \mathbb{N}_\perp & &\text{(the flat cpo of natural numbers),} \\
D^{\tau_1 \to \tau_2} &= (D^{\tau_1} \to D^{\tau_2}) & &\text{(the cpo of continuous functions).}
\end{aligned}$$

The constants are given their usual interpretation. In particular $[\![Y_\tau]\!]$ is the fixed point operator on D^τ and $[\![cond]\!]$ is the sequential conditional, defined by

$$[\![cond]\!]\, b\, d\, e = \begin{cases} d & \text{if } b = 0, \\ e & \text{if } b \in \mathbb{N} \setminus \{0\}, \\ \perp & \text{if } b = \perp. \end{cases}$$

The meaning $[\![M]\!]\rho$ of an expression M in an environment ρ is defined as usual. If M is closed, then we abbreviate $[\![M]\!]\rho$ to $[\![M]\!]$, because it doesn't depend on ρ. An element $d \in D^\tau$ is called (PCF-)*definable* if there is a closed PCF-expression M such that $[\![M]\!] = d$.

We will now define the notions of computational adequacy and full abstraction. Let '$\to$' be the one step transition relation between (closed) PCF-expressions as defined in [Plo77] (the exact definition of '$\to$' is not important for our purpose). Then the *observable behavior* of a PCF-program P can be defined by

$$beh(P) = \{n \mid P \xrightarrow{*} \underline{n}\}.$$

This set is either empty, if P diverges, or a singleton $\{n\}$, if P terminates with normal form $\underline{n}$. A denotational semantics is called *computationally adequate*

---

[1]From now on we will use the type superscript τ only for the binding occurence of a variable.

if it correctly describes the observable behavior of programs. For PCF this is made precise in the following theorem.

Theorem 2.1 *Let* $[\,]$ *be the standard semantics of cpo's and continuous functions. Then*

$$[\![P]\!] = n \quad \text{iff} \quad beh(P) = \{n\} \quad (\text{iff} \quad P \overset{*}{\to} \underline{n}).$$

for every PCF-program P and every $n \in \mathbb{N}$.

An immediate consequence is, that for any two programs P and Q:

$$[\![P]\!] = [\![Q]\!] \quad \text{iff} \quad beh(P) = beh(Q).$$

We take Theorem 2.1 for granted; its proof can be found in [Plo77].

Definition 2.2

(a) *Two PCF-expressions M and N of the same type are called* observationally congruent *(notation:* $M \approx N$*) if* $beh(C[M]) = beh(C[N])$ *for every program context* $C[\,]$.

(b) *A denotational semantics is called* fully abstract, *if observational congruence coincides with semantic equality, i. e. if for any two expressions M and N:*

$$M \approx N \quad \text{iff} \quad [\![M]\!] = [\![N]\!].$$

Note that—in the light of computational adequacy—observational congruence and full abstraction can be defined purely in terms of the denotational semantics: $M \approx N$ holds iff $[\![C[M]]\!] = [\![C[N]]\!]$ for every program context $C[\,]$. Hence full abstraction means that $[\![M]\!] = [\![N]\!]$ iff $[\![C[M]]\!] = [\![C[N]]\!]$ for all program contexts $C[\,]$. One direction of this equivalence is usually trivial; if the denotational semantics is defined by induction on the structure of expressions, then $[\![M]\!] = [\![N]\!]$ implies $[\![C[M]]\!] = [\![C[N]]\!]$. Hence a proof of full abstraction only consists of the other direction: If $[\![M]\!] \neq [\![N]\!]$, then one must find some program context $C[\,]$ such that $[\![C[M]]\!] \neq [\![C[N]]\!]$.

For proving observational congruence of two closed expressions, it is sufficient to consider *applicative* contexts. This is made precise by the following theorem, known as the *Context Lemma*.

Theorem 2.3 *Two closed PCF-expressions* $M, N : \sigma_1 \to \ldots \to \sigma_n \to \iota$ *are observationally congruent iff* $[\![MP_1 \ldots P_n]\!] = [\![NP_1 \ldots P_n]\!]$ *for all closed expressions* $P_1 : \sigma_1, \ldots, P_n : \sigma_n$.

3. Logical Relations

It is well-known why full abstraction fails for PCF, but we repeat the argument here, because we want to illustrate the use of logical relations. Consider the following three equations for a function f of type $\iota \to \iota \to \iota$:

$$
\begin{aligned}
f \quad 0 \quad \perp &= 0 \\
f \quad \perp \quad 0 &= 0 \\
f \quad 1 \quad 1 &= 1
\end{aligned}
\tag{1}
$$

Such a function f exists in the model, namely the *parallel or* operator *por*, defined by

$$
por\, d\, e = \begin{cases} 0 & \text{if } d = 0 \text{ or } e = 0, \\ 1 & \text{if } d, e \in \mathbb{N} \setminus \{0\}, \\ \perp & \text{otherwise.} \end{cases}
$$

On the other hand, there is no PCF-definable function f which satisfies (1). In [Plo77] this was proved with the aid of a so-called *Activity lemma*; here we will use logical relations for the proof. But let us first see how full abstraction fails. The point is that it can be tested with closed PCF-expressions whether a function f satisfies (1). For $i = 1, 2$, let $PORTEST_i$ of type $(\iota \to \iota \to \iota) \to \iota$ be defined by[2]

$$PORTEST_i \equiv \lambda y^{\iota \to \iota \to \iota}. \text{if } y\, \underline{0}\, \Omega_\iota = \underline{0} \wedge y\, \Omega_\iota\, \underline{0} = \underline{0} \wedge y\, \underline{1}\, \underline{1} = \underline{1} \text{ then } \underline{i} \text{ else } \Omega_\iota \text{ fi.}$$

Then

$$
[\![PORTEST_i]\!]\, f = \begin{cases} i & \text{if } f \text{ satisfies (1)}, \\ \perp & \text{otherwise.} \end{cases}
$$

This means that $[\![PORTEST_1]\!] \neq [\![PORTEST_2]\!]$, because they differ on the function *por*, but $PORTEST_1 \approx PORTEST_2$ by Theorem 2.3, because $[\![PORTEST_1 M]\!] = [\![PORTEST_2 M]\!]$ for every closed PCF-expression M.

We now introduce logical relations.

Definition 3.1

(a) *An n–ary logical relation R on a λ-model $(D^\tau)_{\tau \in Type}$ is a family of relations $R^\tau \subseteq (D^\tau)^n$ such that for all types σ, τ and $f_1, \ldots, f_n \in D^{\sigma \to \tau}$*

$$R^{\sigma \to \tau}(f_1, \ldots, f_n) \Leftrightarrow \forall d_1, \ldots, d_n. R^\sigma(d_1, \ldots, d_n) \Rightarrow R^\tau(f_1 d_1, \ldots, f_n d_n).$$

(b) *An element $d \in D^\tau$ is called* invariant *under R if $R^\tau(d, \ldots, d)$ holds.*

Note that a logical relation R is uniquely determined by its definition on ground type(s). The importance of logical relations is expressed by the following theorem, known as the *Main Lemma for Logical Relations* ([Plo80]).

---

[2]We freely use functions like $=, \wedge, \leq : \iota \to \iota \to \iota$ which are obviously definable.

Theorem 3.2 *Let R be a logical relation on a λ-model such that $[\![c]\!]$ is invariant under R for every constant c. Then $[\![M]\!]$ is invariant under R for every closed λ-term M over these constants.*

We are now ready to prove with mathematical rigor that no PCF-definable function f can satisfy (1). Let R be the 3-ary logical relation defined by

$$R^\iota(d_1, d_2, d_3) \Leftrightarrow (d_1 = \bot) \vee (d_2 = \bot) \vee (d_1 = d_2 = d_3).$$

It is easy to see (and follows from Theorem 3.4) that $[\![c]\!]$ is invariant under R for each PCF-constant c. Hence $[\![M]\!]$ is invariant under R for every closed PCF-expression M. But then (1) cannot hold for $f = [\![M]\!]$, because R^ι holds for the argument columns $(0, \bot, 1)$ and $(\bot, 0, 1)$, but *not* for the result column $(0, 0, 1)$.

The example shows us how we can rule out elements like *por* (which destroy full abstraction) with the aid of logical relations. Two questions naturally arise:

- Is there a semantic characterization of all logical relations under which the PCF-constants are invariant?

- Are logical relations sufficient to rule out *all* elements which destroy full abstraction?

For the first question we will immediately give a positive answer; the second question will be investigated in section 4.

Definition 3.3

(a) *For each $n \geq 0$ and each pair of sets $A \subseteq B \subseteq \{1, \ldots, n\}$ let $S^n_{A,B} \subseteq (D^\iota)^n$ be defined by*

$$S^n_{A,B}(d_1, \ldots, d_n) \Leftrightarrow (\exists i \in A. \, d_i = \bot) \vee (\forall i, j \in B. \, d_i = d_j).$$

An n-ary logical relation R is called a sequentiality relation if R^ι is an intersection of relations of the form $S^n_{A,B}$.

(b) *An element $d \in D^\tau$ is called logically sequential if it is invariant under all sequentiality relations.*

Note that the logical relation R which we have used to rule out *por*, was a sequentiality relation, namely $R^\iota = S^3_{\{1,2\},\{1,2,3\}}$.

Theorem 3.4 *The sequentiality relations are the* only *logical relations under which the meanings of all PCF-constants are invariant.*

Proof :

(a) Let R be a sequentiality relation.

The meaning of a ground type constant is invariant under R, because $R^\iota(d, \ldots, d)$ holds for all $d \in D^\iota$.

For a first order constant c, it is sufficient to prove that $[\![c]\!]$ is invariant under the relations $S_{A,B}^n$. This is obvious for the functions $[\![pred]\!]$ and $[\![succ]\!]$, because they are strict. For $[\![cond]\!]$ assume that $(b_1, \ldots, b_n), (d_1, \ldots, d_n)$ and $(e_1, \ldots, e_n)$ are in $S_{A,B}^n$ and let $v_i = [\![cond]\!]\, b_i\, d_i\, e_i$ for $i = 1, \ldots, n$. If $b_i = \bot$ for some $i \in A$, then $v_i = \bot$. Otherwise $b_i = b_j$ for all $i, j \in B$, hence either $v_i = d_i$ for all $i \in B$ or $v_i = e_i$ for all $i \in B$. In all three cases it follows that $(v_1, \ldots, v_n)$ is in $S_{A,B}^n$.

In order to see that the fixed point operators are invariant under R, first note that $R^\iota(\bot_\iota, \ldots, \bot_\iota)$ and hence $R^\tau(\bot_\tau, \ldots, \bot_\tau)$ by induction on τ. Moreover, each R^τ is closed under lubs of directed sets (also by induction on τ). Hence, if $R^{\tau \to \tau}(f, \ldots, f)$ for some $f \in D^{\tau \to \tau}$, then we first obtain $R^\tau(f^n \bot_\tau, \ldots, f^n \bot_\tau)$ for all $n \in \mathbb{N}$ and this implies $R^\tau([\![Y_\tau]\!]\, f, \ldots, [\![Y_\tau]\!]\, f)$.

(b) Let R be a logical relation, under which the meanings of all PCF-constants—and hence, by Theorem 3.2, the meanings of all closed PCF-expressions—are invariant. Let S be the intersection of all $S_{A,B}^n$, which contain R^ι. If we can prove that $S \subseteq R^\iota$, then $S = R^\iota$ and this means that R is a sequentiality relation.

Let $d = (d_1, \ldots, d_n) \in S$. By induction on the size of $C \subseteq \{1, \ldots, n\}$ we show that there is some vector $e = (e_1, \ldots, e_n) \in R^\iota$ with

$$e_i = d_i \text{ for all } i \in C. \tag{2}$$

Setting $C = \{1, \ldots, n\}$ we obtain the desired result that $d \in R^\iota$.

For $|C| \leq 1$ we can choose some constant vector e; hence let $|C| \geq 2$.

Case 1: $d_i \neq \bot$ for all $i \in C$.

Let $T = \{v \in (D^\iota)^n \mid v_i \neq \bot \text{ for all } i \in C\}$. Each $v \in T$ defines an equivalence relation $\sim_v$ on C by: $i \sim_v j \Leftrightarrow v_i = v_j$. Let $u \in R^\iota \cap T$ be such that $\sim_u$ is minimal in $\{\sim_v \mid v \in R^\iota \cap T\}$, i.e. no other vector $v \in R^\iota \cap T$ is 'finer' than u (note that $R^\iota \cap T \neq \emptyset$, because $R^\iota(d, \ldots, d)$ for all $d \in D^\iota$). We will prove $R^\iota \subseteq S_{B,B}^n$ for each equivalence class B of $\sim_u$.

Let $i \in C$, let $B = \{j \in C \mid u_j = u_i\}$ be the equivalence class of i, and assume that $v \in R^\iota \setminus S_{B,B}^n$. Then $v_j \neq \bot$ for all $j \in B$ and $v_j \neq v_k$ for some pair of indices $j, k \in B$. This allows us to find a vector $w \in R^\iota \cap T$ which is finer than u, namely: Let $u_0 \neq u_l$ for all $l \in C$ and define

$$M \equiv \lambda x^\iota. \lambda y^\iota. \text{ if } x = \underline{u_i} \text{ then if } y = \underline{v_j} \text{ then } \underline{u_0} \text{ else } x \text{ fi else } x \text{ fi}.$$

Then $w = (\llbracket M \rrbracket u_1 v_1, \ldots, \llbracket M \rrbracket u_n v_n) \in R^\iota \cap T$ and it is easy to see that $\sim_w \subset \sim_u$ (namely the equivalence class B of $\sim_u$ is split into two equivalence classes of $\sim_w$). This contradicts the minimality of $\sim_u$, hence the assumption was wrong and $R^\iota \subseteq S_{B,B}^n$ is proved.

It follows that $S \subseteq S_{B,B}^n$ and hence $d \in S_{B,B}^n$ for all equivalence classes B of $\sim_u$. This implies $\sim_u \subseteq \sim_d$, because $d_i \neq \bot$ for all $i \in C$. But then it is easy to find a closed PCF-expression N such that $d_i = \llbracket N \rrbracket u_i$ for all $i \in C$, i.e. $e = (\llbracket N \rrbracket u_1, \ldots, \llbracket N \rrbracket u_n) \in R^\iota$ satisfies (2).

Case 2: $d_i = \bot$ for some $i \in C$.

If $R^\iota \not\subseteq S_{C \setminus \{i\}, C}^n$, then there is some $u \in R^\iota$ such that $u_j \neq \bot$ for *all* $j \in C \setminus \{i\}$ and $u_k \neq u_i$ for *some* $k \in C \setminus \{i\}$. By induction hypothesis there are vectors $v, w \in R^\iota$ with

$$v_j = d_j \text{ for all } j \in C \setminus \{i\} \quad \text{and} \quad w_j = d_j \text{ for all } j \in C \setminus \{k\}.$$

These two vectors can now be 'merged' by a closed PCF-expression. Let

$$M \equiv \lambda x^\iota. \lambda y^\iota. \lambda z^\iota. \text{ if } x = \underline{u_k} \text{ then } y \text{ else } z \text{ fi}$$

and let $e = (\llbracket M \rrbracket u_1 v_1 w_1, \ldots, \llbracket M \rrbracket u_n v_n w_n)$. Then e is in R^ι and a straightforward calculation shows that it satisfies (2).

If $R^\iota \subseteq S_{C \setminus \{i\}, C}^n$, then $S \subseteq S_{C \setminus \{i\}, C}^n$ and hence $d \in S_{C \setminus \{i\}, C}^n$. As $d_i = \bot$, this implies $d_k = \bot$ for some $k \in C \setminus \{i\}$. Let $v, w \in R^\iota$ be defined as above (wrt the new index k). If $v_i = \bot$, then we can take $e = v$, otherwise let

$$M \equiv \lambda x^\iota. \lambda y^\iota. \text{ if } x = \underline{v_i} \text{ then } y \text{ else } x \text{ fi}$$

and take $e = (\llbracket M \rrbracket v_1 w_1, \ldots, \llbracket M \rrbracket v_n w_n)$. In both cases e is in R^ι and satisfies (2). □

We conclude this section with a sophisticated example of 'reasoning about sequential functions'. It is essentially the last example on p. 269 of [Cur86]. We use the familiar notation for threshold functions ([Plo77]), defined by

$$(u_1 \Rightarrow \ldots \Rightarrow u_n \Rightarrow v) \, d_1 \ldots d_n = \begin{cases} v & \text{if } d_i \sqsupseteq u_i \text{ for } i = 1, \ldots, n, \\ \bot & \text{otherwise}. \end{cases}$$

Example 3.5 *Let* $g_1 = (\bot \Rightarrow 1 \Rightarrow 0) \sqcup (1 \Rightarrow 0 \Rightarrow 0)$, $g_2 = (0 \Rightarrow \bot \Rightarrow 0)$, $g_3 = (\bot \Rightarrow 0 \Rightarrow 0)$ *and* $g_4 = \bot$. *Then there is no PCF-definable function* f *of type* $(\iota \to \iota \to \iota) \to \iota$, *which satisfies the equations*

$$
\begin{aligned}
f\ g_1 &= 0 \\
f\ g_2 &= 0 \\
f\ g_3 &= 0 \\
f\ g_4 &= \bot
\end{aligned}
$$

Proof :   Let R be the sequentiality relation defined by

$$R^\iota = S^4_{\{1,2\},\{1,2\}} \cap S^4_{\{1,3\},\{1,3\}} \cap S^4_{\{1,2,3\},\{1,2,3,4\}}.$$

The result column $(0,0,0,\bot)$ is *not* in $S^4_{\{1,2,3\},\{1,2,3,4\}} \supseteq R^\iota$, hence we must only convince ourselves that $(g_1,\ldots,g_4)$ is in $R^{\iota\to\iota\to\iota}$. Assume that $g_i\, d_{i1}\, d_{i2} = e_i$ for $i = 1,\ldots,4$, where $(e_1,\ldots,e_4) \notin R^\iota$. This is only possible for $(e_1,\ldots,e_4) = (0,0,0,\bot)$, and then one of the following two cases must occur:

$$
\begin{aligned}
(d_{11},\ldots,d_{41}) &\sqsupseteq (\bot,0,\bot,\bot) \quad \text{and} \quad (d_{12},\ldots,d_{42}) \sqsupseteq (1,\bot,0,\bot) \\
(d_{11},\ldots,d_{41}) &\sqsupseteq (1,0,\bot,\bot) \quad \text{and} \quad (d_{12},\ldots,d_{42}) \sqsupseteq (0,\bot,0,\bot).
\end{aligned}
$$

In the first case $(d_{12},\ldots,d_{42}) \notin S^4_{\{1,3\},\{1,3\}} \supseteq R^\iota$ and in the second case $(d_{11},\ldots,d_{41}) \notin S^4_{\{1,2\},\{1,2\}} \supseteq R^\iota$. Thus we have proved that $(g_1,\ldots,g_4)$ is in $R^{\iota\to\iota\to\iota}$. $\square$

Of course Example 3.5 can be continued in the same sense as the *parallel or-*example: There is a function f in the model which satisfies the 4 equations and it is easy to construct PCF-expressions of type $((\iota \to \iota \to \iota) \to \iota) \to \iota$, which differ in f iff f satisfies these equations.

4.   A Full Abstraction Result

Theorem 4.1 *Let* $\tau = \sigma_1 \to \ldots \to \sigma_n \to \iota$ *be a type of order 1 or 2 and let* $f \in D^\tau$ *be a logically sequential function. Moreover, let* $(d_{i1},\ldots d_{in}) \in D^{\sigma_1} \times \ldots \times D^{\sigma_n}$ *for* $i = 1,\ldots,m$. *Then there is a closed PCF-expression* M *of type* τ *such that*

$$f\, d_{i1}\ldots d_{in} = [\![M]\!]\, d_{i1}\ldots d_{in} \quad for\ i = 1,\ldots,m.$$

Proof :   Let R be the logical relation defined by

$$R^\iota(e_1,\ldots,e_m) \Leftrightarrow \exists\ closed\ M : \tau.\, \forall i \in \{1,\ldots,m\}.\, e_i = [\![M]\!]\, d_{i1},\ldots,d_{in}.$$

We must prove that $R^\iota(f\, d_{11}\ldots d_{1n},\ldots,f\, d_{m1}\ldots d_{mn})$ holds.

(a) First we prove that R is a sequentiality relation, i.e. that all PCF-constants are invariant under R. This is obvious for a ground type constant c: $R^\iota(\llbracket c \rrbracket, \ldots, \llbracket c \rrbracket)$ holds, because $\llbracket \lambda x_1 \ldots \lambda x_n . c \rrbracket d_{i1} \ldots d_{in} = \llbracket c \rrbracket$ for $i = 1, \ldots, m$. Now let c be a first order constant, say $c : \iota \to \iota \to \iota$. If $R^\iota(u_1, \ldots, u_m)$ and $R^\iota(v_1, \ldots, v_m)$, then there are closed PCF-expressions U and V such that for $i = 1, \ldots, m$

$$u_i = \llbracket U \rrbracket d_{i1} \ldots d_{in} \quad \text{and} \quad v_i = \llbracket V \rrbracket d_{i1} \ldots d_{in}$$

and hence

$$\llbracket c \rrbracket u_i v_i = \llbracket \lambda x_1 . \ldots \lambda x_n . c \, (U \, x_1 \ldots x_n)(V \, x_1 \ldots x_n) \rrbracket d_{i1} \ldots d_{in}.$$

This means $R^\iota(\llbracket c \rrbracket u_1 v_1, \ldots, \llbracket c \rrbracket u_m v_m)$, hence $R^{\iota \to \iota \to \iota}(\llbracket c \rrbracket, \ldots, \llbracket c \rrbracket)$ is proved.

(b) Now we prove that $R^{\sigma_j}(d_{1j}, \ldots, d_{mj})$ holds for $j = 1, \ldots, n$. The proof is similar as in (a). If σ_j is the ground type ι, then $R^\iota(d_{1j}, \ldots, d_{mj})$ holds because $d_{ij} = \llbracket \lambda x_1 . \ldots \lambda x_n . x_j \rrbracket d_{i1} \ldots d_{in}$ for $i = 1, \ldots, m$. If σ_j is a first order type, say $\sigma_j = \iota \to \iota \to \iota$, and $(u_1, \ldots, u_m), (v_1, \ldots, v_m)$ are as above, then

$$d_{ij} u_i v_i = \llbracket \lambda x_1 . \ldots \lambda x_n . x_j \, (U \, x_1 \ldots x_n)(V \, x_1 \ldots x_n) \rrbracket d_{i1} \ldots d_{in}$$

for $i = 1, \ldots m$, and this proves $R^{\iota \to \iota \to \iota}(d_{1j}, \ldots, d_{mj})$.

From (a) it follows that $R^\tau(f, \ldots, f)$ holds, and together with (b) this yields the desired result $R^\iota(f \, d_{11} \ldots d_{1n}, \ldots, f \, d_{m1} \ldots d_{mn})$.   □

Theorem 4.1 tells us that for each logically sequential function f there is always a definable function which coincides with f on finitely many given arguments. But we need a stronger result: f must be the *least upper bound* of definable elements. The clue for proving this result is the definability of finite projections.

Definition 4.2 *Let D be a cpo. A function $p : D \to D$ is called a* projection, *if $p \circ p = p$ and $p \sqsubseteq id_D$ (the identity on D). A projection is called* finite *if $p(D)$ is a finite set. D is called an* SFP-domain, *if there is a sequence of finite projections $p_1 \sqsubseteq p_2 \sqsubseteq \ldots$ such that $id_D = \bigsqcup_{i \in \mathbb{N}} p_i$.*

It is well-known that the cpo's D^τ are SFP-domains with the projections p_i^τ defined as follows:

$$p_i^\iota d = \begin{cases} d & \text{if } d \in \{0, \ldots, i\}, \\ \bot & \text{otherwise,} \end{cases}$$
$$p^{\sigma \to \tau} f = p^\tau \circ f \circ p^\sigma.$$

These projections are definable by closed PCF-expression P_i^τ, defined by:

$$P_i^\iota \equiv \lambda x^\iota . \text{if } x \leq \underline{i} \text{ then } x \text{ else } \Omega_\iota \text{ fi}$$
$$P_i^{\sigma \to \tau} \equiv \lambda y^{\sigma \to \tau} . \lambda x^\sigma . P_i^\tau (y \, (P_i^\sigma x))$$

Theorem 4.3 *Let $\tau = \sigma_1 \to \ldots \sigma_n \to \iota$ be a type of order 1 or 2 and let $f \in D^\tau$ be a logically sequential function. Then f is the lub of a directed set of definable elements.*

Proof: We know that $f = \bigsqcup_{i \in \mathbb{N}} p_i^\tau f$, where the p_i^τ are the finite projections. Hence it is sufficient to prove that $p_i^\tau f$ is definable for each $i \in \mathbb{N}$. Now note that $p_i^\tau g \, d_1 \ldots d_n = p_i^\iota (g \, (p_i^{\sigma_1} d_1) \ldots (p_i^{\sigma_n} d_n))$ for all $g \in D^\tau$, $d_1 \in D^{\sigma_1}, \ldots, d_n \in D^{\sigma_n}$. The set $p_i^{\sigma_1}(D^{\sigma_1}) \times \ldots \times p_i^{\sigma_n}(D^{\sigma_n})$ is finite, hence—by Theorem 4.1—there is a closed PCF-expression M such that

$$[\![M]\!] \, (p_i^{\sigma_1} d_1) \ldots (p_i^{\sigma_n} d_n) = f \, (p_i^{\sigma_n} d_1) \ldots (p_i^{\sigma_n} d_n)$$

for all $(d_1, \ldots, d_n) \in D^{\sigma_1} \times \ldots \times D^{\sigma_n}$ and this finally implies

$$p_i^\tau f = p_i^\tau \, [\![M]\!] = [\![P_i^\tau M]\!]. \qquad \square$$

Theorem 4.3 implies that sequentiality relations are sufficient to prove all observational congruences between closed expressions of order ≤ 3:

Theorem 4.4 *For each type τ let L^τ be the set of logically sequential elements of D^τ. If $\tau = \sigma_1 \to \ldots \to \sigma_n \to \iota$ is a type of order ≤ 3 and M, N are closed PCF-terms of type τ, then*

$$M \approx N \Leftrightarrow [\![M]\!] \, d_1 \ldots d_n = [\![N]\!] \, d_1 \ldots d_n \text{ for all } (d_1, \ldots, d_n) \in L^{\sigma_1} \times \ldots \times L^{\sigma_n}.$$

Proof :

'$\Rightarrow$': Assume that $[\![M]\!] \, d_1 \ldots d_n \neq [\![N]\!] \, d_1 \ldots d_n$ for some logically sequential elements $d_1, \ldots, d_n$. By Theorem 4.3, $d_1, \ldots, d_n$ can be approximated by definable elements. Hence the continuity of M and N implies that there are closed PCF-expressions $P_1, \ldots, P_n$ such that $[\![MP_1 \ldots P_n]\!] \neq [\![NP_1 \ldots P_n]\!]$, i.e. $M \not\approx N$.

'$\Leftarrow$': Follows immediately from the Context Lemma. $\qquad \square$

Theorem 4.4 nearly looks like a full abstraction result. A technical difficulty is that $L = (L^\tau)_{\tau \in Type}$ is not a λ–model because it is not extensional. But if $\hat{L}$ is the extensional collapse of L, then we obtain:

Theorem 4.5 *For all closed PCF–terms M, N of order ≤ 3*

$$M \approx N \quad \text{iff} \quad [\![M]\!] = [\![N]\!] \text{ in the model } \hat{L}.$$

Hence the model $\hat{L}$ is fully abstract for the sublanguage of closed expressions of order $\leq$ 3, (and this immediately generalizes to arbitrary expressions of order $\leq$ 3 whose only free variables are of order $\leq$ 2).

Unfortunately we do not know what happens at order $>$ 3. The proof of Theorem 4.1 heavily relies on the fact that the functions are of order 1 or 2, hence it doesn't give us any hint how to generalize the result to order $>$ 2. A similar difficulty appeared in [Plo80], where logical relations are used to characterize functions of order $\leq$ 2, which are definable in the *pure* λ-calculus. Plotkin succeeded to generalize the result to order $>$ 2 with the aid of *Kripke* logical relations. But his proof seems to be closely tailored to the pure λ-calculus, so we couldn't see how a similar idea might work in our case.

5. Conclusion

Of course this is not the first approach to characterize sequential functions.

(a) A model which is fully abstract for sequential PCF has been defined in [Mul86]. This model is generally criticized because of its 'syntactic flavor', which does not provide any new insight into the *logic* of the language.

(b) Models of *stable* and—most recently—*strongly stable* functions have been developed (cf. [BCL85, BE91, Cur86]). Stability rules out the *parallel or* operator, but still allows some first order functions which are not sequential (cf. [Cur86]). Strong stability fills this gap, but hasn't made much progress on second order functions (cf. the conclusion of [BE91]).

(c) An alternative approach are *sequential algorithms* ([BCL85, Cur86]), which lead to a model which is *not extensional* (and hence can't be fully abstract for PCF). Most recently, a variant of sequential algorithms has been used in [CF91] to obtain a fully abstract model for a new 'sequential' *extension* of PCF.

How does our result relate to these approaches?

Technically, there is some similarity with (a), but our semantic characterization of the relations (Theorem 3.4) *does* give us an insight into the logic of PCF: It provides a method for proving observational congruences and we know that this method is in some sense complete. On the other hand we have achieved more than (b), because we could completely characterize second order functions. Finally, (c) is of course a result about a different language.

References

[BCL85] G. Berry, P.-L. Curien, and J.-J. Lévy. Full abstraction for sequential languages: The state of the art. In M. Nivat and J. C. Reynolds, editors, *Algebraic Methods in Semantics*, pages 89–132. Cambridge Univ. Press, 1985.

[BE91] A. Bucciarelli and T. Ehrhard. Sequentiality and strong stability. In 5th *Symposium on Logic in Computer Science*, pages 138–145. IEEE, 1991.

[CF91] R. Cartwright and M. Felleisen. Observable sequentiality and full abstraction. Draft, Dept. of Comp. Sc., Rice University, Houston TX, 1991.

[Cur86] P.-L. Curien. *Categorical Combinators, Sequential Algorithms and Functional Programs*. Research Notes in Theoretical Computer Science. Pitman, Wiley, 1986.

[Mul86] K. Mulmuley. *Full Abstraction and Semantic Equivalences*. ACM Doctoral Dissertation Award. MIT Press, 1986.

[Plo77] G. D. Plotkin. LCF considered as a programming language. *Theoretical Computer Science*, 5:223–256, 1977.

[Plo80] G. D. Plotkin. Lambda-definability in the full type hierarchy. In J. Seldin and J. Hindley, editors, *To H. B. Curry: Essays on Combinatory Logic, Lambda Calculus and Formalism*, pages 363–374. Academic Press, 1980.

I-Categories and Duality

M.B.Smyth
Dept.of Computing
Imperial College
London SW7 2BZ

Introduction

In recent joint work with A. Edalat[ES91] we developed a general approach
to the solution of domain equations, based on information system ideas. The
basis of the work was an axiomatization of the notion of a category of information systems, yielding what we may call an "information category", or
I-category for short. We begin this paper with an exposition of the I-category
work. In the remainder of the paper we consider duality in I-categories, as
the setting for studying initial algebra/final algebra coincidence. We then
look at induction and coinduction principles in the light of these ideas.

To amplify the preceding a little, we note that the existing treatments of information systems, following [Sco82] and [LW84], make use of a global ordering
of the objects of the category (the information systems) in order to "solve"
domain equations by the ordinary cpo fixed point theorem. In the I-category
approach, an initial algebra characterization of the solutions is obtained, by
making use of a global ordering $\sqsubseteq$ of *morphisms* in addition to the ordering
$\trianglelefteq$ of objects. (In the usual cases, where morphisms are "approximable relations" between tokens, the global ordering is essentially set inclusion; more
precisely, we have that $(f : A \to B) \sqsubseteq (f' : A' \to B')$ if $A \trianglelefteq A'$, $B \trianglelefteq B'$
and $f \subseteq f'$.) Moreover the axiomatic formulation enables us to handle many
examples besides the usual categories of domains, in a unified manner: for
example, "domain equations" over Stone spaces, via Boolean algebras as information systems. In the present exposition, we attempt to clarify the relation between the I-category approach and an established method of domain
equation solution, namely the O-category method, using the Basic Lemma of
[SP82] as a key.

The initial algebra-final coalgebra coincidence has been studied recently by
Freyd [Fre91]. He treats the coincidence as an independent property of cat-

egories: a category is *algebraically compact* if every (covariant) endofunctor
has an initial algebra and a final coalgebra and these are canonically isomorphic. Our viewpoint is somewhat different. In the cases in which we
are interested, the initial-final algebra coincidence is simply a by-product of
the colimit-limit coincidence well-known in domain theory [Sco72], and studied in an abstract setting in [SP82] . This means that we are satisfied if
the initial-final algebra coincidence obtains only under the sort of conditions
which prevail in O-category theory: e.g. that the functors involved be continuous (in the appropriate sense) and strictness-preserving. Our problem
then is whether anything can be done with those categories which do *not*
have colimit-limit coincidence, but which are amenable to I-category treatment, such as Stone spaces or, for that matter, sets. The answer we suggest,
following [Smy85] (but now in an I-category setting), is that such categories
may be extendible, by addition of new morphisms, to categories in which
colimit-limit coincidence does hold. In the case of Stone spaces, the added
morphisms are certain many-valued maps, of a kind familiar from studies of
non-determinism.

As a (rather speculative) application we consider induction and coinduction.
The idea is that, if significant recursively defined types are both initial and
final (as algebra/coalgebra respectively, and when looked at in the right category), both induction and coinduction principles should be available for them.
By way of illustrating the ideas about extension of I-categories, we consider
two non-domain examples: natural numbers, $\mathbb{N}$, and Cantor space, 2^ω. The
case of $\mathbb{N}$ is straightforward, but some obscurity remains with respect to induction for Cantor space. What is clear is that much remains to be done.

Conventions, Notations. By a *(ω -)dcpo* we understand a poset having
lubs of increasing sequences; a *cpo* is a dcpo with a least element. We say
that a partial function $f : D \to E$ (D, E dcpo's) is *continuous* if, for any
$x_0 \sqsubseteq x_1 \sqsubseteq \ldots \in D$ such that $f(x_i)$ is defined for all i, $f(\bigsqcup x_i)$ is defined and is
equal to $\bigsqcup f(x_i)$.

We use ; for composition of morphisms: $f; g$ is $\cdot \xrightarrow{f} \cdot \xrightarrow{g} \cdot$. The
classes of objects and of morphisms of a category $\mathbf{K}$ are denoted $\mathrm{Obj}(\mathbf{K})$,
$\mathrm{Mor}(\mathbf{K})$.

1. I-categories

We begin by recalling the Basic Lemma of [SP82] .

Notation.   If $\perp$ is an initial object of a category $\mathbf{K}$, we denote by $\perp_A$ the morphism from $\perp$ to object A. If $\Delta = D_0 \xrightarrow{f_0} D_1 \xrightarrow{f_1} \cdots$ is an ω-chain in $\mathbf{K}$ and $\mu : \Delta \rightarrow A$ a (co)cone from Δ to A, then Δ^- is the ω-chain $D_1 \xrightarrow{f_1} D_2 \xrightarrow{f_2} \cdots$ and $\mu^- : \Delta^- \rightarrow A$ is the cone $\langle \mu_{n+1} \rangle_{n \in \omega}$; further, if $F : \mathbf{K} \rightarrow \mathbf{L}$ is a functor, then $F\Delta$ is the ω-chain $FD_0 \xrightarrow{Ff_0} FD_1 \xrightarrow{Ff_1} \cdots$; similarly for $F\mu$.

Lemma 1.1 ("Basic Lemma") . *Let* $\mathbf{K}$ *be a category with initial object* $\perp$, *and* F *an endofunctor of* $\mathbf{K}$. *Let* Δ *be the* ω-*chain* $\langle F^n(\perp), F^n(\perp_{F\perp}) \rangle_{n \in \omega}$:

$$\Delta : \perp \xrightarrow{\perp_{F\perp}} F(\perp) \xrightarrow{F(\perp_{F\perp})} F^2(\perp) \cdots$$

Suppose that both $\mu : \Delta \rightarrow A$ *and* $F\mu : F\Delta \rightarrow FA$ *are colimiting cones. Then an initial* F-*algebra exists and can be taken as* (A, α), *where* $\alpha : FA \rightarrow A$ *is the mediating morphism from* $F\mu$ *to* μ^-.

Our working hypothesis is that initial solutions of domain equations arise as instantiations of the Basic Lemma. Thus we will build up the defining conditions for an I-category step-by-step, by seeing what is required for an instantiation of the Basic Lemma.

In accordance with the information system approach, we are assuming that we have available (in the ambient category $\mathbf{I}$) a distinguished class Inc of *inclusion* morphisms which form a subcategory of $\mathbf{I}$ (that is, identities are inclusions, and Inc is closed under composition), such that there is at most one inclusion in each hom-set of $\mathbf{I}$. Thus Inc determines a pre-order $\trianglelefteq$ of $\mathbf{Obj(I)}$; we assume that $(\mathbf{Obj(I)}, \trianglelefteq)$ is actually a partial order, indeed a cpo. For our purposes we additionally assume a partial order $\sqsubseteq$ of $\mathbf{Mor(I)}$, the maps **dom**, **cod** being monotonic. To solve a domain equation $D = FD$ in this setting, one requires a functor F which can be regarded as a continuous (self-)map of the cpo $\mathbf{Obj(I)}$. Ordinarily, one now simply invokes the cpo fixed-point theorem to complete the "solution". However, we are looking for an initial algebra characterization, via the Basic Lemma (BL); for this, we need in particular to be able to take the increasing sequence $\perp \trianglelefteq F\perp \trianglelefteq \cdots$ (where $\perp$ is given as the least element of $\mathbf{Obj(I)}$) with its lub as an instance of μ in the statement of BL.

But now the questions arise: why should $\perp$ (given as initial with respect to Inc) be initial in $\mathbf{I}$? Why should

$$\Delta' : \perp \xrightarrow{\text{in}} F\perp \xrightarrow{\text{in}} F^2\perp \longrightarrow \cdots$$

(where $in(A, B)$, or simply in, denotes the inclusion morphism from A to B) be an instance of Δ as in BL ? And, why should $\mu' : \Delta' \to \bigsqcup F^n(\bot)$, where μ'_n is the inclusion, be colimiting in $\mathbf{I}$?

As to the first question, there is in fact no reason why $\bot$ should be initial in $\mathbf{I}$. What we have to do is cut $\mathbf{I}$ down to its subcategory $\mathbf{I}_S$ of strict morphisms, where a morphism $f : A \to B$ is said to be *strict* if the diagram

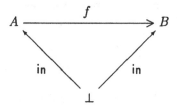

commutes. Clearly, inclusions are strict, the strict morphisms give a subcategory $(\mathbf{I}_S)$, and $\bot$ is initial in $\mathbf{I}_S$. Thus we will work in $\mathbf{I}_S$ to get an initial algebra.

The second question is easily disposed of: Let us require that the functor F preserve inclusions. Then $F^n(\bot_{F\bot})$ is $in(F^n\bot, F^{n+1}\bot)$ for each n, and Δ' is an instance of Δ.

The third question needs more work. What we would really like to show is that if

$$\Gamma = C_0 \xrightarrow{\text{in}} C_1 \xrightarrow{\text{in}} \cdots$$

and ν is the cone of inclusions from Γ to $C = \bigsqcup C_i$, then Γ is colimiting in $\mathbf{I}_S$. Suppose that $\rho : \Gamma \to D$ is another cone from Γ. We need a mediating morphism h from ν to ρ. For h to be mediating means:

$$\rho_i = in(C_i, C); h \qquad\qquad \text{(all } i\text{)}.$$

At this point we would like to invoke the global ordering of morphisms, and eliminate the $in(C_i, C)$ by taking lub's:

$$
\begin{aligned}
\bigsqcup \rho_i &= \bigsqcup in(C_i, C); h \\
&= in(C, C); h \\
&= h
\end{aligned}
$$

Now, for this calculation to make sense, we need to know (among other things) that $\langle in(C_i, C)\rangle_i$, is an increasing sequence. This is achieved most economically by postulating that

$$A \trianglelefteq B \quad\Rightarrow\quad in(A, E) \sqsubseteq in(B, E);$$

that is, in is monotonic (in its first argument). Notice that this will also
ensure that $\langle \rho_i \rangle$ is increasing, since we will have:

$$\rho_i = \text{in}(C_i, C_{i+1}); \rho_{i+1}$$
$$\sqsubseteq \text{in}(C_{i+1}, C_{i+1}); \rho_{i+1} = \rho_{i+1}$$

– provided we have also monotonicity of ; (1st. argument). If we now
strengthen these conditions to *continuity* of in and of ; (in the first argument),
we will have

$$\bigsqcup \text{in}(C_i, C) = \text{in}(\bigsqcup C_i, C)$$
$$= \text{in}(C, C)(= \text{id}_C),$$

and the calculation evidently goes through (establishing that, if h is mediat-
ing, it must be $\bigsqcup \rho_i$). To show that $\bigsqcup \rho_i$ *is* mediating, we expect to argue:

$$\rho_i = \text{in}(C_i, C_{i+1}); \rho_{i+1} = \text{in}(C_i, C_{i+2}); \rho_{i+2} = \cdots$$

so

$$\rho_i = \bigsqcup_k (\text{in}(C_i, C_{i+k}); \rho_{i+k})$$
$$= \text{in}(C_i, C); \bigsqcup \rho_i$$

Here we needed to presuppose that in and ; are continuous in the *second*
argument. (Notice that these assumptions also yield, by an easy argument,
that the least upper bound of a chain of strict morphisms is strict, from which
we can conclude that $\bigsqcup \rho_i$ is in $\mathbf{I}_S$.)

Bringing together all that we have needed to assume about $\mathbf{I}$, we arrive at:

Definition 1.1 A *(complete) I-category* is a 5-tuple $(\mathbf{I}, \text{Inc}, \trianglelefteq, \sqsubseteq, \perp)$ where $\mathbf{I}$
is a category, Inc is a distinguished class of morphisms of $\mathbf{I}$ called *inclusions*,
$\trianglelefteq, \sqsubseteq$ orderings of $\text{Obj}(\mathbf{I})$, $\text{Mor}(\mathbf{I})$ respectively, and $\perp$ a distinguished object,
satisfying:

- there is at most one inclusion, $\text{in}(A, B): A \to B$, for each $A, B \in \text{Obj}(\mathbf{I})$
  (and we consider in to be a partial map from $\text{Obj}(\mathbf{I})^2$ to $\text{Mor}(\mathbf{I})$);

- $\text{in}(A, B)$ exists iff $A \trianglelefteq B$;

- Inc (taken with $\text{Obj}(\mathbf{I})$) is a subcategory of $\mathbf{I}$;

- $(\text{Obj}(\mathbf{I}), \trianglelefteq)$ is a cpo with least element $\perp$, and $(\text{Mor}(\mathbf{I}), \sqsubseteq)$ is a dcpo;

- the maps dom and cod are monotonic, while the partial maps in and ; are continuous.

I-categories that are not complete can also be considered: see the remarks at the end of this section. In this paper we consider in detail only the complete case, and I-categories mentioned should therefore be assumed complete, even if this is not mentioned explicitly.

The subcategory I_S of *strict* morphisms of an I-category I is defined as previously. We say that a functor of I-categories is *standard* if it preserves inclusions, and $\omega - continuous$ if it is continuous w.r.t. the ordering of objects (equivalently, if it preserves ω-colimits of inclusions). An easy verification establishes the following:

Proposition 1.1 *A standard functor preserves strictness of morphisms (that is, any standard $F : I \to J$ cuts down to $F : I_S \to J_S$).* $\square$

The preceding analysis yields:

Theorem 1.2 (Initial Algebra Theorem.) *Let F be a standard, $\omega - con$-tinuous endofunctor of the complete I-category I. Then F has an initial algebra over I_S, namely $\mathrm{id}_{\mu F}$, whre μF is the least fixed point of F in the cpo $\mathrm{Obj}(I)$.* $\square$

To get more of a feel for Definition 1.1, it is useful to look at a few basic facts (assume given a complete I-category $(I, \mathsf{Inc}, \trianglelefteq, \sqsubseteq, \bot)$):

Proposition 1.2 *The maps dom and cod are continuous.*

Proof Let $< f_i : A_i \to B_i >$ be an increasing sequence of morphisms. By monotonicity of dom, $< A_i >_i$ is an increasing sequence. By continuity of in, $\bigsqcup \mathrm{id}_{A_i}$ (that is, $\bigsqcup \mathrm{in}(A_i, A_i)$) is id_A, where $A = \bigsqcup A_i$. Then by continuity of composition, $\mathrm{id}_A; \bigsqcup f_i$ is defined (and equals $\bigsqcup(\mathrm{id}_A; f_i)$). Thus $\mathrm{dom}(\bigsqcup f_i) = A$; cod is, of course, similar. $\square$

Proposition 1.3 $A \trianglelefteq B$ *iff* $\mathrm{id}_A \sqsubseteq \mathrm{id}_B$.

Proof IF: Monotonicity of dom (or cod).
ONLY IF: Monotonicity of in. $\square$

Proposition 1.4 $f : A \to B \ \sqsubseteq \ g : C \to D$ *if and only if:* $A \trianglelefteq C, B \trianglelefteq D$ *and the diagram*

$$
\begin{array}{ccc}
A & \xrightarrow{\text{in}} & C \\
f \downarrow & \sqsubseteq & \downarrow g \\
B & \xrightarrow{\text{in}} & D
\end{array}
$$

weakly commutes (that is, $f ; \mathrm{in}(B, D) \sqsubseteq \mathrm{in}(A, C) ; g$ *).*

Proof ONLY IF: Suppose $f \sqsubseteq g$. Then $A \trianglelefteq C, B \trianglelefteq D$ (by monotonicity of dom, cod); and

$$\mathrm{id}_A ; f ; \mathrm{in}(A, B) \ \sqsubseteq \ \mathrm{in}(A, C) ; g ; \mathrm{id}_D \quad \text{(monotonicity of in and of ;)}$$

so the diagram weakly commutes.

IF:The calculation is:

$$
\begin{aligned}
f \ = \ & f ; \mathrm{id}_B & \sqsubseteq \ & f ; \mathrm{in}(B, D) \\
\sqsubseteq \ & \mathrm{in}(A, C) ; g & \sqsubseteq \ & \mathrm{id}_C ; g \qquad = \ g
\end{aligned}
$$

$\square$

Clearly, the set of primitives in our definition of an I-category is not the most economical possible; in particular, $\trianglelefteq$ could be dispensed with. Also, Proposition 4 shows that the global ordering of morphisms can be defined in terms of the (local) ordering of hom-sets, together with $\trianglelefteq$; this is the approach taken in [ES91]. As compared with what is done here, the definition of a complete I-category in [ES91] adds two extra conditions: a requirement that each morphism of the form $\mathrm{in}(\bot, A)$ is least in its hom-set; and an axiom which implies that inclusions are monomorphisms (Axioms 1(ii) and 1(iii) of [ES91]). These are not useful for our purposes here. On the other hand, it will prove convenient to have $\trianglelefteq$, $\sqsubseteq$, and Inc explicit in the structure of an I-category when we come to consider situations in which $\sqsubseteq$, or Inc, is varied, while $\trianglelefteq$ remains fixed.

Example 1.1 The I-category of *sets* is (**Set**, Inc, $\subseteq$, $\sqsubseteq$, $\emptyset$) where Inc consists of ordinary set inclusions and, for morphisms $f : A \to B$, $g : C \to D$ we have: $f \sqsubseteq g$ iff $A \subseteq C$, $B \subseteq D$ and $f \subseteq g$. The conditions for an I-category are satisfied trivially. Product and coproduct can straightforwardly be introduced as standard, ω–continuous functors. The same is true for the *finite* power-set functor (which maps each set to the collection of its finite subsets), while the unrestricted power-set functor is clearly not ω–continuous. A (severely) restricted exponentiation is available: specifically, given a fixed finite set A, the functor which takes a set S to S^A is standard and ω–continuous.

Example 1.2   Boolean algebras.   Similarly, we can define an   I-category $(\mathbf{BA^+}, \mathsf{Inc}, \trianglelefteq, \sqsubseteq, \mathbf{2})$ of Boolean algebras, with $\trianglelefteq$ as the subalgebra relation. Notice that the one-element algebra is incomparable (under $\trianglelefteq$) with all the other algebras; hence we fix the 2-element algebra $\mathbf{2} = (\{0,1\}; \wedge, \vee, \neg)$ as the least algebra to be considered, taking $\mathbf{BA^+}$ as the category of Booleans algebras $\mathbf{B}$ such that $\mathbf{2} \trianglelefteq \mathbf{B}$ (with Boolean homomorphisms). Our primary interest in Boolean algebras lies in viewing them as information systems for Stone spaces (via Stone duality); since we exclude the one-element algebra, we actually capture the *non-empty* Stone spaces.

It is easy to define, for example, product and coproduct as ω−continuous functors over $\mathbf{BA^+}$. But there is a slight technical difficulty regarding standardness of the functors. We would expect to construct, say, the coproduct $A + B$ (representing the product of the corresponding Stone spaces) as the algebra of terms built from the elements of A and B, subject to certain equations. Specifically, the terms can be disjunctions of pairs (a,b), $a \in A$, $b \in B$, with equations such as

$$\neg(a,b) = (\neg a, 1) \wedge (1, \neg b)$$

(as well as commutativity, etc). Now a construction of the coproduct via equivalence classes of terms is unlikely to give us (assuming $A \trianglelefteq A'$, $B \trianglelefteq B'$) $A + B$ as literally a subset of $A' + B'$; the equivalence classes will typically be enlarged in going from $A + B$ to $A' + B'$. This problem disappears if, in the spirit (and usual practice) of information system work, we present Boolean algebras as *pre*-ordered sets with associated Boolean operations, thus obviating the need to consider equivalence classes. We refer to [ES91] for the details of this,as also of the treatment of morphisms in the pre-ordered algebra setting. On the morphisms as "approximable relations", see further Section 2.3. below.

Remark 1.1 It is easy to see that I-categories are necessarily O-categories. But this does not at all mean that I-category theory is subsumed by O-category theory, since inclusions, with which we solve domain equations in I-categories, need not be embeddings (that is, need not have adjoints).

Remark 1.2 Complete I-categories made their appearance (sketchily) in [Smy89], and some details were added in the M.Sc. dissertation [Eda89] of the author's student A. Edalat. The much more comprehensive account [ES91] extends the comparison between I-categories and information orderings in various ways: notably, the distinction is made between I-categories and complete I-categories (analogous to that between posets and cpo's), leading to notions of *ideal completion* of I-categories and *algebraic* I-categories.

Concrete I-categories are also introduced, as having objects which are structured sets of "tokens"; we take these as our preferred notion of *information category*. Finally, extensive examples are provided.

2. Duality

2.1. Colimit/limit coincidence. Dual I-categories.

In the O-category setting, the coincidence of initial and final algebras is an immediate consequence of the colimit-limit coincidence, the elucidation of which was one of the main objects of [SP82]. Suppose that $\mathbf{K}$ is an O-category having the terminal object $\perp$ which is also initial with respect to $\mathbf{KE}$ (and therefore $\mathbf{KS}$), cf. Theorem 1 of [SP82], and let $F : \mathbf{K} \to \mathbf{K}$ be a functor satisfying an appropriate continuity condition. Replacing the embeddings occurring in the standard construction of the initial algebra $\alpha : FA \to A$ by the corresponding projections, we get a construction (in terms of limiting diagrams of projections) of $\alpha^{-1} : A \to FA$ as final coalgebra. Notice that the limit diagrams arising here are colimit diagrams of $\mathbf{K}^{op}$, so that we have an instance of the Basic Lemma again; but not only that, they are diagrams of *embeddings* of $\mathbf{K}^{op}$, since the embeddings of $\mathbf{K}^{op}$ are the same thing as projections of $\mathbf{K}$ ([SP82],Section 3).

To achieve something similar in the context of an I-category, we will have to postulate, for each inclusion, a corresponding projection, which we shall call a "subsumption". More precisely, we say that an I-category $\mathbf{K}$ *has subsumptions* if for every inclusion $e : A \to B$ there is a morphism $p : B \to A$ such that (e, p) is a projection pair; the class of subsumptions (that is, of projections which thus correspond to inclusions) will be denoted sub. Concerning this notion we have the following result:

Theorem 2.1 *Suppose that the complete I-category* $(\mathbf{K}, \mathsf{Inc}, \trianglelefteq, \sqsubseteq, \perp)$ *has subsumptions. Then* $(\mathbf{K}^{op}, \mathsf{sub}, \trianglelefteq, \sqsubseteq, \perp)$ *is also an I-category (which will be called the* dual *of* $\mathbf{K}$*).*

Proof What has to be shown is the continuity of the partial function sub, where $\mathsf{sub}(A, B)$ is the subsumption from B to A (defined just in case $A \trianglelefteq B$). Consider the monotonicity of sub. To begin with, we have the special case $X \trianglelefteq Y \Rightarrow \mathsf{id}_X \sqsubseteq \mathsf{sub}(X, Y)$, since:

$$\mathsf{id}_X = \mathsf{in}(X, Y); \mathsf{sub}(X, Y) \sqsubseteq \mathsf{id}_Y; \mathsf{sub}(X, Y).$$

Similarly, $X \trianglelefteq Y \Rightarrow \mathsf{sub}(X, Y) \sqsubseteq \mathsf{id}_Y$. Then for the general case, given $A \trianglelefteq B \trianglelefteq D$ and $A \trianglelefteq C \trianglelefteq D$, we have:

$$\mathsf{sub}(A, B) \quad \sqsubseteq \quad \mathsf{sub}(B, D); \mathsf{sub}(A, B)$$

$$= \mathsf{sub}(A, D)$$
$$= \mathsf{sub}(C, D); \mathsf{sub}(A, C)$$
$$\sqsubseteq \mathsf{sub}(C, D)$$

[Note: we have been taking ; in the sense of $\mathbf{K}$, not $\mathbf{K}^{\mathrm{op}}$.] To complete the proof of continuity, one then shows that, if $A_0 \trianglelefteq A_1 \trianglelefteq \dots$ and $B_0 \trianglelefteq B_1 \trianglelefteq \dots$, where $A_n \trianglelefteq B_n$ (all n), then $\bigsqcup_n \mathsf{sub}(A_n, B_n)$ is the subsumption corresponding to the inclusion $\bigsqcup_n \mathsf{in}(A_n, B_n)$; this is an easy verification.   □

2.2.   Initial/final algebras.

Given an I-category with subsumptions, we can obtain final coalgebras (for suitable functors) by constructing initial algebras over the dual category. We need the notion of "costrict " morphisms (= strict in the dual category): a morphism $f : A \to B$ is *costrict* if $f; \mathsf{sub}(\Delta, B) = \mathsf{sub}(\Delta, A)$. Of course, in the usual categories of (information systems for) domains, all morphisms are costrict. An I-category $\mathbf{K}$ and its dual $\mathbf{K}^{\mathrm{op}}$ are identical as cpo's; thus an endofunctor F of $\mathbf{K}$ is continuous iff it is continuous when considered as an endofunctor of $\mathbf{K}^{\mathrm{op}}$. Also, given a suitable endofunctor F, the construction of the initial algebra over $\mathbf{K}$ (with respect to strict morphisms of $\mathbf{K}$) is identical, as a cpo construction, with that of the initial algebra over $\mathbf{K}^{\mathrm{op}}$ (with respect to strict morphisms of $\mathbf{K}^{\mathrm{op}}$). We deduce:

Theorem 2.2 *Let $\mathbf{K}$ be a complete I-category which has subsumptions, and F an ω-continuous endofunctor of $\mathbf{K}$ which preserves both inclusions and subsumptions. Then F has both an initial algebra (with respect to strict morphisms) and a final coalgebra (with respect to costrict morphisms), which coincide.*   □

Regarding the conditions on $\mathbf{F}$ in this Theorem, we may note that a standard functor which is $\sqsubseteq$-monotonic necessarily preserves subsumptions, so that preservation of subsumptions might be replaced by the rather natural $\sqsubseteq$-monotonicity (it is perhaps curious that this condition played no role previously). One could then think of going further and replacing both this and ω-continuity by O-continuity, or continuity on hom-sets (cf. [SP82]). However, we want here to pursue the enquiry in another direction.

2.3.   Conservative extensions.

On the face of it, Theorem 2 has nothing to tell us about the many interesting I-categories that do not have subsumptions. However, we shall see that it may be possible to obtain some of the benefits of the initial/final algebra coincidence even for an I-category $\mathbf{K}$ which lacks subsumptions, by embedding $\mathbf{K}$ in an I-cateory which has them. To obtain useful information about $\mathbf{K}$ in this way, we require an extension which "adds" the necessary subsumptions to $\mathbf{K}$, but otherwise disturbs $\mathbf{K}$ as little as possible.

Definition 2.1 Let $\mathbf{K},\mathbf{K}'$ be (complete) I-categories. We say that $\mathbf{K}'$ is a *conservative extension* of $\mathbf{K}$ if $\mathsf{Obj}(\mathbf{K}) = \mathsf{Obj}(\mathbf{K}'), \mathsf{Inc}_K = \mathsf{Inc}_{K'}, \perp_K = \perp_{K'}$, while $\mathsf{Mor}\mathbf{K}$ is a sub-dcpo of $\mathsf{Mor}\mathbf{K}'$.

Proposition 2.1 *Suppose that $\mathbf{K}'$ is a conservative extension of $\mathbf{K}$. Let F' be a standard, ω-continuous endofunctor of $\mathbf{K}'$ which restricts to an endofunctor of $\mathbf{K}$. Then F, F' have the same initial algebra.*

Proof Evident, since the construction of the initial algebra involves only ingredients which are common to the two categories.   □

While, in this sense, a conservative extension does not provide us with new initial algebras, the position is quite different for final (co)algebras.

Example 2.1 Let the I-category $\mathbf{Set}_P$ be defined as $\mathbf{Set}$, except that $\mathbf{Set}_P$ admits partial maps as morphisms. Clearly, $\mathbf{Set}_P$ is a conservative extension of $\mathbf{Set}$, and $\mathbf{Set}_P$ has subsumptions (which are partial identity maps). $\mathbf{Set}_P$ thus exhibits coincidence of initial and final algebras. As a simple example, we may consider how the canonical solution of

$$X = 1 + X$$

is final in $\mathbf{Set}_P$, but not in $\mathbf{Set}$. The canonical solution, written as a coalgebra, takes the form

$$\mathsf{pred} : \mathbb{N} \to 1 + \mathbb{N}$$

(predecessor). Let $g : S \to 1 + S$ be another coalgebra (assume for simplicity that g is total); it will be convenient to view 1 in this context as the error set. Then it is easy to see that a homomorphism h from g to pred is obtained if we put $h(x) = n$, where n is the least $k \in \mathbb{N}$ such that $g^k(x) = \mathbf{error}$. Clearly, we have to allow that $h(x)$ is undefined in the case that $g^k(S) \subseteq S$ for all k; thus we have to work in $\mathbf{Set}_P$ rather than $\mathbf{Set}$.

The next example is perhaps rather less familiar:

Example 2.2 Boolean algebras/Stone spaces. Consider Cantor space 2^ω, given by

$$\begin{aligned} C &= \mathbf{2} \times C \\ \text{or}\quad C &= C + C, \end{aligned} \tag{1}$$

in terms of Stone spaces, or its Stone dual given by

$$\begin{aligned} B &= E \times B \\ \text{or}\quad B &= B + B, \end{aligned} \tag{2}$$

where E is the four-element Boolean algebra. Straightforwardly, we have an initial algebra solution of (2) over $\mathbf{BA^+}$, hence by Stone duality a final coalgebra solution of (1) (over non-empty Stone spaces). What we do not have is a final solution of (1) or an initial solution of (2); we may attribute these deficiencies to the lack of subsumptions in $\mathbf{BA^+}$.

How might we extend $\mathbf{BA^+}$ to an I-category which has subsumptions? Suppose that B_1, B_2 are Boolean algebras, with $B_1 \trianglelefteq B_2$; let e be the injection of B_1 in B_2. In order to find a "projection" corresponding to e, it will be convenient to take the ideal completions $\overline{B}_1$, $\overline{B}_2$ of B_1, B_2, and extend e (by continuity) to an embedding $\overline{e} : \overline{B}_1 \to \overline{B}_2$ of domains. (Thus we may regard $\overline{B}_i, i = 1, 2$, as the lattice, or frame, of open sets of the Stone space of B_i.) Let $\overline{p} : \overline{B}_2 \to \overline{B}_1$ be the projection adjoint to $\overline{e}$. As with any projection, $\overline{p}$ preserves meets (we shall only be concerned with finite meets) and *directed* joins. Mappings between frames which preserve finite meets and directed (rather than arbitrary) joins are of great interest: they are dual to certain many-valued spatial maps, namely the upper semi-continuous (usc) maps. Here we may take a many-valued map $f : X \to Y$, where X, Y are (say) Hausdorff spaces, to be *usc* provided:

(i) $\{x \in X \mid f(x) \subseteq O\}$ is open, for O open in Y

(ii) $f(x)$ is a compact subset of Y, for all $x \in X$.

(Many-valued usc maps are discussed, for example, by Kuratowski [Kur61, §39] and Berge [Ber59]. For further information on the duality and its computational significance, see [Smy83, Smy85]. Let us just remark that the usc maps may be understood as the topological version of Dijkstra's boundedly non-deterministic functions [Dij76], and the corresponding mappings of frame as "healthy predicate transformers".) We can with advantage restrict a little further the class of multimaps we consider, by imposing

(iii) $f(x)$ is non-empty, for all $x \in X$.

The corresponding requirement for mappings between frames (predicate transformers) is strictness, that is, preservation of 0 (or $\bot$). These frame mappings
are, in effect what we are taking for the conservative extension of **BA+**. To fit
into the information system framework, however, we need to formulate them
as "approximable relations" between Booleans algebras. The details of this
are unimportant here, but, to be specific, we can take relations $R \subseteq B \times B'$
between Booleans algebras B, B' satisfying:

(i) $0R0$

(ii) $a \geq b, bRb', b' \geq c' \quad \Rightarrow \quad aRc'$

(iii) $aRb', aRc' \quad \Rightarrow \quad aRb' \vee c'$

(iv) $1R1$

(v) $aRa', bRb' \quad \Rightarrow \quad a \wedge bRa' \wedge b'$

(vi) $0Rb' \quad \Rightarrow \quad b' = 0$

Conditions (i)-(iii) here say that we have a Scott-continuous map from $\overline{B}$
to $\overline{B'}$, (iv)-(v) gives preservation of finite meets, and (vi) gives strictness. In
this way we get an I-category **BA*** which may be considered as a conservative
extension of **BA+**, and which has subsumptions. The detailed formulation
of this is not too important here, as when we come to the application (in-
duction/coinduction principles) we will work with Stone spaces and usc maps
rather than Boolean algebras and relations satisfying (i)-(vi).

3. Induction and Coinduction

We start with the abstract Induction Principle for initial algebras (over a
category K):

> Suppose that $\alpha : FA \to A$ is an initial algebra for $F : K \to$
> K. Then any $F-$algebra morphism $m : B \to A$ which is monic
> (as a morphism of K) is an isomorphism. Diagrammatically, the
> commuting of:

$$
\begin{array}{ccc}
FB & \longrightarrow & B \\
\downarrow {\scriptstyle Fm} & & \downarrow {\scriptstyle m} \\
FA & \xrightarrow{\;\alpha\;} & A
\end{array}
$$

where α is an initial $F-$algebra, implies that m is an isomorphism.

As a concrete instance, if α is the natural numbers initial algebra (over **Set**),
this reduces to the statement that, if B is a subset of $\mathbb{N}$ which contains 0
and is closed under successor, then $B = \mathbb{N}$. Details may be found in [LS81].
The above formulation was suggested by Plotkin in his capacity as referee

of [LS81], as a replacement for a previous version due to the present author. (That version, which may be found in [Smy79],was formulated in terms of characteristic functions rather than subobjects, and appears to be much more restricted in scope than the above.)

Dually, we have the Coinduction Principle for final algebras:

Suppose that the diagram

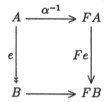

commutes, where α^{-1} is a final $F-$coalgebra, and e is epic. Then e is an isomorphism.

Again taking the natural numbers (now over $\mathbf{Set^P}$), with α^{-1} as pred, it is not difficult to derive the following concrete version:

Suppose that R is an equivalence relation over $\mathbb{N}$ such that

$$xRy \Rightarrow x = y = 0 \text{ or } \mathsf{pred}(x)R\mathsf{pred}(y) \qquad (3)$$

(where the second disjunct on the right is taken to imply that neither x nor y is 0). Then R is the identity relation.

This concrete principle may be put in a more convenient (and familiar) form, in which the condition that R be an equivalence relation is dropped. To see this, one verifies, first, that if an arbitrary binary relation R satisfies (3), then so does its reflexive symmetric transitive closure R^{RST}; and, secondly, that if $R^{\mathrm{RST}} = I$ (the identity relation), then $R \subseteq I$. This means that (3) can be reformulated:

Suppose that R is a binary relation over $\mathbb{N}$ satisfying (3). Then:

$$xRy \Rightarrow x = y.$$

In a completely analogous way, we can derive the Coinduction Principle for Cantor space:

Suppose that R is a binary relation over 2^ω such that:

$$xRy \quad \Rightarrow \quad \mathsf{hd}(x) = \mathsf{hd}(y) \ \& \ \mathsf{tl}(x)R\mathsf{tl}(y)$$

Then: $xRy \Rightarrow x = y$.

The theme in which we are mainly interested here is the idea that *both* induction and coinduction principles are available for typical "inductive" types (as a consequence of initial/final algebra coincidence). We therefore turn to the question of *induction* for Cantor space. The following is a slight rearrangement of an induction principle ("Baseless Induction") proposed by Iain Stewart [Ste89] (quite independently of generalities about initial algebras and the like):

Let X be a closed, non-empty subset of 2^ω such that

$$0.X \subseteq X \ \& \ 1.X \subseteq X$$

that is: $\quad x \in X \quad \Rightarrow \quad 0x \in X \ \& \ 1x \in X \qquad (4)$

Then: $X = 2^\omega$

It may be worth noting that a further reformulation of Baseless Induction yields:

Any non-empty subset X of 2^ω satisfying

$$0.X \subseteq X \ \& \ 1.X \subseteq X$$

is dense

Thus, if the identification of liveness properties with dense subsets of the space of state sequences [AS85] is accepted (it can hardly be said that there is general agreement on this yet), Baseless Induction may be of use in formulating liveness properties.

On a first inspection, Baseless Induction seems to conform well with the initial algebra account of induction. For a monomorphism m of a suitable kind (an embedding of spaces) from a non-empty Stone space into 2^ω may be identified with a subset that is closed and non-empty, and (4) says that m is a morphism of algebras, that is, we have commuting of

$$
\begin{array}{ccc}
2 \times X & \longrightarrow & X \\
{\scriptstyle 2 \times m} \downarrow & & \downarrow {\scriptstyle m} \\
2 \times 2^\omega & \xrightarrow{\ \alpha\ } & 2^\omega
\end{array}
$$

But a curious difficulty now appears. We have so far omitted to determine the class of usc maps with respect to which initial F–algebras can be found (by the methods of this paper). These maps are the spatial duals of the costrict morphisms of **BA***. It turns out that they are the surjective usc maps (where $f : X \to Y$ is surjective if $f(X) = Y$). We shall not stop to prove this here, but just point out that the one-point space is obviously not initial w.r.t the arbitrary usc maps, while it is w.r.t the surjective usc maps (the unique surjective many-valued map from the one-point space to a given non-empty Stone space is trivially usc). Now the injection of a closed subset X into 2^ω is not, in general, surjective usc. Thus the derivation of a simple, concrete induction principle from the initial algebra principle for 2^ω is in some doubt. (The point is not, of course, that we do not know how to prove Baseless Induction: it is very easy to give a direct elementary proof of that. The point, or rather the question, is whether we can systematically explain paired induction/coinduction principles as arising from initial/final algebra coincidences.)

Some concluding remarks about the choice of morphisms are appropriate. The concrete versions of the induction/coinduction principles that we derive are clearly rather sensitive to such choice(s). In addition to the choice of morphisms with respect to which initiality or finality of algebras is proved, another possible source of variation (no more than hinted at in this paper) lies in the choice of exactly which mono's (or epi's) to consider: it may, for example, be appropriate to consider extremal mono's rather than arbitrary mono's. Such considerations may be relevant to resolving the difficulty about Baseless Induction.

Finally, a remark on what may seem to be the artificiality of going to usc maps to make certain principles "work". In fact, these maps are in many situations a computationally very natural choice of morphism. Dijkstra's work has already been mentioned; for an example of a very different kind, see [Vic88] .

References

[AS85]   B. Alpern and F. B. Schneider. Defining liveness. *Information Processing Letters*, 21:181–185, 1985.

[Ber59]   C. Berge. *Espaces Topologiques: Fonctions Multivoques*. Dunod, 1959.

[Dij76]   Edsger W. Dijkstra. *A Discipline of Programming*. Prentice-Hall, Englewood Cliffs, New Jersey, 1976.

[Eda89] A. Edalat. Categories of information systems. Master's thesis, Imperial College, 1989.

[ES91] A. Edalat and M. Smyth. Categories of information systems. Technical Report Doc-91-21, Imperial College, 1991.

[Fre91] P. Freyd. Algebraically complete categories. 1991.

[Kur61] C. Kuratowski. *Topologie*. Warszawa, 1961.

[LS81] D. Lehmann and M. Smyth. Algebraic specification of data types: a synthetic approach. *Math. Systems Theory*, 14:97–139, 1981.

[LW84] K. G. Larsen and G. Winskel. Using information systems to solve recursive domain equations effectively. In D. B. MacQueen G. Kahn and G. Plotkin, editors, *Semantics of Data Types*, pages 109–130, Berlin, 1984. Springer-Verlag. Lecture Notes in Computer Science Vol. 173.

[Sco72] D. S. Scott. Continuous lattices. In E. Lawvere, editor, *Toposes, Algebraic Geometry and Logic*, pages 97–136. Springer-Verlag, Berlin, 1972. Lecture Note in Mathematics 274.

[Sco82] D. S. Scott. Domains for denotational semantics. In M. Nielson and E. M. Schmidt, editors, *Automata, Languages and Programming: Proceedings 1982*. Springer-Verlag, Berlin, 1982. Lecture Notes in Computer Science **140**.

[Smy79] M. Smyth. Effectively given domains. D.Phil. Thesis, 1979. Oxford.

[Smy83] M. B. Smyth. Powerdomains and predicate transformers: a topological view. In J. Diaz, editor, *Automata, Languages and Programming*, pages 662–675, Berlin, 1983. Springer-Verlag. Lecture Notes in Computer Science Vol. 154.

[Smy85] M. B. Smyth. Finite approximation of spaces. In D. Pitt, S. Abramsky, A. Poigné, and D. Rydeheard, editors, *Category Theory and Computer Programming*, volume 240 of *Lecture Notes in Computer Science*, pages 225–241. Springer Verlag, 1985.

[Smy89] M. Smyth. Axioms for a "category of information systems". manuscript, May 1989.

[SP82] M. B. Smyth and G. D. Plotkin. The category-theoretic solution of recursive domain equations. *SIAM J. Computing*, 11:761–783, 1982.

[Ste89] I. Stewart. Notes on baseless induction. manuscript, Imperial College, 1989.

[Vic88] S. Vickers. An algorithmic approach to p-adic integers. In *3rd Workshop MFPLS*, volume 298 of *LNCS*, pages 599–614. Springer, 1988.

Geometric Theories and Databases

Steven Vickers *

Abstract

Domain theoretic understanding of databases as elements of powerdomains is modified to allow multisets of records instead of sets. This is related to geometric theories and classifying toposes, and it is shown that algebraic base domains lead to algebraic categories of models in two cases analogous to the lower (Hoare) powerdomain and Gunter's mixed powerdomain.

Terminology

Throughout this paper, "domain" means algebraic poset – not necessarily with bottom, nor second countable. The information system theoretic account of algebraic posets fits very neatly with powerdomain constructions. Following Vickers [90], it may be that essentially the same methods work for continuous posets; but we defer treating those until we have a better understanding of the necessary generalizations to topos theory.

More concretely, a domain is a preorder (information system) $(D, \subseteq)$ of *tokens,* and associated with it are an algebraic poset pt D of *points* (ideals of D; one would normally think of pt D as the domain), and a frame ΩD of *opens* (upper closed subsets of D; ΩD is isomorphic to the Scott topology on pt D).

"Topos" always means "Grothendieck topos", and not "elementary topos"; morphisms between toposes are understood to be geometric morphisms.

S, italicized, denotes the category of sets.

We shall follow, usually without comment, the notation of Vickers [89], which can be taken as our standard reference for the topological and localic notions used here.

---

* Department of Computing, Imperial College, London, England; email sjv@doc.ic.ac.uk

Acknowledgements

It has taken me a long time to understand toposes well enough to be able to handle them at all mathematically. I think others will share my experience, for even now there is no introductory text that treats in depth the relation between toposes and *geometric* logic (as opposed to intuitionistic logic), this being the connection that is relevant to Grothendieck's original intuition of a topos as a generalized topological space. My thanks are therefore due to all the people who have ever told me anything about toposes. I should like in particular to mention: Martin Hyland, for his unfailing helpfulness; Mike Fourman, for his clear talk (reported as Fourman and Vickers [85]) on classifying toposes; Samson Abramsky, Axel Poigné and Mike Smyth, whose participation in a reading group at Imperial in 1985 helped me gain a toehold of understanding of Johnstone [77]; and Peter Johnstone and other participants at the Durham Symposium for their advice on technical matters.

I should also like to acknowledge the influence of Carl Gunter, whose elegant database account of powerdomains gave me both the intuitions (databases) and the mathematical context (generalized powerdomain theory) that I needed for a natural application of topos theory.

The work was made possible by the financial support of the British Science and Engineering Research Council for two research projects at Imperial: Formal Methods for Declarative Languages, and Foundational Structures for Computer Science.

1. Introduction

Gunter [89] gives a powerdomain theoretic account of some aspects of databases (or, perhaps more fairly, a database account of powerdomains). Let us first summarize the intuitions in the case of the lower (or Hoare) powerdomain.

The account starts with a domain D in which each real-world object considered by the database has a semantic value as a point. A *record* is then a token of the domain. It can be thought of in two ways:

(i)   The token represents a compact point, and hence a possible finite approximation to the semantic values of objects (which may be more infinite in some sense).

(ii)   The token represents a completely coprime open, and hence an observable property of points (namely, that they are approximated by the token's corresponding compact point; so if t is the token, we shall write ↑t for this property). Every other open is then a disjunction of those represented by tokens.

In other words, the token lives both in a straightforward domain theoretic world as a point, and in a logical world as a property. The logic is an observational logic, and it is *geometric* – that is to say, in this propositional case, its connectives are finite meets and arbitrary joins. It has been argued enough elsewhere (see Vickers [89]) that these connectives, unlike ¬ and →, have a direct observational content.

The real world contains more than one object to be described, and in Gunter's account a database has a finite *set* of records to represent a set of objects. One can refine a database in two ways, either by refining the records (how to do this is described by the base domain), or by adding new records. These are encapsulated in the *lower* preorder on finite sets of tokens,

$$X \sqsubseteq_L Y \quad \Leftrightarrow \quad \forall s \in X. \exists t \in Y. s \sqsubseteq t$$

These finite sets of tokens are themselves the tokens for a new domain, the lower powerdomain $P_L D$. Such a set, X, can be seen either as a finite approximation to a reality comprising possibly infinitely many, possibly non-compact points, or as a property of that reality – that for each token s in X, there is a point in reality satisfying it (in the modal logic of powerdomains, X represents $\bigwedge_{s \in X} \Diamond \uparrow s$).

There are curiosities here that arise from the use of sets. For instance, suppose a database has two records that are only partial:

{forename = "John"}
{surname = "Smith"}

In other words, the database has records of – apparently – two people, John and Smith. If it now turns out that Smith's forename is John, that record gets refined to

{forename = "John", surname = "Smith"}

But this record now subsumes the old record for John. The powerdomain semantics implies that the two databases

{forename = "John"}
{forename = "John", surname = "Smith"}

and

{forename = "John", surname = "Smith"}

are equivalent, so the computer might as well save space by dropping the record about John. But that's odd, because the original John was in fact John *Smythe,* a completely different person.

The proposal to be developed here is that the world is full of Johns, all different, and the database should be prepared to contain distinct copies of the record {forename = "John"}. In other words, the database is not a *set* of records, but a *bag.* (Many databases do indeed work this way.)

Definition 1.1 Let S be a set. A *bag* in S is a set X (the *base* set) equipped with a function from X to S, written x ↦ |x|, the *value* of x.

It is also possible to define a bag by stating, for each element of S, what its multiplicity (possibly infinite) is in the bag. But the set X in the definition here makes concrete the "underlying distinct identities" of the elements of the bag and makes it easier to define bag morphisms.

Having accepted that a database should be a finite bag of records (tokens from the base domain), we should now like to extend the domain theory by defining a "bagdomain"; moreover, this will be a category rather than a domain. One possible definition of "bagdomain" ("categorical powerdomain") has been given by Lehmann [76] and developed by Abramsky [83], but we define a different notion. The relation between Lehmann's bagdomain and ours is roughly that between the upper and lower powerdomains.

Let us look carefully at how one bag, Y, can refine another, X. As before, we want the addition of extra elements to refine the bag. We also want to be able to refine a bag by refining its elements, so for each x in X we want some y in Y with |x| ⊑ |y|. But remember also that the distinct elements of the bag were supposed to represent distinct objects in reality; so in making the refinement perhaps we should keep track of which element is which.

Consider –

$X =$	{surname = "Smith"}	(object x)
$Y =$	{surname = "Smith", age = 0}	(object y_1)
	{surname = "Smith"}	(object y_2)

Obviously Y can refine X, and it is very reasonable to suppose that it is by mapping x to y_2, with y_1 new. But it could equally well be by mapping x to y_1 with y_2 new, and these are really two different ways of making the refinement.

Definition 1.2 Let D be a domain, and let X and Y be bags in D (i.e. bags of *tokens*, representing *compact* points). A *refinement* from X to Y is a function f: X → Y such that ∀x∈X. |x| ⊑ |f(x)|.

Note a subtle observational implication of this definition. If f is not onto, then we have refined X by adding new elements. This has already been mentioned. But if f is not 1-1, then the refinement says that two elements that were thought to be distinct have now been found to be the same. The idea of making this kind of observation is quite plausible, but it must be understood that it is a physical assumption about the systems we are trying to model. Another way of understanding it is that there must be an observational, intensional meaning to equality (physical identity of objects, not equality of their values). We shall see later that this helps to extend the observational content of propositional geometric logic to the predicate case.

Definition 1.3 Let D be a domain. Then the *lower bagdomain* over D, $B_L(D)$, is the category for which –

- objects are finite bags of tokens from D
- morphisms are refinements.

It is not hard to show that this actually is a category. It is also essentially small, for it is equivalent to its full subcategory in which the base sets of the bags are restricted to be finite subsets of the natural numbers, or indeed of any standard countable alphabet. We shall tacitly replace $B_L(D)$ by this small category equivalent to it, and use as the "standard countable alphabet" the same one as supplies the stock of logical variables; this will ease some technical proofs later on.

Our aim now is to show that this is a good categorical extension of powerdomain theory. The analogy is as follows.

- The category $B_L(D)$ is a categorical information system. Its objects are tokens and its morphisms are refinements.
- The categorical "ideal completion" (the analogy of pt D) is the *ind-completion* (see Johnstone [82]), whose objects are filtered diagrams in the base category. We shall show that for $B_L(D)$, the objects of the ind-completion are the arbitrary bags of possibly non-compact points of D, with refinements as morphisms.
- The observational theory for $P_L(D)$ is a *propositional* geometric theory, generated by propositions (nullary predicates) $\diamond a$ ($a \in \Omega D$). For $B_L(D)$ this is replaced by a *predicate* geometric theory with unary predicates a(x) ($a \in \Omega D$). Abstracting away from its presentation, a propositional geometric theory can be identified with a frame, its geometric Lindenbaum algebra; but for a predicate theory one must instead use the more complicated *classifying topos*. (For a propositional theory, the

classifying topos is the topos of sheaves over the corresponding locale.) We shall show that the classifying topos for the theory here is equivalent to the functor category $S^{BL(D)}$ (where we write S for the category of sets). This is analogous to the localic proof with $P_L(D)$ that the locale whose frame is presented with generators $\diamond a$ (i.e. given by a theory presentation with those as primitive propositions) is homeomorphic to the algebraic poset whose compact points are represented by finite sets of compact points of D under the lower preorder.

2. Geometric logic

We summarize here the principal ideas of geometric logic and topos theory that we shall use. For fuller references, see Johnstone [77] and Makkai and Reyes [77]; for another introduction for computer scientists see Fourman and Vickers [85].

A geometric theory is presented by –

- sorts
- primitive predicate and function symbols, each with an arity specifying the number and sorts of its arguments and (for a function) the sort of its result
- axioms of the form $\phi \vdash \psi$, where ϕ and ψ are geometric formulas, constructed using the primitive symbols, sorted variables and the geometric connectives $\wedge$ (and **true**), $\vee$ (including infinite disjunctions), $=$ (sorted) and $\exists$.

Note the special case of a propositional theory, i.e. one with no sorts. The only primitive symbols are the nullary predicates (propositions), there can be no variables, and $=$ and $\exists$ have no role to play. Such a presentation is exactly a presentation of a frame.

Formulas as sets

The crucial intuition is that a *formula* is a parametrized *set*, the parameter being a model. In other words, for a given theory there is a pairing Models × Formulas → Sets. It pairs a model M with a formula ϕ to give the *extent* of ϕ in M, the set of ways in which the free variables of ϕ can be instantiated in M to make ϕ true.

We write this extent as $\{M|\phi\}$ (the notation was suggested by Dirac's [47] bras and kets). This can be thought of as $\{x \in M \mid \phi(x)\}$, though x is really a vector of all the free variables in ϕ, so that $\{M|\phi\}$ is a set of tuples of elements of the appropriate carriers of M. (Remember that the theory may be many-sorted.)

A formula with *no* free variables has an extent that is a subset of M^0, which is a singleton set $\{*\}$. Classically, either M satisfies the formula, in which case the extent is all of $\{*\}$, or it doesn't, and the extent is $\emptyset$. Also, a sort can be represented by the formula x=x; $\{M|x=x\}$ is the carrier of M for the sort of x.

For both models and formulas, we have a notion of morphism and the pairing $\{M|\phi\}$ is functorial in both arguments.

For models, the morphisms are homomorphisms in an obvious sense (with functions mapping carriers to carriers, and preserving the operations and predicates). Then for a homomorphism from M to N we get for each ϕ a corresponding function from $\{M|\phi\}$ to $\{N|\phi\}$ *because ϕ is geometric.*[*] (To see an example where this breaks down for non-geometric formulas, consider $\phi \equiv \forall y.(x \cdot y = y \cdot x)$ in the theory of monoids. This states that x is central (commutes with all other elements), so $\{M|\phi\}$ is the centre of M; but homomorphisms of monoids do not necessarily map centres to centres.)

For formulas, once we have agreed that they are sets, the morphisms should be functions; the idea is to use their graphs. If $\phi(x)$ and $\psi(y)$ are formulas (x and y here may be vectors of variables), then a function from ϕ to ψ is a formula $\theta(x,y)$ satisfying –

- $\theta(x, y) \vdash \phi(x) \wedge \psi(y)$ (ϕ and ψ are the source and target of θ)
- $\theta(x, y) \wedge \theta(x, y') \vdash y = y'$ (single-valuedness)
- $\phi(x) \vdash \exists y. \theta(x, y)$ (totality)

When we parametrize by (take extents in) a model M, these conditions ensure that $\{M|\theta\}$ is the graph of a function from $\{M|\phi\}$ to $\{M|\psi\}$.

The category of formulas, the *syntactic category* of the theory, has as objects the formulas (actually, formulas modulo relabelling of variables) and as morphisms the function formulas (actually modulo provable equivalence).

Because the formulas are parametrized sets, one can imagine applying set constructions to them, such as products, equalizers, unions, etc. Some of these can be done just using the logical connectives. For instance, Cartesian product and intersection can both be constructed using conjunction:

$$\{M|\phi(x)\} \cap \{M|\phi(x)\} = \{M|\phi(x) \wedge \phi(x)\}$$
$$\{M|\phi(x)\} \times \{M|\phi(x)\} = \{M|\phi(x) \wedge \phi(y)\}$$

---

[*] This also holds when infinite conjunctions are used in the construction of ϕ, though we shan't use this.

Others can not. Some, such as complements and function spaces, are essentially non-geometric. But some, notably disjoint union, are geometric in flavour and can be added. In the next section, we shall describe more precisely, though in categorical terms, what are "geometric set constructions".

Although the syntactic category does not contain all the geometric set theory, its logical structure is very convenient for reasoning with. This is illustrated in our main Theorems, 3.1 and 4.2, where in order to interpret certain geometric theories in terms of other ones, it suffices to define functors into the syntactic category.

"Giraud frames"

Makkai and Reyes [77] use the phrase "Giraud toposes" temporarily as a description of categories E satisfying the following conditions (see Johnstone [77] p. 17 for more details) –

- E has all finite limits
- E has all small colimits, and they are universal (preserved under pullback)
- Coproducts in E are disjoint, i.e. the injections pull back pairwise to the initial object $\emptyset$
- Epimorphisms out of an object are equivalent to "equivalence relations" on that object

- E is locally small, i.e. each "hom-set" is indeed a set
- E has a set of generators, i.e. a set G of objects such that if f and g are distinct morphisms from X to Y, then there is some h: $G \rightarrow X$ with G in G such that h;f $\neq$ h;g

The last two conditions are size conditions, needed to construct a small site for which the Giraud topos can be the category of sheaves. The first four are the categorical embodiment of "geometric set theory": so the constructions wanted are finite limits (which can already be done by logic, in the syntactic category), and all small colimits (which cannot).

Makkai and Reyes use their term "Giraud topos" only while they are proving Giraud's theorem, that Giraud toposes and (Grothendieck) toposes are the same – the corollary to this, that every Giraud topos is an elementary topos, is somewhat remarkable, for nowhere in the definition are mentioned Cartesian closedness or subobject classifiers.

Let us try to give a more permanent usefulness to the notion of Giraud topos by making the same distinction as there is between frames and locales: the morphisms go in opposite directions. We therefore rename Giraud toposes as *Giraud frames:* categories E satisfying the conditions given above. A homomorphism of Giraud frames will then be a functor that preserves this geometric structure of finite limits and all colimits; or, rather, let us express the preservation of colimits by the possession of a right adjoint, so that a homomorphism of Giraud frames from E to F is an adjoint pair (f^*, f_*), $f^*: E \rightarrow F$ and $f_*: F \rightarrow E$ such that f^* preserves finite limits. f^* (the *inverse image*) is the primary part, the structure-preserving functor, and in fact we shall often talk as though f^* *is* the Giraud frame homomorphism.

Note that these homomorphisms do not necessarily preserve other structure of elementary toposes, such as exponentials and subobject classifiers. An exact analogue (the propositional case) is with frames: frames are, as it happens, complete Heyting algebras, but frame homomorphisms are only required to preserve joins and finite meets and do not necessarily preserve the Heyting arrow.

Now a locale is a frame "pretending to be a topological space", and this pretence is maintained in part by the morphism reversal. The same can be done with toposes. Grothendieck says that a topos is a generalized topological space. More precisely, a topos is a generalized *locale,* and in a directly generalized way it too pretends to be a space, its "space" of models – to aid the pretence, the set-theoretic models are called *points* of the topos. The reason that this can only be a pretence is that we do not identify toposes on the mere grounds that they have the same space of points – for instance, some non-trivial toposes don't have any points at all. Again, to maintain the pretence, the appropriate morphisms between toposes, the *geometric morphisms,* are defined to be homomorphisms of Giraud frames, but in the opposite direction. We can now say with precision that toposes (and geometric morphisms) are to locales as Giraud frames (and homomorphisms) are to frames.

The classifying topos

Let us first recall the definition of the classifying topos, and then try to elucidate it.

Let $\mathbb{T}$ be a geometric theory. A topos $S[\mathbb{T}]$ is a *classifying topos* for $\mathbb{T}$ iff –

- $S[\mathbb{T}]$ has a specified model of $\mathbb{T}$, the *generic* model, and
- if E is any other topos with a model of $\mathbb{T}$, then there is a unique (up to equivalence) geometric morphism f from E to $S[\mathbb{T}]$ such that f^* maps the generic model to the given model in E.

Fact *Every geometric theory has a classifying topos (unique up to equivalence).*

Proof See Makkai and Reyes [77]. (Johnstone [77] gives a proof for coherent theories, i.e. geometric theories in which no infinite disjunctions are used in the axioms).   ]

$S[\mathbb{T}]$ is to be the Giraud frame freely generated by the primitives of $\mathbb{T}$ (as sets and functions) subject to the relations expressed in the axioms in $\mathbb{T}$. To say that we have interpreted the primitives in $S[\mathbb{T}]$ (sorts as objects, predicates as subobjects of products of sort objects, functions as morphisms) and satisfied the axioms is precisely to say that we have a model of $\mathbb{T}$ in $S[\mathbb{T}]$. To say that this is done freely is to imply a universal property, that if we have a model of $\mathbb{T}$ in any other Giraud frame E, then there is a unique (up to equivalence here) homomorphism from $S[\mathbb{T}]$ to E that preserves the generators, and in the language of toposes and geometric morphisms this is as stated in the Definition.

2.1 Flat functors and Diaconescu's Theorem

Consider a domain D. A point is an ideal of D, but it can be viewed another way: it is a function f from D to $\mathbf{2} = \wp(\{*\})$, mapping s to $\{*\}$ iff s is in the ideal, and this function is antitone and satisfies

$$\{*\} = \bigcup\{f(u): u \in D\}$$
$$f(s) \cap f(t) = \bigcup\{f(u): s \sqsubseteq u, t \sqsubseteq u\} \qquad (s, t \in D)$$

The first of these says that there is some u such that $* \in f(u)$, i.e. the ideal is non-empty. The second says that if $* \in f(s)$ and $* \in f(t)$, i.e. s and t are both in the ideal, then they have some upper bound in the ideal: in other words, the ideal is directed.

These translate directly into the localic presentation of ΩD (see Vickers [90]; ub(S) is the set of upper bounds of S):

$$\Omega D = \text{Fr} \langle \uparrow\{s\}\ (s \in D) \mid \bigwedge_{s \in S} \uparrow\{s\} = \bigvee_{u \in \text{ub}(S)} \uparrow\{u\}\ (S \subseteq_{\text{fin}} D) \rangle$$

so that the points of the locale are exactly the ideals of D.

Our aim now is very quickly to present the categorical version of this idea. Throughout this discussion, let D be a small category.

First, the *ind-completion,* Ind-D, is the analogue of the ideal completion of a poset. A very good account is given in Johnstone [82]. Its objects are filtered diagrams in D, considered as formal representatives of their colimits. The morphisms are defined between the diagrams in such a way as to represent morphisms between colimits.

Now the Yoneda embedding from D into $S^{D^{op}}$ (where S is the category of sets) extends to Ind-D and in fact gives an equivalence between Ind-D and a full subcategory of $S^{D^{op}}$. Johnstone describes this as the category of "filtered colimits of representable functors", but that is for the sake of technical simplicity. There is an alternative characterization (see Johnstone [77]) of these functors as *flat*.

Definition 2.1

(i)   Let G be a category. G is *filtered* iff every finite diagram in G has a cocone over it, i.e.

 • G is nonempty.
 • Given any two objects X and Y in G, there can be found a third object Z and morphisms f: $Y \to Z$, g: $Y \to Z$.
 • Given two morphisms f, g: $X \to Y$, there can be found a third morphism h: $Y \to Z$ with f;h = g;h.

(ii)   Let D be a category, and let F: $D^{op} \to S$ be a functor.
The *Grothendieck construction* on F, Groth F, is the category in which

 • an object is a pair (x, X) where X is an object of D and $x \in F(X)$
 • a morphism from (x, X) to (y, Y) is a morphism f: $X \to Y$ in D such that x = F(f)(y)

F is *flat* iff Groth F is filtered.

Example 2.2 Let D be a poset. Then flat functors F: $D^{op} \to S$ are equivalent to ideals of D.
Proof Let F be flat. First, note that F(X) is a singleton for all X in D, for suppose x, y $\in$ F(X). Consider the objects (x, X) and (y, X) in Groth F. By flatness, there is an object (z, Z) and morphisms f and g from (x, X) and (y, X) respectively to (z, Z). But we must have f = g in D, so then x = F(f)(z) = F(g)(z) = y. Let I be the set of X in D such that F(X) $\neq \emptyset$. Again by flatness, it follows that I is an ideal.]

The next step is to observe that *flatness is geometric*.

Definition 2.3 The geometric theory Flat(D) has –

sorts: for each object X of D, a corresponding sort
functions: for each morphism f: $X \to Y$ in D, a corresponding function
              f: $Y \to X$
axioms: $\vdash Id(x) = x$
              $\vdash (f;g)(z) = f(g(z))$ (f: $X \to Y$, g: $Y \to Z$ in D)
              $\vdash \bigvee_{X \in D} \exists x:X.\ x = x$
              $\vdash \bigvee_{Z \in D} \bigvee \{\exists z:Z.\ (x = f(z) \land y = g(z)): f: X \to Z, g: Y \to Z$ in D$\}$
              $f(y) = g(y) \vdash \bigvee_{Z \in D} \bigvee \{\exists z:Z.\ (y = h(z)): h: Y \to Z, f;h = g;h\}$
                                            (f, g: $X \to Y$ in D)

A model of Flat(D) in S is precisely a flat functor from D^{op} to S. The object part is described by the carriers, and the morphism part by the interpretation of the functions. The first two axioms say that it is a (contravariant) functor, and the remaining three that it is flat.

Theorem 2.4 (Diaconescu's Theorem)

Let D be a small category. Then the classifying topos for Flat(D) is S^D.

The generic model is the Yoneda embedding of D contravariantly into S^D.

Proof In other words, for any topos F there is an equivalence between models of Flat(D) in F and geometric morphisms from F to S^D. This is Theorem 4.34 in Johnstone [77]. To see why, first note that Johnstone covers not just Grothendieck toposes but more generally toposes "defined over" an elementary topos E. We are interested in the case where E = S, the category of sets. The geometric morphism f: F → S has for its inverse image part the functor f*: S → E that maps a set X to the coproduct in F of |X| copies of 1; f* maps the small category D, an internal category in S, to an internal category f*D in F. Now Johnstone shows that geometric morphisms from F to S^D are equivalent to "flat internal presheaves on f*D", so there is a gap: we ought to show that flat internal presheaves on f*D, which are defined categorically, are equivalent to the models in F of Flat(D), which are defined more in logical terms.

This is a gap of exposition rather than of real mathematics. A development of topos theory using the Mitchell-Bénabou language would naturally define the flat internal presheaves in terms similar to Definition 2.3. Johnstone took a deliberately categorical approach, and presented the "obvious" categorical formalization of the logical ideas. Nonetheless, once the treatments – logical and categorical – have diverged it seems a non-trivial exercise to show that corresponding points on them are genuinely equivalent.

The basic trick here is that because coproducts are preserved by pullback in a topos, and because we are interested in flat internal presheaves over the category f*D whose object and morphism families are copowers of 1, the objects used in the course of defining "flat internal presheaves" can be decomposed as coproducts indexed by structures from the external category D. (Recall also that in a Giraud frame colimits are universal, i.e. preserved by pullback.)   ∎

2.2 Geometric logic as observational logic

In Abramsky [87] and Vickers [89], it is argued that disjunction and finitary conjunction, the connectives of propositional geometric logic, have observational content in that observability of properties is preserved by these connectives. The notion of "observability" is not formally defined, but it is intended to capture these two ideas:

- *positivity:* if an observable property holds, then it is possible to observe it; but if it fails, there need not be any way of ever discovering that.
- *serendipity:* one is told how to know in retrospect when one has observed something, but not any method that's guaranteed to result in the observation whenever possible.

(Serendipity perhaps represents the distinction between this notion of observability and semi-decidability.) We wish here to extend these intuitions to full predicate geometric logic. Although this section is necessarily informal, we hope that it helps to answer the question "What use is geometric logic in the real world?"

In the framework outlined above, $\{M|\phi\}$ (when ϕ is a proposition) is a subset of the singleton $\{*\}$. Hence observing ϕ is equivalent to discovering an element of this subset. We should like to extend this to predicate logic by the notion of "apprehending" elements of the sets $\{M|\phi\}$ (for general ϕ), i.e. observing the existence of elements and moreover getting some kind of grasp on them so that they can be related to others. We also want to be able to observe equality between elements, though this is an *intensional* equality – two twin elements may be indistinguishable in all respects, but still not equal.

We therefore propose the idea of "observability set", specified by two methods (both interpreted positively and serendipitously):

- how to apprehend elements of it
- how to observe equality between two elements of it

Equality (i.e. possibility of observing equality) must be an equivalence relation.

This idea is actually not very different from Bishop's (Bishop and Bridges [85]) account of sets as "the totality of all mathematical objects constructed in accordance with certain requirements ... endowed with a binary relation = of *equality*," though in generalizing "construction" to "apprehension" (one can plausibly apprehend an element by constructing it, though not necessarily the other way round) we are trying to be more neutral about the physical reality.

Similarly, we drop Bishop's constructional or operational import from the notion of function.* A function f: $X \rightarrow Y$ is simply its graph, a subset of $X \times Y$, a retrospective way of observing that for two elements x and y we have "y=f(x)". (To apprehend an element of f you apprehend a pair (x,y) and observe that "y=f(x)"; to observe that (x,y) = (x',y') you observe that x = x' and y = y'. We shall preserve the quotes round "y=f(x)" to make it plain that this is not an

---

* For this reason, we believe that geometric logic has a role to play in *specification* of computer programs.

instance of the the equality observation.) Of course, this method must be extensional (with respect to =), single-valued and total, though these are matters for proof rather than observation.

It is now possible to justify the first four conditions in the definition of Giraud frame in terms of observability sets, to reason informally that observability sets are preserved under finite limits and small colimits and that they satisfy the properties relating these.

For instance, consider how one could construct the equalizer of two functions f, g: $X \to Y$. We must describe the equalizer E, the function h: $E \to X$, and the mediating functions as observability sets, and we must also prove that these satisfy the right properties.

To apprehend an element of E, apprehend elements x and y of X and Y, and observe that "y=f(x)" and "y=g(x)". To observe that $(x,y) = (x',y')$, just observe that $x = x'$ (though it will follow from single-valuedness of f and g that it is then possible to observe that $y = y'$).

To observe (having apprehended x', x and y) that "x'=h(x, y)", observe that x' = x. One can then prove that h is a function. Moreover, h;f = h;g: for suppose it is possible to observe "y'=(h;f)(x,y)". The construction of composite functions is that to observe "c=(ϕ;ψ)(a)" you must apprehend some b and observe that "c=ψ(b)" and "b=ϕ(a)". In our case, therefore, it is possible to apprehend some x' and observe that "y'=f(x')" and x' = x. But $(x,y) \in E$, so we already know that "y=f(x)" and "y=g(x)". By extensionality and single-valuedness of f, it is possible to observe that y' = y and hence by extensionality of g that "y'=g(x')". It follows that it is possible to observe that "y'=(h;g)(x,y)". By symmetry it follows that it's possible to observe "y'=(h;f)(x,y)" iff it's possible to observe "y'=(h;g)(x,y)", i.e. h;f = h;g.

Now suppose we have a function k: $Z \to X$ such that k;f = k;g. We want to construct the mediating function k': $Z \to E$ such that k = k';h. (Note that these equalities between functions are not themselves observable, since they relate sets, not elements.) To observe that "(x,y)=k'(z)", observe that "x=k(z)". Again, one can prove that k' is a function, that k = k';h, and that k' is the unique such.

We shall not give the details of the rest of the proofs needed, but restrict ourselves to describing the constructions of products, coproducts and coequalizers.

Products: To apprehend an element of X×Y, apprehend a pair (x,y) of elements from X and Y. To observe that $(x,y) = (x',y')$, observe that x = x' and y = y'. The projections and mediating functions are all obvious. (Note also the nullary product 1: to apprehend an element of it or to observe two elements equal you need do nothing. Hence it has a single element that exists of its own accord.)

Coproducts, $X = \coprod_{\lambda \in \Lambda} X_\lambda$: Note that Λ is not an observability set, but an ordinary discrete set in which equality and inequality are perfectly known. To apprehend an element (x,λ) of X, apprehend an element x of X_λ for some λ. To observe $(x,\lambda) = (y,\lambda)$, observe that $x = y$ in X_λ; if $\lambda \neq \mu$ then to observe that $(x,\lambda) = (y,\mu)$ is impossible.

Coequalizers: Let f, g: $X \rightarrow Y$ be two functions. We wish to construct the coequalizer h: $Y \rightarrow C$. To apprehend an element of C, just apprehend an element of Y. To observe that $y = y'$ in C, apprehend elements y_0, y_1, ..., y_n of Y and elements x_1, ..., x_n of X, observe that $y = y_0$ and $y' = y_n$, and for each i $(1 \leq i \leq n)$ observe either that "$y_{i-1}=f(x_i)$" and "$y_i=g(x_i)$" or that "$y_{i-1}=g(x_i)$" and "$y_i=f(x_i)$".

The thesis, then, is that geometric theories can be satisfactorily used to formalize informal (or real-world) structures that seem to contain the ingredients of observability sets. Each sort will then represent (or be interpreted in a real-world model as) an observability set, each predicate symbol an observability subset of a product of sort sets, and each function symbol a function as described above. The axioms – including axioms to specify functionhood of functions – will be constraints on the admissible models, justifiable on physical grounds for the physical models that we have in mind. They are not in themselves observable.

The classifying topos will contain not only the observability sets specified in the theory presentation (the generic model), but also all others that can be derived using limits and colimits as described above.

The remainder of this paper attempts to apply these intuitions to databases. The idea is that *people* in the real world can be considered an observability set. To apprehend a person, you take a firm grasp of his or her collar and say "'Allo, 'allo, 'allo, what's going on 'ere, then?" To show that two apprehended people are equal, you try to knock their heads together and discover that you can't. In the examples dealt with, the only other primitive observability sets are subsets of this, and propositional observations. There are no observations other than equality to relate different people. This is rather a simple case, but we hope that the same intuitions can be used in other contexts, and in particular for specifying software systems.

3. The lower bagdomain

Theorem 3.1 *Let D be a domain. Then the theory Flat ($B_L(D)$) is equivalent to the single-sorted theory T presented by:*

predicates: $a(x)$ $(a \in \Omega D)$

axioms: $a(x) \vdash b(x)$ $(a \leq b \text{ in } \Omega D)$

                     $\bigwedge_{a \in S} a(x) \vdash (\bigwedge_{a \in S} a)(x)$ $(S \subseteq_{fin} \Omega D)$

                     $(\bigvee_{a \in S} a)(x) \vdash \bigvee_{a \in S} a(x)$ $(S \subseteq \Omega D)$

Proof We must show that $S^{BL(D)}$ (the classifying topos of Flat $(BL(D))$, by Diaconescu) is equivalent to $S[T]$, finding a model of each theory in the other's classifying topos in order to describe the geometric morphisms between them.

First, we describe a flat functor F from $B_L(D)^{op}$ to $S[T]$. This automatically extends to a Giraud frame homomorphism F: $S^{BL(D)} \to S[T]$ (i.e. a geometric morphism from $S[T]$ to $S^{BL(D)}$; F is the inverse image part).

Let X be an object of $B_L(D)$, a finite bag of tokens of D. Then

$$F(X) = \bigwedge_{x \in X} (\uparrow|x|)(x)$$

We are assuming here that the stock of elements from which the base sets of the bags are formed is actually *the same as* the stock of variables used in the logic. If $x \in X$ then $|x|$ is a token of D, and $\uparrow|x|$ is the corresponding open. Note that because the objects of the syntactic category (formulas) are defined only up to renaming of variables, we may always rename the base elements of bags to ensure that base sets are disjoint.

Let f: $X \to Y$ be a morphism in $B_L(D)$, i.e. a refinement; by renaming, we can assume that X and Y are disjoint. Then F(f) is the function formula

$$F(X) \wedge F(Y) \wedge \bigwedge_{x \in X} (x = f(x))$$

Note that "f(x)" here is not a composite term, but an actual variable. The refinement f maps variables to variables. Also, because $|x| \sqsubseteq |f(x)|$, this formula is equivalent to $F(Y) \wedge \bigwedge_{x \in X} (x = f(x))$.

The identity morphism is mapped to the identity function. As for composition, suppose also that g: $Y \to Z$ in $B_L(D)$. Then $F(f) \circ F(g)$ is $\exists y.(F(f) \wedge F(g))$; when this is expanded as a logical formula, it is easily shown to be equivalent to $F(f;g)$.

Next, we must prove flatness. The first flatness axiom, follows from the easily verified $\vdash \exists x: F(\emptyset). \; \mathbf{x} = \mathbf{x}$ where $\emptyset$ is the empty bag. (The bold face $\mathbf{x}$ represents the – empty – vector of free variables in $F(\emptyset)$.)

For the second flatness axiom, let X and Y be bags of tokens with disjoint base sets and let Z be a disjoint union of X and Y with injections f: $X \to Z$, g: $Y \to Z$. We can assume Z is disjoint from X and Y and write x' = f(x), y' = g(y). Then it suffices to show that in $S[T]$,

$$F(X) \wedge F(Y) \vdash \exists z: F(Z). \; (\mathbf{x} = f(\mathbf{z}) \wedge \mathbf{y} = g(\mathbf{z}))$$

For the third flatness axiom, let X and Y be disjoint bags of tokens, with refinements f, g: X $\to$ Y. The left hand side of the axiom is interpreted as $\exists x.(F(f) \wedge F(g))$, which is equivalent to $F(Y) \wedge \bigwedge_{x \in X} f(x) = g(x)$. Let $\sim$ be the equivalence relation on Y generated by the pairs $f(x) \sim g(x)$ ($x \in X$), and let Z_0 be $Y/\sim$: as a set, this is the coequalizer of f and g. Choose a representative element of each equivalence class, and let Y_0 be the set of chosen representatives. Then our formula is equivalent to

$$\bigwedge_{y \in Y_0} [(\bigwedge_{y' \sim y} \uparrow|y'|)(y) \wedge \bigwedge_{y' \sim y} (y' = y)]$$

$$\dashv\vdash \quad \bigwedge_{y \in Y_0} [(\bigvee\{\uparrow t: \forall y'. (y' \sim y \to |y'| \sqsubseteq t)\})(y) \wedge \bigwedge_{y' \sim y} (y' = y)]$$

$$\dashv\vdash \quad \bigvee\{\bigwedge_{y \in Y_0} [(\uparrow t_y)(y) \wedge \bigwedge_{y' \sim y} (y' = y)]:$$
$$\forall y \in Y_0. \ \forall y' \in Y. \ (y' \sim y \to |y'| \sqsubseteq t_y)\}$$

$$\dashv\vdash \quad \bigvee\{\bigwedge_{y \in Y} (\uparrow t_{[y]})(y) \wedge \bigwedge_{y' \sim y} (y' = y): \forall y \in Y. \ |y| \sqsubseteq t_{[y]})\}$$

where we are writing [y] for the equivalence class of y, an element of Z_0. Given a family of tokens t_z ($z \in Z_0$) with all $|y| \sqsubseteq t_{[y]}$, we can define a bag Z whose underlying set is Z_0 and with $|z| = t_z$, and then the projection h: Y $\to$ Z is a refinement with f;h = g;h. Therefore, the LHS of the axiom is a disjunction of formulas of the form

$$\bigwedge_{y \in Y} (\uparrow|h(y)|)(y) \wedge \bigwedge_{y' \sim y} (y' = y)$$

which is equivalent to $\exists z: F(Z). \ y = h(z)$.

Next, we describe a model for T in $S^{BL(D)}$, giving a Giraud frame homomorphism G from $S[T]$ to $S^{BL(D)}$. The single sort is carried by the functor (object of $S^{BL(D)}$) that maps each bag X to its base set X, and each refinement to its underlying function. If $a \in \Omega D$, then G(a(x)) is the subfunctor that takes X to $\{x \in X: |x| \vDash a\}$. If $a \le b$, then clearly a(x) gives a subfunctor of b(x).

Limits and colimits in $S^{BL(D)}$ are computed argumentwise, so it is clear that the subfunctor corresponding to $\bigwedge_{a \in S} a(x)$ (the intersection) is $\{x \in X: \forall a \in S. |x| \vDash a\}$, i.e. $\{x \in X: |x| \vDash \bigwedge_{a \in S} a\}$, i.e. the subfunctor corresponding to $(\bigwedge_{a \in S} a)(x)$. The other axiom is similar.

We must now show that the two geometric morphisms thus constructed are the two parts of an equivalence.

First, let X be an object of $B_L(D)$. Then GF(X) $\cong$ X, for

$$GF(X)(Z) = G(\bigwedge_{x \in X} (\uparrow|x|)(x))(Z) = \prod_{x \in X} G((\uparrow|x|)(x))(Z)$$
$$= \prod_{x \in X} \{z \in Z: |x| \sqsubseteq |z|\} \cong B_L(D)(X, Z) = X(Z)$$

Next, let f: X $\to$ Y be a morphism in $B_L(D)$. GF(f)(Z) is a subset of $GF(Y)(Z) \times GF(X)(Z) \cong B_L(D)(Y, Z) \times B_L(D)(X, Z)$, and this subset is the graph of a function. We must show that the function is the same as f(Z), i.e. that (g, h) ∈ GF(f)(Z) iff h = f;g. (Some of the notation is a bit condensed here. When we write X(Z) to mean $B_L(D)(X, Z)$, we are identifying X with its image under the Yoneda embedding; f(Z) is similar.)

$$GF(f)(Z) = G(F(X) \wedge F(Y) \wedge \bigwedge_{x \in X} (x = f(x)))(Z)$$
$$= \{z \in Z^Y \times Z^X: |x| \sqsubseteq |z_x|, |y| \sqsubseteq |z_y|, z_x = z_{f(x)}\}$$
$$\cong \{(g, h) \in Y(Z) \times X(Z): \forall x.\ h(x) = g(f(x))\}$$
$$= \{(g, h) \in Y(Z) \times X(Z): h = f;g\}$$

Hence, F;G is equivalent to the identity on $\mathcal{S}^{BL(D)}$.

Let us now consider G;F. First, note that G(a(x)) is the *colimit* (not just the union) of the subfunctors G((↑s)(x)) for tokens s ⊢ a. For suppose we have a functor U ∈ $\mathcal{S}^{BL(D)}$, and natural transformations f^s: G((↑s)(x)) → U (s ⊢ a). If z ∈ G(a(x))(Z), i.e. z ∈ Z and |z| ⊢ a = $\bigvee_{s \vdash a}$ ↑s, then |z| ⊒ s for some s ⊢ a and z ∈ G((↑s)(x))(Z). We wish to define $f_Z(z) = f^s_Z(z)$, and in fact this is unambiguous. For if also |z| ⊒ s' ⊢ a then

$$|z| \vdash \uparrow s \wedge \uparrow s' = \bigvee \{\uparrow t: t \sqsupseteq s, t \sqsupseteq s'\}$$

so |z| ⊒ some such t, and $f^s_Z(z) = f^t_Z(z) = f^{s'}_Z(z)$.

Next, G((↑s)(x)) is isomorphic to the (image in $\mathcal{S}^{BL(D)}$ of the) singleton bag $X_s = \{x\}$ with |x| = s. F preserves colimits, so F(G(a(x))) = A (say) is the colimit of the objects F(X_s) = (↑s)(x), s ⊢ a, and it remains to show that this colimit is isomorphic to a(x). The inclusions (↑s)(x) → a(x) (s ⊢ a) form a cocone over the diagram; let f: A → a(x) be the mediating morphism. f is epi, because a(x) is the union of its subobjects (↑s)(x). Let R, a subobject of A×A, be the corresponding equivalence relation. A×A is the colimit of objects (↑s)(x) × (↑t)(x), and pulling back to R we get that R is the colimit of objects (↑s)(x) ∩ (↑t)(x) ≅ (↑s ∧ ↑t)(x), the intersection being taken in a(x). Hence R is a colimit of objects (↑u)(x). It follows from this that we can find an inverse to the reflexivity monic A → R, and hence that the equivalence relation R is equality and f is an isomorphism.

Hence G;F is equivalent to the identity on $\mathcal{S}[T]$.   ⟧

In view of this Theorem, we can consider the presentations of T and Flat $B_L(D)$ to be different presentations of the same theory, which we shall write $B_L(D)$ (the *lower bagdomain* theory over D).

Corollary 3.2 *The category of models of $B_L(D)$ is equivalent to*

(i)   *The ind-completion of the category $B_L(D)$;*

(ii)   *The category whose objects are bags (possibly infinite) of points (possibly non-compact) of D, and whose morphisms are the refinements.*

Proof

(i) follows because the theory is Flat $B_L(D)$.

(ii) A model X for the theory as presented in Theorem 3.1 has a carrier (let's call that X too) equipped with subsets $\{X|a\}$, and the axioms ensure that for each $\xi \in$ X the map $|\xi|$: $\Omega D \to \mathbf{2}$, $|\xi|(a) = \mathbf{true}$ iff $\xi \in \{X|a\}$, is a frame homomorphism. In other words, $|\xi|$ is a point of the locale D; so X is a bag of points of D.

If X and Y are two models, then a homomorphism from X to Y is a function f: X $\to$ Y that maps each $\{X|a\}$ to within $\{Y|a\}$. In other words, $\forall \xi \in X.\ \forall a \in \Omega D.$ $(|\xi| \vDash a \to |f(\xi)| \vDash a)$, i.e. $\forall \xi \in X.\ |\xi| \subseteq |f(\xi)|$, i.e. f is a refinement.   ]

The models are *topological systems* in the sense of Vickers [89], those systems E for which $\Omega E = \Omega D$. However, refinements are not necessarily the point parts of continuous maps.

Proposition 3.3 B_L *is functorial: if f: D $\to$ E is continuous, then there is a geometric morphism $B_L(f)$: $B_L(D) \to B_L(E)$ (i.e. from the classifying topos of the theory $B_L(D)$ to that of $B_L(E)$).*

Proof It suffices to find a model for $B_L(E)$ in $S^{B_L(D)}$; we interpret b(x) (b $\in \Omega E$) as $(\Omega f(b))(x)$. Functoriality is clear.]

Proposition 3.4 *There is a natural transformation from B_L to P_L, mapping – on models – the bag X to Cl $\{|\xi|: \xi \in X\}$.*

Proof To give a geometric morphism from $B_L(D)$ to $P_L(D)$, we find a model for $P_L(D)$ in $B_L(D)$. It maps $\diamond a$ (a $\in \Omega D$) to $\exists x.\ a(x)$. Let F be its inverse image part.

Given a model X, let Y be the corresponding point of $P_L(D)$; it is defined by $\{Y|\phi\} = \{X|F(\phi)\}$ for $\phi \in \Omega P_L(D)$. (One could modify Dirac's notation and say $\{Y|\phi\} = \{X|F|\phi\}$, or $\{Y| = \{X|F.\ )$ By the standard theory of the lower powerdomain, Y can be identified with a closed set of points of D, namely the complement of $\vee \{a \in \Omega D: Y \nvDash \diamond a\}$. But

$Y \vDash \diamond a \Leftrightarrow \{X|\exists x.\ a(x)\} = \{Y|\diamond a\} = \{*\} \Leftrightarrow$ extent(a) $\cap \{|x|: x \in X\} \neq \emptyset$

$Y \nvDash \diamond a \Leftrightarrow \{|x|: x \in X\} \subseteq$ extent(a)^c

and the result follows from this.   ]

4. The mixed bagdomain

Gunter [89] presents a scheme in which databases have not only the records as we have described them above, representing actual objects, but also a different kind of record describing background assumptions about the objects of the form "all objects are such and such". The distinction represents the two modalities in the localic theory of the Vietoris (Plotkin) powerdomain: $\Diamond$a says "there is some object satisfying a", whereas $\Box$a says "all objects satisfy a".

This new kind of database has two sets of records, called the "flat" part* and the "sharp" part. If the flat records (tokens) are $r_1, ..., r_m$, and the sharp records are $s_1, ..., s_n$, then the database as a whole represents the property

$$\Diamond\!\uparrow\!r_1 \wedge ... \wedge \Diamond\!\uparrow\!r_m \wedge \Box(\uparrow\!s_1 \vee ... \vee \uparrow\!s_n\})$$

This says that for each r_i there is something in reality that it approximates, and everything in reality refines at least one of the s_j's.

Although these modalities are best known in connection with the Vietoris powerdomain, one of the axioms turns out to be inappropriate. It is:

$$\Box(a\vee b) \leq \Box a \vee \Diamond b$$

The reason is that it ties the sharp part of the database to the contingencies of what objects we have found, whereas we should actually prefer to have our background assumptions holding regardless.

Gunter therefore develops a new powerdomain, the "mixed" powerdomain, presented locally by

$$\Omega P_M(D) = Fr \langle\ \Diamond a,\ \Box a\ (a \in \Omega D)\ |$$
$$\Diamond \text{ preserves all joins}$$
$$\Box \text{ preserves finite meets and directed joins}$$
$$\Diamond a \wedge \Box b \leq \Diamond(a\wedge b)\ \ \rangle$$

He shows that the usual methods of powerdomain theory work with this one, giving an information system theoretic construction (i.e. the tokens for the powerdomain are databases with flat and sharp parts, the flat parts having to conform with the sharp parts) and an algebraic characterization.

Our aim now is to revise this theory by replacing $\Diamond a$ by $a(x)$ as before. We do not change $\Box a$; this remains as a proposition. The mixed axiom is replaced by $a(x)\wedge\Box b \vdash (a\wedge b)(x)$, which is slightly stronger. The old version says that if all

---

* This use of the word "flat" is by musical analogy, and is quite unconnected with the notion of flat functor.

objects satisfy b, and some object satisfies a, then some object satisfies a∧b. The new one says that if all objects satisfy b, and some object satisfies a, then *that very same object* satisfies a∧b.

Definition 4.1 Let D be a domain. (As usual, we think of D concretely as the set of *tokens*.) The *mixed bagdomain* over D, $B_M(D)$, is the category for which –

- objects are pairs $X = (X^b, X^\#)$ where X^b is a finite bag in D, $X^\#$ is a finite subset of D, and $X^\# \subseteq_U \{|x| : x \in X^b\}$ (i.e. for every $x \in X^b$ there is some $t \in X^\#$ such that $|x| \sqsupseteq t$).
- morphisms from X to Y are refinements from X^b to Y^b, provided that $X^\# \subseteq_U Y^\#$.

In Gunter's [89] construction, X^b is a *set*. Just as with B_L, we shall assume that the base sets for our bags X^b are finite subsets of some standard countable alphabet used also for logical variables.

Theorem 4.2 *Let D be a domain. Then the theory Flat $(B_M(D))$ is equivalent to the single-sorted theory T presented by:*

predicates:	$a(x)$, $\Box a$	$(a \in \Omega D)$
axioms:	$a(x) \vdash b(x)$	$(a \leq b \text{ in } \Omega D)$
	$\bigwedge_{a \in S} a(x) \vdash (\bigwedge_{a \in S} a)(x)$	$(S \subseteq_{fin} \Omega D)$
	$(\bigvee_{a \in S} a)(x) \vdash \bigvee_{a \in S} a(x)$	$(S \subseteq \Omega D)$
	$\Box a \vdash \Box b$	$(a \leq b \text{ in } \Omega D)$
	$\bigwedge_{a \in S} \Box a \vdash \Box(\bigwedge_{a \in S} a)$	$(S \subseteq_{fin} \Omega D)$
	$\Box(\bigvee^{\uparrow}_{a \in S} a) \vdash \bigvee^{\uparrow}_{a \in S} \Box a$	$(S \subseteq \Omega D \text{ directed})$
	$a(x) \land \Box b \vdash (a \land b)(x)$	$(a, b \in \Omega D)$

Proof First, we describe a flat functor F from $B_M(D)$ to $S[T]$, extending to a Giraud frame homomorphism from $S^{B_M(D)}$ to $S[T]$.

Let X be an object of $B_M(D)$. Then

$$F(X) = \bigwedge_{x \in X^b} (\uparrow|x|)(x) \land \Box(\uparrow X^\#)$$

Let $f: X \to Y$ be a morphism in $B_M(D)$; by renaming, we can assume that X^b and Y^b are disjoint. Then F(f) is the function formula

$$F(X) \land F(Y) \land \bigwedge_{x \in X^b} (x = f(x))$$

which is equivalent to $F(Y) \land \bigwedge_{x \in X^b} (x = f(x))$.

The functorial axioms of Flat($B_M(D)$) are satisfied much as in Theorem 3.1.

The first flatness axiom follows from the fact that in ΩD,

true $= \uparrow D = \vee^{\uparrow} \{\uparrow X^{\#}: X^{\#} \subseteq_{fin} D\}$

Hence in T,

true $\dashv\vdash \square$**true** $\dashv\vdash \vee^{\uparrow} \{\square\uparrow X^{\#}: X^{\#} \subseteq_{fin} D\} \vdash \vee_{X \in B_M(D)} \exists x: F(X). \, x = x$

by considering those X for which $X^b = \emptyset$.

For the second flatness axiom, let X and Y be objects of $B_M(D)$, with X^b and Y^b are disjoint. Let Z_0 be their union, and f: $X^b \to Z_0$, g: $Y^b \to Z_0$ the injections, both refinements. Then in $S[T]$,

$$F(X) \wedge F(Y) \dashv\vdash \wedge_{z \in Z_0} (\uparrow|z|)(z) \wedge \square(\uparrow X^{\#} \wedge \uparrow Y^{\#})$$
$$\dashv\vdash \vee^{\uparrow} \{\wedge_{z \in Z_0} (\uparrow|z|)(z) \wedge \square(\uparrow Z^{\#}): Z^{\#} \supseteq_U X^{\#}, Z^{\#} \supseteq_U Y^{\#}\}$$
$$(\text{because in } \Omega D \uparrow X^{\#} \wedge \uparrow Y^{\#} = \vee^{\uparrow} \{\uparrow Z^{\#}: Z^{\#} \supseteq_U X^{\#}, Z^{\#} \supseteq_U Y^{\#}\})$$

Now

$$(\uparrow|z|)(z) \wedge \square(\uparrow Z^{\#}) \dashv\vdash (\uparrow|z| \wedge \uparrow Z^{\#})(z) \wedge \square(\uparrow Z^{\#})$$
$$\dashv\vdash \vee \{(\uparrow t_z)(z) \wedge \square(\uparrow Z^{\#}): t_z \supseteq |z|, t_z \in \uparrow Z^{\#}\}$$

so what we have is a join of formulas of the form $\wedge_{z \in Z_0} (\uparrow t_z)(z) \wedge \square(\uparrow Z^{\#})$, where $t_z \supseteq |z|$, $t_z \in \uparrow Z^{\#}$, $Z^{\#} \supseteq_U X^{\#}$ and $Z^{\#} \supseteq_U Y^{\#}$. Let us define Z by taking $Z^b = Z_0$ with $|z|_Z = t_z$, and $Z^{\#}$ as we have it already. Then Z is an object in $B_M(D)$, and f: $X \to Z$, g: $Y \to Z$ are morphisms. This is enough to entail the right hand side of the axiom just as in Theorem 3.1.

For the third flatness axiom, let X and Y be objects of $B_M(D)$, with morphisms f, g: $X \to Y$. The argument is the same as in Theorem 3.1, the sharp parts $Z^{\#}$ all being $Y^{\#}$.

Next, we describe a model for T in $S^{B_M(D)}$, giving a Giraud frame homomorphism G from $S[T]$ to $S^{B_M(D)}$. The single sort and the predicates are interpreted as functors that operate on objects Z by applying the method of Theorem 3.1 to Z^b. For instance, $G(x=x)(Z)$ is the set Z^b. The axioms just involving those predicates are respected as in Theorem 3.1.

The proposition $\square a$ is interpreted as the functor

$$G(\square a)(Z) = \begin{cases} 1 & \text{if } \forall z \in Z^{\#}. \, |z| \vDash a \\ \emptyset & \text{otherwise} \end{cases}$$

$\square$ preserves finite meets and directed joins for the same reason as in upper (Smyth) powerdomain theory, and using finiteness of $Z^{\#}$.

As for the mixed axiom,

$$G(a(x) \wedge \square b)(Z) = \{z \in Z^b: |z| \vDash a \wedge \forall z' \in Z^{\#}. \, z' \vDash b\}$$

$$= \{z \in Z^b : |z| \vDash (a \wedge b) \wedge \forall z' \in Z^\#. \; z' \vDash b\}$$

$$\text{(because if } z \in Z^b \text{ then } z \sqsupseteq \text{ some } z' \in Z^\#)$$

$$\subseteq \{z \in Z^b : |z| \vDash (a \wedge b)\} = G((a \wedge b)(x))(Z)$$

We must now show that the two geometric morphisms thus constructed are the two parts of an equivalence.

First, let X be an object of $B_M(D)$. Then $GF(X) \cong X$, for

$$GF(X)(Z) = G(\wedge_{x \in X^b} (\uparrow|x|)(x) \wedge \square(\uparrow X^\#))(Z)$$

$$\cong \begin{cases} \{\text{refinements from } X^b \text{ to } Z^b\} & \text{if } X^\# \sqsubseteq_U Z^\# \\ \varnothing & \text{otherwise} \end{cases}$$

$$= B_M(D)(X, Z) = X(Z)$$

The argument for morphisms is somewhat as in Theorem 3.1, and so F;G is equivalent to the identity on $S^{BL(D)}$.

Let us now consider G;F. Just as in 3.1, $G(a(x))$ is the colimit of the subfunctors $G((\uparrow s)(x))$ for tokens $s \vDash a$, and so $F(G(a(x)))$ is the colimit of the $F(G((\uparrow s)(x)))$'s. $G((\uparrow s)(x))$ is the colimit in $S^{BM(D)}$ of the objects X from $B_M(D)$ for which X^b is the singleton bag $X_s = \{x\}$ with $|x| = s$, and $X^\# \sqsubseteq_U \{s\}$; for

$$G((\uparrow s)(x))(Z) \cong \{\text{refinements from } X_s \text{ to } Z^b\} \cong B_M(D)((X_s, Z^\# \cup \{s\}), Z)$$

Hence $F(G((\uparrow s)(x)))$ is the colimit of the objects $F(X) = (\uparrow s)(x) \wedge \square(\uparrow X^\#)$, and because

$$\vee \{\square(\uparrow X^\#) : X^\# \sqsubseteq_U \{s\}\} = \square (\vee \{\uparrow X^\# : X^\# \sqsubseteq_U \{s\}\}) = \square\text{true} = \textbf{true}$$

(the join is directed), it follows that this colimit is $(\uparrow s)(x)$. Hence $F(G(a(x)))$ is the colimit of the objects $(\uparrow s)(x)$, and much as in Theorem 3.1 this is $a(x)$. The proof that $F(G(\square a)) \cong \square a$ is somewhat similar.   ]

Again, what we now have are two presentations of a single theory. Let us call it $B_M(D)$ (the *mixed bagdomain* theory over D).

Corollary 4.3 *The category of models of $B_M(D)$ is equivalent to*

(i)   *The ind-completion of the category $B_M(D)$;*

(ii)   *The category whose objects are pairs $X = (X^b, X^\#)$ where $X^\#$ is a compact saturated set of points of D and X^b is a bag of points from $X^\#$; and whose morphisms from X to Y are refinements from X^b to Y^b provided that $Y^\# \subseteq X^\#$.*

Proof (i) follows because the theory is Flat $B_M(D)$. For (ii), the X^b part of the model arises as in 3.2, while $X^\#$ is a model for the propositions $\square a$ (by the standard theory for the upper, or Smyth powerdomain, using the Hofmann-Mislove theorem). The mixed axiom specifies that all the values of elements of X^b lie in $X^\#$.]

Proposition 4.4 B_M *is functorial.*
Proof Given a continuous map f: D → E, we map b(x) to $(\Omega f(b))(x)$ and □b to $\square(\Omega f(b))$ (b ∈ ΩE).]

Proposition 4.5 *There is a natural transformation from B_M to P_M. On models, it maps $(X^b, X^\#)$ to $(Cl \{|x|: x \in X^b\}, X^\#)$.*
Proof Much as for Proposition 3.4. ◇a ↦ ∃x. a(x), □a ↦ □a.   ]

5. Further work

5.1 Categorical generalizations
The idea in $B_L(D)$ of tacking arguments on the propositions generalizes to non-propositional theories, tacking an extra argument on all the predicates; this has now been done by Johnstone [91]. Moreover, rather than starting from this syntactic construction as a definition (which would leave open the question of presentation independence), he characterizes it universally as a "partial product". He also defines a monad on the category of toposes whose functor part is somewhat analogous to the construction given in Theorem 3.1, and shows – as in Theorem 3.1 – that in the case of algebraic toposes (i.e. those of the form S^C for some small category C) it is equivalent to one analogous to the finite bagdomain construction.

5.2 Continuous domains
In Vickers [90] it is shown that the method of information systems can also be applied to continuous posets, not just algebraic ones. A continuous information system as presented there has a token set with an order < that is transitive and *interpolative:* if s < u, then s < t < u for some t. This generalizes the reflexive axiom of posets. The construction of points and opens out of these tokens is then really very similar to that of the poset case, but now giving a continuous poset.

I conjecture that the bagdomains preserve continuity just as they do algebraicity. (The appropriate notion of continuous categories has been investigated by Johnstone and Joyal [82].)

5.3 More restricted information systems

The information systems used above are of the simplest kind, describing algebraic posets. By putting extra conditions on, one obtains more restricted kinds of information system that give more restricted kinds of domains: for instance, spectral algebraic, strongly algebraic (SFP), Scott domains.

Spectral algebraic domains can be rationalized as being governed by the desire to have a *coherent* theory, i.e. a geometric theory in which infinite disjunctions are not used in the presentation. This idea still makes sense in the predicate case, and in fact it is easy to see that if D is spectral algebraic then $B_L(D)$ and $B_M(D)$ are coherent. For instance, it is not hard to show that $B_M(D)$ can be presented using predicates a(x) and $\Box a$ where a is a *compact* open.

Flat D is equivalent to a coherent theory if, for instance, D satisfies certain conditions analogous to the "2/3 SFP" conditions for spectral algebraic information systems, namely that for every finite diagram Δ in D there is a finite set S of cocones over Δ such that every cocone over Δ factors via one in S. Such D's would be a categorical generalization of spectral algebraic domains. There may be similar generalizations of SFP and Scott domains, expressible by conditions on information systems, and if so, one can ask whether they are closed under B_L and B_M.

5.4 Other uses of predicate geometric logic

Although the database ideas here are rather specialized, the methods may turn out to be useful models for other applications of predicate geometric logic, using the intuitions described in Section 2.

One interesting possibility that shows some promise is the use of geometric logic in specification languages. One reason for this may be the *ex post facto* nature of the intuition for observations of functions: the specifications tell you how to observe (or check) when you've got the right answer, but not how to calculate it. More concrete evidence is provided by some recent work of Hodges [90], who shows that specifications using the Z language can in practice be restricted to a logic (which he calls Σ_1^+) that is essentially geometric.

5.5 Dynamic predicate geometric logic

Another feature of the database ideas that may be intrinsic at this stage of development of the theory is that the databases are describing a *static universe*. The only process of change is that by which one refines one's knowledge of this universe by refining one's knowledge of the objects, by discovering new objects, or by observing equalities. The whole theory falls far short of real databases in its failure to account for change in the world.

It has been argued in Abramsky and Vickers [90] that in the propositional case, to take account of change one must introduce a dynamic observational logic in which the observations form a quantale instead of a frame. The essential difference is that conjunction a∧b (observe a and b) must be replaced by a much more precise temporal composition a·b (observe a then b), with the understanding that there may have been a change in the object in connection with the observation a.

It therefore seems quite likely that in order to model dynamic databases we should use a theory that combines the dynamic features of quantale logic with the multiplicity of objects in the predicate logic: we need a predicate quantale logic. I do not yet understand what this theory would have to be.

References

S. Abramsky [83], "On semantic foundations of applicative multiprogramming", *Automata, Languages and Programming*, Lecture Notes in Computer Science **154**, (Springer-Verlag 1983), 1-14.

S. Abramsky [87], *Domain Theory and the Logic of Observable Properties*, PhD Thesis, Queen Mary College, University of London, 1987.

S. Abramsky [88], "Domain theory in logical form", *Annals of Pure and Applied Logic* **51** (1991), 1-77.

S. Abramsky and S.J. Vickers [90], *Quantales, Observational Logic and Process Semantics*, Report DOC 90/1, Department of Computing, Imperial College, London, 1990.

Errett Bishop and Douglas Bridges [85], *Constructive Analysis*, Grundlehren der mathematischen Wissenschaften **279**, Springer-Verlag (1985).

P. A. M. Dirac [47], *The Principles of Quantum Mechanics* (3rd edition), Oxford University Press (1947).

Michael P. Fourman and Steven Vickers [85], "Theories as categories", in David Pitt, Samson Abramsky, Axel Poigné and David Rydeheard (eds) *Category Theory and Computer Programming*, Lecture Notes in Computer Science **240** (Springer-Verlag 1986), 434-448.

Carl A. Gunter [89], "The mixed powerdomain", to appear in *Theoretical Computer Science*.

Wilfrid Hodges [90], "Another Semantics for Z", notes.

Peter Johnstone [77], *Topos theory,* Academic Press, London (1977).

Peter Johnstone [82], *Stone Spaces,* Cambridge University Press (1982).

Peter Johnstone [91], "Partial products, bagdomains and hyperlocal toposes", this volume.

Peter Johnstone and André Joyal [82], "Continuous categories and exponentiable toposes", *J. Pure and Appl. Algebra* **25**, 255-296.

D. Lehmann [76], "Categories for fixed point semantics", FOCS 17.

Michael Makkai and Gonzalo E. Reyes [77], *First Order Categorical Logic,* Lecture Notes in Mathematics **611** (Springer-Verlag 1977).

Steven Vickers [89], *Topology via Logic,* Cambridge University Press (1989).

Steven Vickers [90], "Continuous information systems", submitted for publication.

Partial Products, Bagdomains and Hyperlocal Toposes

P.T. Johnstone *

1. Introduction

In a recent paper [21], Steven Vickers introduced a 'generalized powerdomain' construction, which he called the (lower) *bagdomain*, for algebraic posets, and argued that it provides a more realistic model than the powerdomain for the theory of databases (cf. Gunter [4]). The basic idea is that our 'partial information' about a possible database should be specified not by a *set* of partial records of individuals, but by an indexed family (or, in Vickers' terminology, a *bag*) of such records: we do not want to be forced to identify two individuals in our database merely because the information that we have about them so far happens to be identical (even though we may, at some later stage, obtain the information that they are in fact the same individual).

There is an obvious problem with this notion. Even if the domain from which we start has only one point, the points of its bagdomain should correspond to arbitrary sets, and the 'refinement ordering' on them to arbitrary functions between sets, so that the bagdomain clearly cannot be a space (or even a locale) as usually understood. However, topos-theorists have long known how to handle 'the space of all sets' as a topos (the *object classifier*, cf. Johnstone and Wraith [15], pp. 175–6), and this is what Vickers constructs: that is, given an algebraic poset D, he constructs a topos $B_L(D)$ whose points are bags of points of D (and in the case when D has just one point, $B_L(D)$ is indeed the object classifier).

The asymmetry of Vickers' construction, however, poses further problems. Ideally, we should like to view $B_L(D)$, like the various powerdomain constructions, as freely generated by D in some suitable sense: that is, we should like to have a monad structure on the functor B_L, together with a concrete description of the algebras for this monad. But this is impossible, since D and $B_L(D)$ live in different categories, and in particular we cannot assign a meaning to $B_L(B_L(D))$. Vickers' work thus raises three interconnected questions:

---

*Department of Pure Mathematics, University of Cambridge, England.

(a) does the functor B_L extend to a functor from (Grothendieck) toposes to toposes? (b) if so, does it carry a monad structure? and (c) if so, can one give an explicit description of the algebras for this monad?

In this paper we show that the answer to all three questions is 'yes'. The layout of the paper is as follows: in section 2 we review some indexed category theory that we shall need later, and then in section 3 we construct the bagdomain functor on the 2-category of Grothendieck toposes. We do so initially by showing that it is a special case of a certain universal construction, the *partial product* construction studied by Dyckhoff and Tholen [2], but we also show how it may be regarded as a 'presentation-independent' construction on geometric theories, generalizing the construction given by Vickers for algebraic posets. Once we have this description, the existence of a monad structure on B_L is more or less immediate. Algebras for the bagdomain monad are studied in section 5; our main result is that a given topos admits (up to natural isomorphism) at most one B_L-algebra structure, and that it does so iff it is a retract of its own bagdomain, iff its categories of ($\mathcal{E}$-valued) points have finite coproducts which are stable under composition. These toposes are closely related to (albeit more special than) the local toposes studied by Johnstone and Moerdijk [13], and we have chosen to call them *hyperlocal*. The relation between local and hyperlocal toposes is explored in section 6, in which we also introduce a 'hyper' version of the localization construction studied in [13]. In section 7 we establish the relationship between the bagdomain and (lower) powerdomain constructions: we show that, if we start from a localic topos, the latter is the localic reflection of the former (so that, if one is determined to stay within the category of locales, the distinction introduced by Vickers between sets and bags of points simply cannot be accommodated). The final section of the paper, which is more speculative, discusses whether an analogous development is possible for the upper and mixed bagdomains.

Before we proceed further, we should make some remarks about notation and terminology. We are interested primarily in Grothendieck toposes (that is, in toposes defined and bounded over the 'classical' topos of sets—cf. [6], section 4.4), but our results are not sensitive to particular properties of the base topos (other than its possession of a natural number object—cf. Blass [1], for the reason why this cannot be dispensed with); that is, they work perfectly well in the 2-category of bounded $\mathcal{S}$-toposes, for any topos $\mathcal{S}$ with a natural number object. However, in order to prove them formally in this setting, we should have to work throughout the paper in the context of $\mathcal{S}$-indexed categories, which would render it rather impenetrable to those who are not thoroughly familiar with indexed category theory. We have therefore chosen to perpetrate the usual abuse of notation, by working over an arbitrary base topos but treating it as if it were the classical category of sets; the experts will know how to translate our proofs into indexed-category language, and those who don't can simply pretend throughout the paper that the topos denoted $\mathcal{S}$ *is* the classical category of sets—they won't lose anything significant by doing so.

A further convention we should mention is that, in this paper, 'geometric morphism' always means 'bounded geometric morphism'; we write Top for the 2-category of toposes and (bounded) geometric morphisms, so that Top/$\mathcal{S}$ denotes the 2-category of 'Grothendieck' toposes. We shall occasionally use the name 'geometric transformation' for one of the 2-cells of Top; we recall from [6] that these are defined to be natural transformations between the *inverse* images of geometric morphisms. And we shall follow the custom of naming an $\mathcal{S}$-topos (that is, a geometric morphism with codomain $\mathcal{S}$) by the name of the topos $\mathcal{E}$ which is its domain, suppressing the name of the geometric morphism itself: this abuse may be justified by the fact that the geometric morphism is determined, up to unique isomorphism, by the knowledge of $\mathcal{E}$ as an $\mathcal{S}$-*indexed* category.

I cannot conclude this Introduction without recording my thanks to three people: to Steve Vickers, for his excellent lecture at the Durham symposium (which started the train of thought that led to this paper), to Martin Hyland, for the use of his Macintosh (not to mention the freedom to occupy his office at all hours), and to Paul Taylor, for his diagram macros.

2.   Indexed Coproducts

Despite the remarks in the antepenultimate paragraph above, we shall not be able to avoid indexed category theory altogether. In particular, when $\mathcal{E}$ and $\mathcal{F}$ are $\mathcal{S}$-toposes, we shall need to consider the category Top/$\mathcal{S}$ $(\mathcal{F}, \mathcal{E})$ of '$\mathcal{F}$-valued points of $\mathcal{E}$' as an $\mathcal{F}$-indexed category: if I is an object of $\mathcal{F}$, then an I-indexed family of objects of this category is a geometric morphism $\mathcal{F}/I \to \mathcal{E}$ (that is—if $\mathcal{E}$ is the classifying topos for a geometric theory T— an 'I-indexed family of T-models in $\mathcal{F}$'), and the 're-indexing' functors are given by composition with the geometric morphisms $\mathcal{F}/J \to \mathcal{F}/I$ induced by morphisms $\alpha \colon J \to I$ in $\mathcal{F}$.

It is well known that, if $\mathcal{C}$ is an ordinary category, we may freely adjoin (set-indexed) coproducts to $\mathcal{C}$ by forming the category Fam($\mathcal{C}$) whose objects are set-indexed families $(C_i \mid i \in I)$ of objects of $\mathcal{C}$, and whose morphisms $(C_i \mid i \in I) \to (D_j \mid j \in J)$ are specified by a function $\alpha \colon I \to J$ together with an I-indexed family of morphisms $(f_i \colon C_i \to D_{\alpha(i)} \mid i \in I)$ in $\mathcal{C}$. (If we merely wish to adjoin finite coproducts, we may do so by forming the full subcategory Fam$_f(\mathcal{C})$ in which the indexing sets I and J are required to be finite.) We shall require the indexed-category version of this construction, which may be described as follows.

We recall that an $\mathcal{F}$-indexed category $\mathcal{C}$ is said to have $\mathcal{F}$-indexed coproducts if its re-indexing functors $\alpha^* : \mathcal{C}^I \to \mathcal{C}^J$ have left adjoints $\Sigma_\alpha : \mathcal{C}^J \to \mathcal{C}^I$, which satisfy the Beck-Chevalley condition for pullback squares in $\mathcal{F}$. Now if $\mathcal{C}$ is an arbitrary $\mathcal{F}$-indexed category, we define $\mathrm{Fam}_{\mathcal{F}}(\mathcal{C})$ (or simply $\mathrm{Fam}(\mathcal{C})$, if $\mathcal{F}$ is obvious from the context) to be the indexed category whose I-indexed families of objects are pairs (α, A) where $\alpha : J \to I$ in $\mathcal{F}$ and A is an object of $\mathcal{C}^J$, and whose morphisms $(\alpha_1, A_1) \to (\alpha_2, A_2)$ are pairs (β, f) where $\beta : J_1 \to J_2$ is a morphism satisfying $\alpha_2 \beta = \alpha_1$ and $f : A_1 \to \beta^* A_2$ in $\mathcal{C}^{J_1}$. (In particular, note that the fibre of $\mathrm{Fam}(\mathcal{C})$ over the terminal object 1 of $\mathcal{F}$ is none other than the total category of the fibration corresponding to $\mathcal{C}$.) The re-indexing functors of $\mathrm{Fam}(\mathcal{C})$ are defined by pullback in $\mathcal{F}$ in an obvious way: if $\beta : K \to I$ in $\mathcal{F}$, then $\beta^*(\alpha, A) = (\pi_1, \pi_2^* A)$ where the π_i are the projections from the pullback $K \times_I J$. And these functors have left adjoints defined by composition (that is, $\Sigma_\beta(\alpha, A) = (\beta\alpha, A)$), which satisfy the Beck-Chevalley condition. Moreover, we have

Proposition 2.1 *There is an $\mathcal{F}$-indexed full embedding $\mathcal{C} \to \mathrm{Fam}(\mathcal{C})$, which is universal among indexed functors from $\mathcal{C}$ to $\mathcal{F}$-indexed categories with $\mathcal{F}$-indexed coproducts.*

Proof :   The full embedding sends an object A of $\mathcal{C}^I$ to $(1_I, A)$. Given an $\mathcal{F}$-indexed functor $F : \mathcal{C} \to \mathcal{D}$, where $\mathcal{D}$ has $\mathcal{F}$-indexed coproducts, its unique extension (up to natural isomorphism) to a coproduct-preserving functor $\mathrm{Fam}(\mathcal{C}) \to \mathcal{D}$ sends an object (α, A) of $\mathrm{Fam}(\mathcal{C})^I$ to $\Sigma_\alpha F^J(A)$. The remaining details are straightforward.

It follows from 2.1 (but is just as easily proved by direct arguments) that the operation $\mathcal{C} \mapsto \mathrm{Fam}(\mathcal{C})$ has the structure of a monad on the 2-category of $\mathcal{F}$-indexed categories: the unit $\mathcal{C} \to \mathrm{Fam}(\mathcal{C})$ is the full embedding defined above, and the multiplication $\mathrm{Fam}(\mathrm{Fam}(\mathcal{C})) \to \mathrm{Fam}(\mathcal{C})$ sends an object $(\alpha, (\beta, A))$ to $(\alpha\beta, A)$.

Proposition 2.2 *Let $\mathcal{C}$ be an $\mathcal{F}$-indexed category, and suppose idempotents split in $\mathcal{C}$. The following are equivalent:*

(i) *$\mathcal{C}$ has $\mathcal{F}$-indexed coproducts.*

(ii) *$\mathcal{C}$ admits an algebra structure for the monad Fam.*

(iii) *The canonical embedding $\mathcal{C} \to \mathrm{Fam}(\mathcal{C})$ has an ($\mathcal{F}$-indexed) left adjoint*

(iv) *The canonical embedding $C \to \mathrm{Fam}(C)$ admits a retraction.*

(v) C *is a retract of some category of the form* $\mathrm{Fam}(\mathcal{D})$.

(vi) C *is a retract of an indexed category having $\mathcal{F}$-indexed coproducts.*

Proof :   (i) $\Rightarrow$ (ii) and (iii): Applying 2.1 to the identity functor of C, we obtain a functor $\mathrm{Fam}(C) \to C$, which in fact sends (α, A) to $\Sigma_\alpha(A)$. It is easy to see that this functor is an algebra structure for Fam, and that it is left adjoint to the embedding.

The implications (ii) $\Rightarrow$ (iv), (iv) $\Rightarrow$ (v) and (v) $\Rightarrow$ (vi) are all trivial; and (iii) also implies (iv) because the embedding $C \to \mathrm{Fam}(C)$ is full and faithful.

(vi) $\Rightarrow$ (i): Suppose we have indexed functors

$$C \xrightarrow{\ \ U\ \ } \mathcal{D} \xrightarrow{\ \ R\ \ } C$$

with $RU \cong 1_C$, where $\mathcal{D}$ has $\mathcal{F}$-indexed coproducts. For any morphism $\alpha \colon J \to I$ in $\mathcal{F}$, we have a diagram

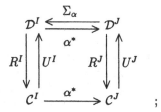

the composite functor $T_\alpha = R^I \Sigma_\alpha U^J \colon C^J \to C^I$ is not in general left adjoint to α^*, but it comes equipped with a canonical idempotent endomorphism, and if we split this endomorphism (pointwise) we do obtain a left adjoint for α^* (see Johnstone [9], 1.5). It is straightforward to verify that these left adjoints inherit the Beck-Chevalley condition from those for $\mathcal{D}$.

For a category C satisfying the hypotheses of 2.2, the retraction in (iv) may not be uniquely determined, but the algebra structure in (ii) is (at least up to canonical isomorphism). Let $T \colon \mathrm{Fam}(C) \to C$ be a retraction; let $\alpha \colon J \to I$ be a morphism of $\mathcal{F}$, and A an object of C^J. Further, let $\Delta \colon J \to J \times_I J$ be the diagonal map, and $\pi_1, \pi_2 \colon J \times_I J \to J$ the two projections. Then the morphism

$$(1_J, A) \xrightarrow{\ (\Delta, 1_A)\ } (\pi_1, \pi_2^* A) = \alpha^*(\alpha, A)$$

in $\text{Fam}(\mathcal{C})^J$ is sent by T^J to a morphism $u\colon A \to \alpha^*T^I(\alpha, A)$ which exhibits $T^I(\alpha, A)$ as a 'weak coproduct' of the α-indexed family A: the image under T^I of

$$(\alpha, A) \xrightarrow{\ (\alpha, u)\ } (1_I, T^I(\alpha, A))$$

is an idempotent endomorphism of $T^I(\alpha, A)$, which we must split in order to obtain the true coproduct. However, in the case when T is a Fam-algebra structure, there is a morphism

$$(\alpha, (1_J, A)) \xrightarrow{\ (\alpha, (\Delta, 1_A))\ } (1_I, (\alpha, A))$$

in $\text{Fam}(\text{Fam}(\mathcal{C}))^I$, whose image under $\text{Fam}(T)^I$ is (α, u) and whose image under the multiplication of the monad is the identity morphism on (α, A); so these two must have the same image under T^I, i.e. the idempotent endomorphism defined above is the identity, which is equivalent to saying that $T^I(\alpha, A)$ is the true coproduct of the α-indexed family. Using this, we may prove that conditions (i), (ii) and (iii) of 2.2 are equivalent even if idempotents do not split in $\mathcal{C}$.

There is a deeper reason for the results noted in the preceding paragraph: namely, that the monad Fam is a *KZ-doctrine* in the sense of Street [18], i.e. the multiplication $\text{Fam}(\text{Fam}(\mathcal{C})) \to \text{Fam}(\mathcal{C})$ is left adjoint to the unit map $\text{Fam}(\mathcal{C}) \to \text{Fam}(\text{Fam}(\mathcal{C}))$ in the 2-category of $\mathcal{F}$-indexed categories— from which it follows that any Fam-algebra structure is left adjoint to the appropriate unit. However, the theory of KZ-doctrines seems not to be widely known; so it seemed worthwhile to include the more specific argument above.

3.   Partial Products

We recall the definition of a partial product, originally introduced by Pasynkov [17] in the category of topological spaces, and later studied by Dyckhoff and Tholen [2] in a more general setting. Let $\mathcal{C}$ be a category with finite limits, A an object of $\mathcal{C}$ and $s\colon B' \to B$ a morphism of $\mathcal{C}$. A *partial product* of A over s is an object $P = P(s, A)$ of $\mathcal{C}$ together with morphisms $p\colon P \to B$ and $e\colon P \times_B B' \to A$ which is universal among such data: i.e., given $p'\colon P' \to B$

and $e' \colon P' \times_B B' \to A$, there is a unique $h \colon P' \to P$ making the diagram

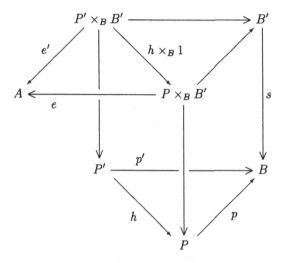

commute.

We note in passing that partial products include both products and exponentials as special cases: if $s = 1_B$, then $P(s, A)$ is simply the product $A \times B$, whereas if B is the terminal object of $\mathcal{C}$ then it is the exponential $A^{B'}$. In general, we have the following criterion for the existence of partial products:

Proposition 3.1 (Dyckhoff and Tholen [2], 2.1) *Given $s \colon B' \to B$, the partial product $P(s, A)$ exists for all objects A of $\mathcal{C}$ iff s is exponentiable as an object of the slice category $\mathcal{C}/B$ (i.e. the product functor*

$$(-) \times s \colon \mathcal{C}/B \to \mathcal{C}/B$$

has a right adjoint).

Under the hypotheses of the Proposition, the operation $A \mapsto P(s, A)$ is (covariantly) functorial, and may be obtained as the composite

$$\mathcal{C} \xrightarrow{\;B^*\;} \mathcal{C}/B \xrightarrow{\;(-)^s\;} \mathcal{C}/B \xrightarrow{\;\Sigma_B\;} \mathcal{C}$$

where Σ_B is, as usual, the forgetful functor $\mathcal{C}/B \to \mathcal{C}$ and B^* is its right adjoint. The functoriality of P in its first variable is a more delicate matter, but the following result is worth noting:

Lemma 3.2 *Let*

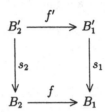

be a pullback square in C, *and suppose the partial product* $P(s_1, A)$ *exists. Then* $P(s_2, A)$ *also exists, and there is a pullback square*

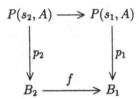

Proof :   A straightforward diagram-chase shows that, if we take the above pullback square as a definition of $P(s_2, A)$, then it has the right universal property.

In what follows, we shall be concerned with partial products in the 2-category $\mathsf{Top}/\mathcal{S}$ of bounded $\mathcal{S}$-toposes, as defined in the Introduction. Of course, the fact that we are working in a 2-category means that we cannot simply adopt the definition given earlier and rely without further comment on results such as 3.1: in order to specify a partial product $\mathcal{P}(s \colon \mathcal{E}' \to \mathcal{E}, \mathcal{F})$ by a universal property in $\mathsf{Top}/\mathcal{S}$, we have to specify the category-valued 2-functor T of which it is to be a representation. Clearly, objects of $T(\mathcal{G})$ should be pairs of geometric morphisms $(p \colon \mathcal{G} \to \mathcal{E}, e \colon \mathcal{G} \times_{\mathcal{E}} \mathcal{E}' \to \mathcal{F})$, but what should be the morphisms between such objects? The problem is that a geometric transformation $\alpha \colon p_1 \to p_2$ between geometric morphisms $\mathcal{G} \to \mathcal{E}$ does not live in the 2-category $\mathsf{Top}/\mathcal{E}$, and so does not in general induce any comparison between the two possible pullbacks of $\mathcal{G}$ along s.

The general problem of defining partial products in a 2-category will be discussed in a subsequent paper [11]; for our present purposes, it will be (more than) enough to specialize to the case when $s \colon \mathcal{E}' \to \mathcal{E}$ is an *algebraic topos* over $\mathcal{E}$—by which we mean that $\mathcal{E}'$ is equivalent to the topos $[\mathbf{C}, \mathcal{E}]$ of diagrams on an internal category $\mathbf{C}$ in $\mathcal{E}$ (this is compatible with Vickers' use of the term 'algebraic poset', but more general than the algebraic toposes considered by Hyland and Pitts [5]). In this case, the pullback of s along $p \colon \mathcal{G} \to \mathcal{E}$ may be taken to be $[p^*\mathbf{C}, \mathcal{G}]$ ([6], 4.35); and a natural transformation $\alpha \colon p_1^* \to p_2^*$ induces an internal functor $\alpha_{\mathbf{C}} \colon p_1^*\mathbf{C} \to p_2^*\mathbf{C}$ in $\mathcal{G}$ and hence a geometric morphism $[p_1^*C, \mathcal{G}] \to [p_2^*C, \mathcal{G}]$ over $\mathcal{G}$. Moreover, although the triangle

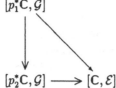

does not commute, there is a canonical geometric transformation which we may insert in it: the inverse image of the diagonal arrow sends a discrete opfibration $F \to C$ in $\mathcal{E}$ to $p_1^*F \to p_1^*C$, the other composite sends it to the pullback of $p_2^*F \to p_2^*C$ along α_C, and the naturality of α yields a morphism from the first of these to the second. We may thus define a morphism $(p_1, e_1) \to (p_2, e_2)$ in our category $T(\mathcal{G})$ to be a pair (α, β) where α is a geometric transformation $p_1 \to p_2$ and β fits into the diagram

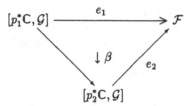

Composition of morphisms in $T(\mathcal{G})$, and the effect of T on morphisms and 2-cells of $\mathsf{Top}/\mathcal{S}$, are defined in the obvious way by pasting. Thus we define a *partial product* $\mathcal{P}(s_C, \mathcal{F})$ of $s_C \colon [C, \mathcal{E}] \to \mathcal{E}$ and $\mathcal{F}$ to be a representing object (in the up-to-equivalence sense) for the category-valued 2-functor T just defined.

By a (very) special case of the main result of Johnstone and Joyal [12], we know that any algebraic $\mathcal{E}$-topos $[C, \mathcal{E}]$ is exponentiable in $\mathsf{Top}/\mathcal{E}$; and if we now follow through the proof of 3.1 in our 2-categorical setting, we duly obtain

Proposition 3.3 *If $\mathcal{E}$ and $\mathcal{F}$ are bounded $\mathcal{S}$-toposes, and C is any internal category in $\mathcal{E}$, then the partial product $\mathcal{P}(s_C, \mathcal{F})$ (as defined above) exists in $\mathsf{Top}/\mathcal{S}$.*

Proposition 3.3 may be obtained as a particular case of a general result which will be established in [11]. In fact it is still more general than we need for the purposes of this paper; we shall only use the case when C is a discrete internal category in $\mathcal{E}$, so that we may identify it with an object X of $\mathcal{E}$ and identify $[C, \mathcal{E}]$ with the slice category $\mathcal{E}/X$. We write s_X for the geometric morphism $\mathcal{E}/X \to \mathcal{E}$ whose inverse image functor is X^*; for convenience, we now state the special case of 3.3 which we need as

Corollary 3.4 *If $\mathcal{E}$ and $\mathcal{F}$ are bounded $\mathcal{S}$-toposes, and X is any object of $\mathcal{E}$, then the partial product $\mathcal{P}(s_X, \mathcal{F})$ exists in* Top$/\mathcal{S}$.

However, there is still some spurious generality in this result. Let $\mathcal{O}$ denote the object classifier in Top$/\mathcal{S}$ (cf. Johnstone and Wraith [15]), and let U be the generic object of $\mathcal{O}$. Then, for any $\mathcal{S}$-topos $\mathcal{E}$ and any object X of $\mathcal{E}$, we have a pullback square

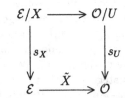

where $\tilde{X}$ is the classifying map of X; so, by 3.2, the existence of $\mathcal{P}(s_X, \mathcal{F})$ for all X and all $\mathcal{F}$ follows from that of $\mathcal{P}(s_U, \mathcal{F})$ for all $\mathcal{F}$—and this is the only case we shall need to consider. (3.3 may similarly be deduced from a special case of itself in which $\mathcal{E}$ is taken to be the classifying topos for the theory of categories.) In passing, we remark that in view of ([6], 4.37(i)), the slice category $\mathcal{O}/U$ is a classifying topos for the theory of pointed objects; hence we may view it as the topos of internal diagrams on the internal category of pointed finite cardinals in $\mathcal{S}$, and indeed s_U is the geometric morphism induced by the (internal) forgetful functor from pointed finite cardinals to finite cardinals—note that the latter is a discrete opfibration.

We next interpret 3.4 in terms of geometric theories. Suppose $\mathcal{F}$ is the classifying topos (over $\mathcal{S}$) for a geometric theory T. Then we see that, for an arbitrary $\mathcal{S}$-topos $\mathcal{G}$, a morphism $\mathcal{G} \to \mathcal{P}(s_U, \mathcal{F})$ in Top$/\mathcal{S}$ corresponds to a morphism $p: \mathcal{G} \to \mathcal{O}$ (equivalently, an object I of $\mathcal{G}$) together with a morphism $e: \mathcal{G}/p^*U \cong \mathcal{G}/I \to \mathcal{F}$ (equivalently, a T-model M in $\mathcal{G}/I$, which we may think of as an I-indexed family of T-models $(M_i \mid i \in I)$ in $\mathcal{G}$). Moreover, it follows from the discussion preceding 3.3 that, if $(M_i \mid i \in I)$ and $(N_j \mid j \in J)$ are two such indexed families, then a geometric transformation between the corresponding geometric morphisms into $\mathcal{P}(s_U, \mathcal{F})$ corresponds to a morphism $\alpha: I \to J$ in $\mathcal{G}$ together with a morphism $M \to \alpha^*N$ of T-models in $\mathcal{G}/I$ (which we may think of as an I-indexed family of T-model morphisms $M_i \to N_{\alpha(i)}$ in $\mathcal{G}$). That is,

Proposition 3.5 *For any two $\mathcal{S}$-toposes $\mathcal{F}$ and $\mathcal{G}$, the $\mathcal{G}$-indexed category* Top$/\mathcal{S}\,(\mathcal{G}, \mathcal{P}(s_U, \mathcal{F}))$ *is equivalent to the category* Fam$_\mathcal{G}$(Top$/\mathcal{S}\,(\mathcal{G}, \mathcal{F})$) *obtained by freely adjoining $\mathcal{G}$-indexed coproducts to* Top$/\mathcal{S}\,(\mathcal{G}, \mathcal{F})$, *as in 2.1.*

Henceforth we shall call $\mathcal{P}(s_U, \mathcal{F})$ the *bagdomain* of $\mathcal{F}$, and denote it by $\mathcal{B}_L(\mathcal{F})$.

Given a presentation for the theory T classified by $\mathcal{F}$, it is easy to construct one for the theory T-bag classified by $\mathcal{B}_L(\mathcal{F})$. For convenience, we suppose that T has been presented using primitive predicate symbols rather than function symbols (i.e. that each n-ary function symbol has been replaced by an $n + 1$-ary predicate symbol together with axioms saying that the interpretation of the predicate is a functional relation). The primitive sorts of T-bag are those of T together with one new sort Y; for each sort X of T we take a unary function symbol $f_X \colon X \to Y$. For each primitive predicate ϕ of T (of signature $(X_1, \ldots, X_n)$, say), we take a predicate $\overline{\phi}$ of signature $(X_1, \ldots, X_n, Y)$, together with the axiom

$$\overline{\phi}(x_1, \ldots, x_n, y) \vdash \bigwedge_i f_{X_i}(x_i) = y$$

which says that $\overline{\phi}$ is 'interpreted fibrewise'; and for each axiom of T we similarly add a conjunction of equations to the premiss to ensure that all the free variables in it live in the same fibre.

Readers who are unhappy with the abstract 2-categorical machinery used in proving 3.3 may, if they prefer, take the construction of T-bag from T as a definition of the bagdomain functor. Of course, in order to do this it is necessary to check that the construction is presentation-independent: that is, that if T_1 and T_2 are geometric theories which are 'Morita-equivalent' in the sense that they have equivalent classifying toposes (aliter, their categories of models in an arbitrary topos are (naturally) equivalent), then T_1-bag and T_2-bag are also Morita-equivalent. However, it is easy to see that this must be so, using the fact that 3.5 holds (more or less by definition) for this construction of the bagdomain.

It is also easy, in the particular case when T is the propositional theory whose models are points of a given algebraic poset D, to identify T-bag (up to Morita equivalence) with the geometric theory described in [21], Theorem 3.1. Thus we see that our construction is indeed a generalization of that introduced by Vickers.

4. Algebraic Bagdomains

In this section we generalize a result proved by Vickers [21], by showing that if $\mathcal{F}$ is an algebraic $\mathcal{S}$-topos (that is, a topos of the form $[\mathbf{C}, \mathcal{S}]$ where $\mathbf{C}$ is an internal category in $\mathcal{S}$), then so is its bagdomain $\mathcal{B}_L(\mathcal{F})$. For simplicity, we shall treat $\mathcal{S}$ as if it were the classical topos of sets, and hence regard $\mathbf{C}$ as an ordinary small category. We write $\mathbf{S}_f$ for (a suitable small skeleton of) the category of finite sets, and $(\mathbf{S}_f)_\bullet$ for the category of finite pointed sets; let $v: (\mathbf{S}_f)_\bullet \to \mathbf{S}_f$ denote the forgetful functor.

Given a small category $\mathbf{C}$, we write $\mathrm{Fam}_f(\mathbf{C})$ for the category of finite families of objects of $\mathbf{C}$, defined as in section 2; again, we regard this as a small category by using a skeletal category of finite sets. Our first result is

Lemma 4.1 $\mathrm{Fam}_f(\mathbf{C})$ *is a partial product* $P(v, \mathbf{C})$ *in the category of small categories.*

Proof :   There is an obvious functor $\mathrm{Fam}_f(\mathbf{C}) \to \mathbf{S}_f$, which sends an indexed family of objects of $\mathbf{C}$ to its indexing set. If we form the pullback

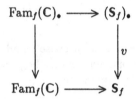

we see that $\mathrm{Fam}_f(\mathbf{C})_\bullet$ may be described as the category of pointed finite families of objects of $\mathbf{C}$, that is finite families of objects with one distinguished member. This category clearly admits a functor to $\mathbf{C}$, sending a pointed family of objects to its distinguished member. Now let $\mathbf{D}$ be a small category equipped with functors $T: \mathbf{D} \to \mathbf{S}_f$ and $U: \mathbf{D}_\bullet = \mathbf{D} \times_{\mathbf{S}_f} (\mathbf{S}_f)_\bullet \to \mathbf{C}$. Given an object D of $\mathbf{D}$, let $F(D) = (C_i \mid i \in I)$, where $I = T(D)$ and $C_i = U(D, (I, i))$ for each $i \in I$; then it is easy to verify that F is a functor $\mathbf{D} \to \mathrm{Fam}_f(\mathbf{C})$, and that it is the unique functor making the appropriate diagrams commute. So $\mathrm{Fam}_f(\mathbf{C})$ has the universal property of a partial product.

In general, we would not expect the functor $\mathbf{cat}(\mathcal{S}) \to \mathrm{Top}/\mathcal{S}$ sending $\mathbf{C}$ to $[\mathbf{C}, \mathcal{S}]$ to preserve partial products—after all, it does not even preserve pullbacks in general (cf. [6], Exercise 6.7). However, it does happen to preserve this particular partial product. First of all, it preserves the pullback diagram defining $\mathrm{Fam}_f(\mathbf{C})_\bullet$; this is because the vertical arrows in this diagram are both discrete opfibrations, corresponding to the inclusion functor $U: \mathbf{S}_f \to \mathcal{S}$ and the composite of this functor with the forgetful functor $\mathrm{Fam}_f(\mathbf{C}) \to \mathbf{S}_f$. Now we have

Proposition 4.2 $\mathcal{B}_L([\mathbf{C}, \mathcal{S}])$ *is equivalent to* $[\mathrm{Fam}_f(\mathbf{C}), \mathcal{S}]$.

Proof :   The argument is essentially that given by Vickers [21] in the case when **C** is a preorder: we show that the theory classified by $[\mathrm{Fam}_f(\mathbf{C}), \mathcal{S}]$ (that is, the theory of flat presheaves on $\mathrm{Fam}_f(\mathbf{C})$) is Morita-equivalent to that classified by $\mathcal{B}_L([\mathbf{C}, \mathcal{S}])$ (that is, the theory of bags of flat presheaves on **C**). For simplicity, we shall describe an equivalence between the categories of $\mathcal{S}$-valued models of these two theories; experienced topos-theorists will know how to 'lift' it to an equivalence between their categories of $\mathcal{G}$-valued models, for any $\mathcal{S}$-topos $\mathcal{G}$.

Given a bag $F = (F_j \mid j \in J)$ of presheaves on **C** (i.e. functors $\mathbf{C}^{\mathrm{op}} \to \mathcal{S}$), we define a functor $\tilde{F} \colon \mathrm{Fam}_f(\mathbf{C})^{\mathrm{op}} \to \mathcal{S}$ by setting $\tilde{F}(C_i \mid i \in I)$ to be the set of pairs

$$(\alpha, (x_i \mid i \in I))$$

where $\alpha \colon I \to J$ is a function and $(x_i \mid i \in I) \in \prod_{i \in I} F_{\alpha(i)}(C_i)$. The action of $\tilde{F}$ on morphisms of $\mathrm{Fam}_f(\mathbf{C})$ is obvious; and it is straightforward to verify that $\tilde{F}$ is a flat presheaf (i.e. that the total category of the corresponding fibration is filtered) if each F_i is flat.

In the opposite direction, we note that the functors expressing $\mathrm{Fam}_f(\mathbf{C})$ as a partial product (together with the observation, already made, that the pullback defining $\mathrm{Fam}_f(\mathbf{C})_\bullet$ is preserved by the functor $[-, \mathcal{S}]$) induce a geometric morphism

$$[\mathrm{Fam}_f(\mathbf{C}), \mathcal{S}] \longrightarrow \mathcal{B}_L([\mathbf{C}, \mathcal{S}]) \,,$$

and composition with this morphism clearly defines a functor from flat presheaves on $\mathrm{Fam}_f(\mathbf{C})$ to bags of flat presheaves on **C**. If we unravel all the definitions involved, it is again straightforward to verify that this functor is inverse (up to natural isomorphism) to the one described above.

As a further generalization of 4.2, it is possible to construct a site of definition for $\mathcal{B}_L(\mathcal{F})$ from one for $\mathcal{F}$. Suppose we are given a Grothendieck topology J on $\mathbf{C}^{\mathrm{op}}$; then we may construct one on $\mathrm{Fam}_f(\mathbf{C})^{\mathrm{op}}$ by saying that, given an object $(C_i \mid i \in I)$, a cocovering family for this object is defined by choosing a J-cocovering family R_i on C_i for each i, and then taking the family of all morphisms of the form $(1_I, (f_i \mid i \in I))$ where $f_i \in R_i$ for each i. Then we have

Proposition 4.3 *If $\mathcal{F}$ is the topos $\mathrm{Sh}(\mathbf{C}^{\mathrm{op}}, J)$ of J-sheaves on $\mathbf{C}^{\mathrm{op}}$, then $\mathcal{B}_L(\mathcal{F})$ may be identified with the topos $\mathrm{Sh}(\mathrm{Fam}_f(\mathbf{C})^{\mathrm{op}}, \mathrm{Fam}_f(J))$, where $\mathrm{Fam}_f(J)$ is the Grothendieck topology defined above on $\mathrm{Fam}_f(\mathbf{C})^{\mathrm{op}}$.*

The proof is straightforward but tedious, and we shall omit it.

5.   The Bagdomain Monad and its Algebras

From the results of section 3, it is easy to impose a monad structure on the functor $\mathcal{F} \mapsto \mathcal{B}_L(\mathcal{F})$, paralleling that on the functor Fam of section 2. The unit map $u: \mathcal{F} \to \mathcal{B}_L(\mathcal{F})$ corresponds to the operation of putting a T-model into a singleton bag; that is, it corresponds to the pair consisting of the classifying map $\hat{1}: \mathcal{F} \to \mathcal{O}$ of the terminal object of $\mathcal{F}$ and the isomorphism $\mathcal{F}/1 \cong \mathcal{F}$. Note that this morphism is actually an inclusion, since we may regard T as a quotient theory of T-bag, obtained by adding axioms which force the interpretation of the indexing type Y to be 1. Similarly, the multiplication $\mathcal{B}_L(\mathcal{B}_L(\mathcal{F})) \to \mathcal{B}_L(\mathcal{F})$ corresponds to the operation of taking a 'bag of bags of T-models' and putting them all into a single bag: formally, the generic model of the theory classified by $\mathcal{B}_L(\mathcal{B}_L(\mathcal{F}))$ consists of an object X of $\mathcal{B}_L(\mathcal{B}_L(\mathcal{F}))$, an object $f: Y \to X$ of $\mathcal{B}_L(\mathcal{B}_L(\mathcal{F}))/X$, and a T-model in $(\mathcal{B}_L(\mathcal{B}_L(\mathcal{F}))/X)/f \cong \mathcal{B}_L(\mathcal{B}_L(\mathcal{F}))/Y$, and we obtain a morphism to $\mathcal{B}_L(\mathcal{F})$ by forgetting X and f but remembering Y. The verification that this is indeed a monad structure on $\mathcal{B}_L$ is straightforward.

The next question, naturally, is to ask what we can say about the toposes which admit algebra structures for this monad. There are many possible answers to this question: the following result collects the most interesting of them.

Theorem 5.1 *Let $\mathcal{F}$ be an $\mathcal{S}$-topos. The following conditions are equivalent:*

(i) *$\mathcal{F}$ admits an algebra structure for the bagdomain monad.*

(ii) *The inclusion $u: \mathcal{F} \to \mathcal{B}_L(\mathcal{F})$ admits a splitting in Top/$\mathcal{S}$.*

(iii) *$\mathcal{F}$ is a retract of some topos of the form $\mathcal{B}_L(\mathcal{E})$.*

(iv) *For each $\mathcal{S}$-topos $\mathcal{G}$, the $\mathcal{G}$-indexed category Top/$\mathcal{S}$ $(\mathcal{G}, \mathcal{F})$ has an algebra structure for the monad Fam$_{\mathcal{G}}$ of section 2, and these structures are 'natural in $\mathcal{G}$' in a suitable sense.*

(v) *For each $\mathcal{S}$-topos $\mathcal{G}$, Top/$\mathcal{S}$ $(\mathcal{G}, \mathcal{F})$ has $\mathcal{G}$-indexed coproducts, which are natural in $\mathcal{G}$.*

(vi) *For each $\mathcal{S}$-topos $\mathcal{G}$, Top/$\mathcal{S}$ $(\mathcal{G}, \mathcal{F})$ has finite coproducts, which are natural in $\mathcal{G}$ (i.e. are preserved under composition with geometric morphisms $\mathcal{G}' \to \mathcal{G}$).*

(vii) *The inclusion $u: \mathcal{F} \to \mathcal{B}_L(\mathcal{F})$ admits a left adjoint in the 2-category Top/$\mathcal{S}$.*

(viii) *For each object I of S, the 'diagonal' geometric morphism $\mathcal{F} \to \mathcal{F}^{S/I}$ (i.e. the exponential transpose of the projection $\mathcal{F} \times S/I \to \mathcal{F}$) has a left adjoint in Top/S.*

(ix) *The structure morphism $\mathcal{F} \to S$, and the diagonal morphism $\mathcal{F} \to \mathcal{F} \times \mathcal{F}$, have left adjoints in Top/S.*

Remark 1. The sense in which the structures described in (iv) and (v) are required to be natural in $\mathcal{G}$ is a little delicate: we shall give the precise definition for (v), and leave (iv) as an exercise for the reader. Given a geometric morphism $f : \mathcal{G}' \to \mathcal{G}$ over S, we may 're-index' the $\mathcal{G}'$-indexed category $\mathsf{Top}/S\,(\mathcal{G}', \mathcal{F})$ over $\mathcal{G}$, by restricting our attention to families indexed by objects of the form f^*I; and this $\mathcal{G}$-indexed category has $\mathcal{G}$-indexed coproducts if the original one had $\mathcal{G}'$-indexed coproducts (note that the preservation of pullbacks by f^* is needed to verify the Beck-Chevalley condition). Composition with f induces a $\mathcal{G}$-indexed functor

$$\mathsf{Top}/S\,(\mathcal{G}, \mathcal{F}) \longrightarrow \mathsf{Top}/S\,(\mathcal{G}', \mathcal{F}) \quad ;$$

what we require is that this functor should preserve $\mathcal{G}$-indexed coproducts (i.e. should commute up to isomorphism with the functors Σ_α). However, readers who are unfamiliar with indexed categories are advised not to worry unduly about this; for all we really require of the naturality condition in (v) is that it should be sufficiently strong to imply that in (vi)—whose meaning is entirely straightforward.

Remark 2. We recall that, because of the reversal of direction involved in passing from a geometric morphism to its inverse image, the assertion that a geometric morphism f has a left adjoint in Top/S is equivalent to saying that f^* has a *right* adjoint which is the inverse image of a geometric morphism over S—equivalently, that the direct image f_* is also the inverse image of a geometric morphism over S. Compare [13], 1.1.

Proof : The implications (i) $\Rightarrow$ (ii) and (ii) $\Rightarrow$ (iii) are trivial. (iii) implies that, for each $\mathcal{G}$, $\mathsf{Top}/S\,(\mathcal{G}, \mathcal{F})$ is a retract (as a $\mathcal{G}$-indexed category) of $\mathsf{Top}/S\,(\mathcal{G}, \mathcal{B}_L(\mathcal{E}))$ in a way which is natural in $\mathcal{G}$. But by 3.5 the latter is equivalent to $\mathrm{Fam}_\mathcal{G}(\mathsf{Top}/S\,(\mathcal{G}, \mathcal{E}))$; and idempotents split in $\mathsf{Top}/S\,(\mathcal{G}, \mathcal{F})$ since it has filtered ($\mathcal{F}$-indexed) colimits. So it follows from 2.2 that (iii) implies (iv) and (v). However, from (v) we can get back to (i), by applying (v) in the generic case: that is, we obtain the structure map $\mathcal{B}_L(\mathcal{F}) \to \mathcal{F}$ by forming the coproduct of the T-models in the generic T-bag-model, and regarding the result as a T-model in $\mathcal{B}_L(\mathcal{F})$, and we show that it is indeed a $\mathcal{B}_L$-algebra structure using the naturality condition which we stated in Remark 1 above.

(v) also implies (vii), by essentially the same construction: if $\sigma: \mathcal{B}_L\mathcal{F} \to \mathcal{F}$ is the morphism classifying the coproduct of the generic T-bag-model, then it follows easily from the definition of morphisms of T-bag-models that, given any topos $\mathcal{G}$ and morphisms $a: \mathcal{G} \to \mathcal{F}$, $b: \mathcal{G} \to \mathcal{B}_L(\mathcal{F})$, we have a bijection between geometric transformations $b \to ua$ and transformations $\sigma b \to a$, which is natural in $\mathcal{G}$. Applying this bijection to the identity transformations on u and σ, we obtain transformations $\sigma u \to 1_{\mathcal{F}}$ and $1_{\mathcal{B}_L\mathcal{F}} \to u\sigma$, which are the counit and unit of an adjunction in Top/$\mathcal{S}$. However, (vii) implies (ii), because u is an inclusion.

A similar argument shows that (v) implies (viii). We may think of the exponential $\mathcal{F}^{S/I}$ as a classifying topos for I-indexed families of T-models, since morphisms $\mathcal{G} \to \mathcal{F}^{S/I}$ correspond to morphisms $\mathcal{G}/I \simeq \mathcal{G} \times \mathcal{S}/I \to \mathcal{F}$ (here we are following the usual custom of identifying the object I of $\mathcal{S}$ with the I-indexed copower of 1 in any $\mathcal{S}$-topos). If we form the coproduct of the generic I-indexed family of T-models, we obtain a morphism $\sigma_I: \mathcal{F}^{S/I} \to \mathcal{F}$, and the argument of the previous paragraph shows that σ_I is left adjoint to the diagonal. (Unfortunately, it does not seem possible to deduce (viii) directly from (vii): the reason is that, although we have an inclusion $\mathcal{F}^{S/I} \to \mathcal{B}_L(\mathcal{F})$, corresponding to the quotient theory of T-bag obtained by adding axioms to force the indexing object to coincide with I, the diagram

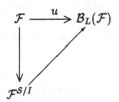

does not commute.)

It is clear that (viii) implies (ix), since the two halves of (ix) may be regarded, up to equivalence, as the particular cases $I = 0$ and $I = 1 + 1$ of (viii). And (ix) implies (vi), since composition with the given left adjoints yields functors $1 \to \text{Top}/\mathcal{S}(\mathcal{G}, \mathcal{F})$ and $\text{Top}/\mathcal{S}(\mathcal{G}, \mathcal{F})^2 \to \text{Top}/\mathcal{S}(\mathcal{G}, \mathcal{F})$ which are respectively left adjoint to the unique functor to 1 and the diagonal functor—that is, they endow $\text{Top}/\mathcal{S}(\mathcal{G}, \mathcal{F})$ with finite coproducts—and which are clearly natural in $\mathcal{G}$. Finally, (vi) implies (v) because, for any $\mathcal{G}$, the category $\text{Top}/\mathcal{S}(\mathcal{G}, \mathcal{F})$ has filtered $\mathcal{G}$-indexed colimits, and composition with a fixed geometric morphism $\mathcal{G}' \to \mathcal{G}$ preserves these colimits; and any object of $\mathcal{G}$ can be expressed as a filtered $\mathcal{G}$-indexed colimit of finite cardinals (cf. [15], 4.4).

We shall call a topos *hyperlocal* if it satisfies the conditions of Theorem 5.1; the reason for this name will be revealed in section 6 below.

Example 5.2 Let C be a small category having finite coproducts. Then the topos $[C, S]$ is hyperlocal: this follows immediately from 4.2, since the finite coproducts endow C with an algebra structure for the monad Fam_f on $\mathrm{cat}(S)$. The converse is false, because the embedding $\mathrm{cat}(S) \to \mathrm{Top}/S$ is not full: an algebra structure $\sigma \colon [\mathrm{Fam}_f(C), S] \to [C, S]$ need not be induced by a functor $\mathrm{Fam}_f(C) \to C$. However, because σ is left adjoint to u, it is necessarily an essential geometric morphism—that is, its inverse image has a left adjoint—and so it carries essential points of $[\mathrm{Fam}_f(C), S]$ into essential points of $[C, S]$; hence if idempotents split in C it is induced by a functor, and C must have finite coproducts. In general, therefore, we can say that $[C, S]$ *is hyperlocal iff the Karoubian completion of* C *has finite coproducts.*

Example 5.3 More generally, let C be a category with finite products, and let J be a Grothendieck topology on C with the properties

 (i) The terminal object 1 of C has no nontrivial covers, and

 (ii) Any cover of a product $C \times D$ is refined by a family of the form $(f \times g \mid f \in R, g \in S)$, where R and S are covers of C and D respectively.

Then the functor $\mathrm{Fam}_f(C^{\mathrm{op}})^{\mathrm{op}} \to C$ sending a finite family of objects of C to its product is cocontinuous (that is, reflects covers) for the topology on its domain defined before 4.3, and hence induces a geometric morphism $\mathcal{B}_L(\mathrm{Sh}(C, J)) \to \mathrm{Sh}(C, J)$. Moreover, the adjunction between the corresponding morphism of presheaf toposes and the inclusion u restricts to an adjunction at the level of sheaves, and so we deduce that *if* (C, J) *is a site satisfying the above hypotheses, then* $\mathrm{Sh}(C, J)$ *is a hyperlocal topos.*

Example 5.4 Let X be a (sober) topological space, and for convenience assume that the spatial product $X \times X$ is also the product in the category of locales (as is well known, local compactness of X is sufficient for this). Then the topos $\mathrm{Sh}(X)$ of sheaves on X is hyperlocal iff there are continuous maps $1 \to X$ and $X \times X \to X$ assigning a least element and binary joins in the poset (under the specialization ordering) of points of X; this follows easily from condition (ix). Since the poset of points of X always has directed joins (Johnstone [10], II 1.9), this implies that it is a complete lattice; conversely, if X is a space which is a complete lattice in its specialization ordering, then $\mathrm{Sh}(X)$ will be hyperlocal provided the binary join map $X \times X \to X$ is continuous. In particular, any continuous lattice equipped with its Scott topology (or more generally, any complete lattice whose Scott topology is itself continuous—cf. [3], II 4.13) defines a hyperlocal topos.

As in the argument at the end of section 2, we may show that the algebra structure in (ii) of 5.1, if it exists, is determined up to canonical isomorphism—that is, it must be left adjoint to u in $\mathsf{Top}/\mathcal{S}$—whereas the retraction in (iii) is merely a 'weak left adjoint' (that is, it comes equipped with geometric transformations satisfying one of the triangular identities, but not necessarily the other) and so need not be unique. (The reason, once again, is that the monad $\mathcal{B}_L$ is a KZ-doctrine on $\mathsf{Top}/\mathcal{S}$.) In particular, we note that the (2-)category of $\mathcal{B}_L$-algebras may be regarded as a sub-2-category of $\mathsf{Top}/\mathcal{S}$; and it is not difficult to determine what its morphisms are:

Proposition 5.5 *The 2-category of $\mathcal{B}_L$-algebras may be identified with the 2-category whose objects are hyperlocal $\mathcal{S}$-toposes, whose morphisms are geometric morphisms over $\mathcal{S}$ 'preserving finite coproducts' (i.e. commuting in the obvious sense with the left adjoints defined in 5.1(ix)), and whose 2-cells are arbitrary geometric transformations (over $\mathcal{S}$) between such morphisms.*

6.   Bagdomains and Scones

The name 'hyperlocal topos' is intended to remind the reader of the local toposes studied by Johnstone and Moerdijk [13]; recall that an $\mathcal{S}$-topos $\mathcal{F}$ is said to be *local* if the structure morphism $\mathcal{F} \to \mathcal{S}$ has an $\mathcal{S}$-indexed left adjoint, and so 5.1(ix) implies that every hyperlocal topos is indeed local. In [13], we established the analogues for local toposes of several of the other conditions of 5.1; see in particular Theorem 1.5 and Corollary 2.10 of that paper. However, we did not provide a description of local toposes as algebras for a monad on $\mathsf{Top}/\mathcal{S}$; since it is possible to do so without much further effort, it seems worthwhile to give the details here.

The functor part of the monad is the *scone* (or Sierpiński cone) construction, also known as the *Freyd cover*: the scone $\hat{\mathcal{F}}$ of an $\mathcal{S}$-topos $\mathcal{F}$ is usually defined as the topos obtained by Artin glueing (cf. [22]) along the global sections functor $\mathcal{F} \to \mathcal{S}$. However, it can also be described as a partial product, emphasizing the analogy with the bagdomain. Recall ([6], 4.37(iii)) that the *Sierpiński topos* $\hat{\mathcal{S}} = [\mathbf{2}, \mathcal{S}]$ is a classifying topos for subobjects of 1 in $\mathcal{S}$-toposes. If U denotes the generic subobject of 1 in $\hat{\mathcal{S}}$ (which is in fact the morphism $(0 \to 1)$ in $\mathcal{S}$), then $\hat{\mathcal{F}}$ may be described as the partial product of $\mathcal{F}$ over $s_U : \hat{\mathcal{S}}/U \to \hat{\mathcal{S}}$. Equivalently, by 3.2, it is the pullback of $\mathcal{B}_L(\mathcal{F})$ along the geometric morphism $\hat{\mathcal{S}} \to \mathcal{O}$ classifying the generic subobject of 1. We note also that, since this latter morphism is an inclusion (it is induced by the obvious full embedding $\mathbf{2} \to \mathbf{S}_f$ in $\mathbf{cat}(\mathcal{S})$), $\hat{\mathcal{F}}$ may be regarded as a subtopos of $\mathcal{B}_L(\mathcal{F})$; the theory it classifies may be described as the quotient of $\mathbf{T}$-**bag** obtained by adding the axiom which forces the interpretation of the indexing sort Y to be a subobject of 1. As such, it is clear that the unit map

$u: \mathcal{F} \rightarrow \mathcal{B}_L(\mathcal{F})$ factors through it, and indeed that the functor $\mathcal{F} \mapsto \hat{\mathcal{F}}$ has the structure of a submonad of $\mathcal{B}_L$. Now we can state the 'local' analogue of 5.1:

Theorem 6.1 *For an $\mathcal{S}$-topos $\mathcal{F}$, the following are equivalent:*

(i) *$\mathcal{F}$ has an algebra structure for the scone monad on $\mathsf{Top}/\mathcal{S}$.*

(ii) *The canonical inclusion $\mathcal{F} \rightarrow \hat{\mathcal{F}}$ has a splitting in $\mathsf{Top}/\mathcal{S}$.*

(iii) *$\mathcal{F}$ is a retract of some $\mathcal{S}$-topos of the form $\hat{\mathcal{E}}$.*

(iv) *The inclusion $\mathcal{F} \rightarrow \hat{\mathcal{F}}$ has a left adjoint in $\mathsf{Top}/\mathcal{S}$.*

(v) *The structure morphism $\mathcal{F} \rightarrow \mathcal{S}$ has a left adjoint in $\mathsf{Top}/\mathcal{S}$.*

(vi) *For every $\mathcal{S}$-topos $\mathcal{G}$, the category $\mathsf{Top}/\mathcal{S}\,(\mathcal{G}, \mathcal{F})$ has an initial object, and these initial objects are preserved under composition with geometric morphisms $\mathcal{G}' \rightarrow \mathcal{G}$.*

Proof : All of this, except (i) and (iv), is contained in 1.5, 2.9 and 2.10 of [13]. However, the construction given there (starting from (vi)) of a retraction for $\mathcal{F} \rightarrow \hat{\mathcal{F}}$ is easily seen to yield a morphism which is both an algebra structure and a left adjoint for the inclusion, by arguments exactly like those in the proof of 5.1.

Once again, the 2-category of algebras for the scone monad on $\mathsf{Top}/\mathcal{S}$ may be explicitly described: it is the 2-category of local toposes, *strict* morphisms (that is, morphisms which preserve their distinguished points) and arbitrary geometric transformations between them. The proof of this is essentially contained in Lemma 3.2 of [13].

Johnstone and Moerdijk also consider the notion of *localization* of a given topos at a given point: that is, the universal solution to the problem of mapping a local topos into the given one, in such a way that its initial point is mapped onto the chosen one. Is there an analogous notion of hyperlocalization? The answer is yes: but we need to be given more data than a single point of our topos to construct it. In fact the initial data for hyperlocalization consists of an $\mathcal{S}$-topos $\mathcal{E}$ together with a 'symmetric monoidal structure' on $\mathcal{E}$ in the 2-category $\mathsf{Top}/\mathcal{S}$; that is, a pair of geometric morphisms $\mathcal{S} \rightarrow \mathcal{E}$ and $\mathcal{E} \times \mathcal{E} \rightarrow \mathcal{E}$, together with invertible geometric transformations which fit into the associativity, symmetry and unit diagrams and satisfy the usual coherence conditions. (Note that this has nothing to do with a symmetric monoidal structure on the underlying category of $\mathcal{E}$; but it induces a symmetric monoidal structure on each category of the form $\mathsf{Top}/\mathcal{S}\,(\mathcal{G},\mathcal{E})$, in a manner which is natural in $\mathcal{G}$.)

It is well known (cf. [14], 4.1) that if we are given a symmetric monoidal structure on a category C and wish to turn it into a cocartesian structure, the thing to do is to pass to the category $\mathbf{CMon}(C)$ of commutative monoids (for the given monoidal structure) in C; this corresponds to the passage from C to a co-slice category $C\backslash C$ in order to make a given object C initial. So we need to find a construction on a topos $\mathcal{E}$ equipped with a symmetric monoidal structure in $\mathsf{Top}/\mathcal{S}$, which will induce this construction on its category of $\mathcal{G}$-valued points, for any $\mathcal{G}$.

Let $\otimes : \mathcal{E} \times \mathcal{E} \to \mathcal{E}$ and $i : \mathcal{S} \to \mathcal{E}$ be the geometric morphisms defining the symmetric monoidal structure. Let $\mathbf{C}$, for the moment, denote (the internalization in $\mathcal{S}$ of) the finite category represented diagrammatically by

$$\bullet \longrightarrow \bullet \longleftarrow \bullet \;,$$

and let $\mathcal{E}'$ denote the exponential $\mathcal{E}^{[\mathbf{C},\mathcal{S}]}$. If $\mathcal{E}$ classifies a geometric theory T, then $\mathcal{E}'$ classifies the theory whose models are diagrams of T-models of shape $\mathbf{C}$, so there is a geometric morphism $u : \mathcal{E}' \to \mathcal{E}^3$ corresponding to the operation which sends such a diagram to the ordered triple of T-models appearing in it. Let $v : \mathcal{E} \to \mathcal{E}^3$ be the geometric morphism whose components are respectively the composite of $\otimes$ with the diagonal, the identity on $\mathcal{E}$, and the composite of i with the structure morphism $\mathcal{E} \to \mathcal{S}$; then if we form the pullback

$$
\begin{array}{ccc}
\mathcal{E}'' & \longrightarrow & \mathcal{E}' \\
\downarrow & & \downarrow{\scriptstyle u} \\
\mathcal{E} & \xrightarrow{\;v\;} & \mathcal{E}^3
\end{array}
$$

we obtain a classifying topos for the theory whose models are diagrams of shape

$$A \otimes A \longrightarrow A \longleftarrow I$$

in the category of T-models. By applying a suitable forcing topology to this topos (cf. Tierney [19]), we obtain a subtopos of $\mathcal{E}''$ which classifies commutative monoids in the category of T-models, as required. We thus obtain

Proposition 6.2 *Let $\mathcal{E}$ be an $\mathcal{S}$-topos and $(\otimes, i)$ a symmetric monoidal structure on $\mathcal{E}$ in $\mathsf{Top}/\mathcal{S}$. Then there exists a hyperlocal topos $\mathcal{F}$, the hyperlocalization of $\mathcal{E}$ at the structure $(\otimes, i)$, equipped with a morphism $f : \mathcal{F} \to \mathcal{E}$ which*

is universal amongst morphisms making

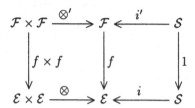

commute, where $(\otimes', i')$ *is the cocartesian monoidal structure on* $\mathcal{F}$ *(i.e. the two left adjoints defined in 5.1(ix)).*

7. Bagdomains and Powerdomains

In Example 5.4, we saw (more or less) that the localic toposes which admit $\mathcal{B}_L$-algebra structures are exactly those whose underlying locales admit algebra structures for the lower (Hoare) powerdomain monad on the category of locales. In order to understand why this should be so, we need to look more closely at the relationship between the bagdomain and the powerdomain.

Given a locale X, we write ΩX for the frame of opens of X, and $P_L(X)$ for the Hoare powerdomain of X, that is the locale whose frame of opens is freely generated by symbols $\lozenge U$, $U \in \Omega X$, subject to the condition that $U \mapsto \lozenge U$ preserves arbitrary joins (cf. Vickers [20], 11.3.1—note that we follow Vickers in *not* imposing the relation which says that $\lozenge X$ is the whole of $P_L(X)$). Our main result is

Theorem 7.1 *For any locale* X, $\mathrm{Sh}(P_L(X))$ *is the localic reflection of the topos* $\mathcal{B}_L(\mathrm{Sh}(X))$.

Proof :   Given the factorization theorem of Johnstone [8], it suffices to construct a hyperconnected geometric morphism

$$\mathcal{B}_L(\mathrm{Sh}(X)) \longrightarrow \mathrm{Sh}(P_L(X)) .$$

To do this, we first observe that there is a functor

$$T \colon \mathrm{Fam}_f(\Omega X^{\mathrm{op}})^{\mathrm{op}} \longrightarrow \Omega P_L(X)$$

sending a finite family $(U_1, \ldots, U_n)$ of opens to $\bigwedge_{i=1}^n \lozenge U_i$. This is functorial because a morphism $(U_1, \ldots, U_n) \to (V_1, \ldots, V_m)$ in $\mathrm{Fam}_f(\Omega X^{\mathrm{op}})^{\mathrm{op}}$ is a function $\alpha \colon \{1, \ldots, m\} \to \{1, \ldots, n\}$ such that $U_{\alpha(j)} \leq V_j$ for each j, and if we have such a function then

$$\bigwedge_{i=1}^n \lozenge U_i \leq \bigwedge_{j=1}^m \lozenge U_{\alpha(j)} \leq \bigwedge_{j=1}^m \lozenge V_j .$$

In the converse direction, if $\bigwedge_{i=1}^{n} \Diamond U_i \le \Diamond V$ in $\Omega P_L(X)$, then we must have $U_i \le V$ for some i; from this it follows easily that T is a full functor. So if we write $BP_L(X)$ for the image of T, i.e. the poset of 'basic opens' of $P_L(X)$ which are expressible as finite meets of opens of the form $\Diamond U$, then it follows from [8], 3.1(ii), that T induces a hyperconnected geometric morphism

$$[\mathrm{Fam}_f(\Omega X^{\mathrm{op}}), \mathcal{S}] \longrightarrow [BP_L(X)^{\mathrm{op}}, \mathcal{S}] \; .$$

Now let J be the Grothendieck topology on $BP_L(X)$ induced from the canonical topology on $\Omega P_L(X)$ (i.e. the largest topology making the inclusion functor continuous): since $BP_L(X)$ is a base for $\Omega P_L(X)$, it follows from the Comparison Lemma that $(BP_L(X), J)$ is a site of definition for $\mathbf{Sh}(P_L(X))$. If we now form the pullback topology J' on $\mathrm{Fam}_f(\Omega X^{\mathrm{op}})^{\mathrm{op}}$ (i.e. the smallest topology making T cocontinuous), it is not hard to identify it with (the case $\mathbf{C} = \Omega X^{\mathrm{op}}$ of) that described before 4.3. Thus we have a pullback diagram

$$\mathcal{B}_L(\mathbf{Sh}(X)) \simeq \mathbf{Sh}(\mathrm{Fam}_f(\Omega X^{\mathrm{op}})^{\mathrm{op}}, J') \longrightarrow \mathbf{Sh}(BP_L(X), J) \simeq \mathbf{Sh}(P_L(X))$$

$$[\mathrm{Fam}_f(\Omega X^{\mathrm{op}}), \mathcal{S}] \longrightarrow [BP_L(X)^{\mathrm{op}}, \mathcal{S}]$$

in $\mathbf{Top}/\mathcal{S}$. But pullbacks of hyperconnected morphisms are hyperconnected ([8], 2.3).

The hyperconnected morphism defined in the proof of 7.1 is readily checked to be natural in X, and to commute with the units and multiplications of the monad structures on $\mathcal{B}_L$ and P_L, in a suitable sense. Now if $\mathcal{F} = \mathbf{Sh}(X)$ is a localic $\mathcal{S}$-topos, then any morphism $\mathcal{B}_L(\mathcal{F}) \to \mathcal{F}$ (in particular, a $\mathcal{B}_L$-algebra structure on $\mathcal{F}$, if there is one) factors uniquely through the localic reflection of its domain, from which it is a mere formality to deduce

Corollary 7.2 *For any locale X, there is a bijection between P_L-algebra structures on X and $\mathcal{B}_L$-algebra structures on $\mathbf{Sh}(X)$.*

(Of course, since there is at most one structure of either type on a given X, this result is less interesting than it appears at first sight.)

8.   Upper and Mixed Bagdomains

Having defined and studied the lower bagdomain, it is natural to ask whether we can also define a bagdomain analogue of the upper (Smyth) powerdomain. As usual, let $\mathcal{F}$ be the classifying topos for a geometric theory $\mathbf{T}$.

Then a first guess at what the upper bagdomain $\mathcal{B}_U(\mathcal{F})$ should be might suggest that its points should be bags of T-models, as before, but that a morphism $(M_i \mid i \in I) \to (N_j \mid j \in J)$ between such bags should consist of a function $\alpha \colon J \to I$ together with a J-indexed family of T-model morphisms $M_{\alpha(j)} \to N_j$. That is, we want $\mathsf{Top}/\mathcal{S}\,(\mathcal{S}, \mathcal{B}_U(\mathcal{F}))$ to be equivalent to the category $\mathsf{Fam}(\mathsf{Top}/\mathcal{S}\,(\mathcal{S}, \mathcal{F})^{\mathrm{op}})^{\mathrm{op}}$ which is obtained by freely adjoining products to $\mathsf{Top}/\mathcal{S}\,(\mathcal{S}, \mathcal{F})$.

However, this is impossible even in the case $\mathcal{F} = \mathcal{S}$; for it would require the category of points of $\mathcal{B}_U(\mathcal{S})$ to be equivalent to $\mathcal{S}^{\mathrm{op}}$. But $\mathcal{S}^{\mathrm{op}}$ is not an accessible category, and so cannot be the category of ($\mathcal{S}$-valued) models of any sketchable theory (cf. Makkai and Paré [16]); a fortiori it is not the category of models of a geometric theory.

A better guess is that, if the category of points of $\mathcal{F}$ is the ind-completion of a small category C, then the category of points of $\mathcal{B}_U(\mathcal{F})$ should be the ind-completion of $\mathsf{Fam}_f(\mathsf{C}^{\mathrm{op}})^{\mathrm{op}}$. This leads in particular to the idea that $\mathcal{B}_U(\mathcal{S})$ should be the classifying topos for the theory of Boolean algebras (cf. [10], VI 3.1), and hence to the idea that $\mathcal{B}_U(\mathcal{F})$ might be the partial product of $\mathcal{F}$ over $\mathcal{E}' \to \mathcal{E}$, where $\mathcal{E}$ is the classifying topos for Boolean algebras and $\mathcal{E}'$, as an $\mathcal{E}$-topos, is (the topos of sheaves on) the 'Stone locale' of the generic Boolean algebra. (As an $\mathcal{S}$-topos, $\mathcal{E}'$ classifies the theory whose models are Boolean algebras equipped with a distinguished prime ideal.) Note that, since Stone locales are compact and regular, $\mathcal{E}'$ is exponentiable in $\mathsf{Top}/\mathcal{E}$ ([12], 5.10), and so 3.1 should assure us that this partial product exists. Thus a point of $\mathcal{B}_U(\mathcal{F})$ will correspond, not to a 'discrete' indexed family of T-models, but to a sheaf of T-models on a Stone space.

Preliminary calculations suggest that this is indeed an appropriate definition of $\mathcal{B}_U$, but a lot of work remains to be done before we can establish the analogues of the results about $\mathcal{B}_L$ proved in this paper.

In his paper [21], Vickers also defines a 'mixed bagdomain' construction for algebraic posets. However, a glance at this construction reveals that it is not so much a mixture of the upper and lower bagdomains, as of the upper powerdomain and the lower bagdomain. The main obstacle to extending this construction to arbitrary $\mathcal{S}$-toposes therefore seems to be the lack of an extension of the powerdomain functor from locales to toposes. (Clearly, further work needs to be done in order to establish whether this is possible; but the final section of the present paper is not the place to embark on it.) On the other hand, if we wish to produce a mixture of the upper and lower bagdomains, it seems likely that the starting-point will be the classifying topos for finite decidable objects (cf. [15], p. 189); it is surely significant that such objects are exactly the discrete Stone locales (cf. [7], Lemma 1.8).

References

[1] A.R. Blass, 'Classifying topoi and the axiom of infinity', *Alg. Universalis* 26 (1989), 341–345.

[2] R. Dyckhoff and W. Tholen, 'Exponentiable morphisms, partial products and pullback complements', *J. Pure Appl. Alg.* 49 (1987), 103–116.

[3] G. Gierz, K.H.Hofmann, K. Keimel, J.D. Lawson, M. Mislove and D.S. Scott, *A Compendium of Continuous Lattices*, Springer-Verlag (1980).

[4] C. Gunter, 'The mixed powerdomain', *Theoret. Comp. Sci.*, to appear.

[5] J.M.E. Hyland and A.M. Pitts, 'The theory of constructions: categorical semantics and topos-theoretic models', in *Categories in Computer Science and Logic*, Contemp. Math. vol. 92 (Amer. Math. Soc., 1989), 137–199.

[6] P.T. Johnstone, *Topos Theory*, Academic Press (1977).

[7] P.T. Johnstone, 'The Gleason cover of a topos, II', *J. Pure Appl. Alg.* 22 (1981), 229–247.

[8] P.T. Johnstone, 'Factorization theorems for geometric morphisms, I', *Cahiers Top. Géom. Diff.* 22 (1981), 3–17.

[9] P.T. Johnstone, 'Injective toposes', in *Continuous Lattices*, Lecture Notes in Math. vol. 871 (Springer-Verlag, 1981), 284–297.

[10] P.T. Johnstone, *Stone Spaces*, Cambridge Univ. Press (1983).

[11] P.T. Johnstone, 'Fibrations and partial products in a 2-category', in preparation.

[12] P.T. Johnstone and A. Joyal, 'Continuous categories and exponentiable toposes', *J. Pure Appl. Alg.* 25 (1982), 255–296.

[13] P.T. Johnstone and I. Moerdijk, 'Local maps of toposes', *Proc. Lond. Math. Soc.* (3) 58 (1989), 281–305.

[14] P.T. Johnstone and S.J. Vickers, 'Preframe presentations present' to appear in *CT '90*, Lecture Notes in Math. (Springer-Verlag).

[15] P.T. Johnstone and G.C. Wraith, 'Algebraic theories in toposes', in *Indexed Categories and Their Applications*, Lecture Notes in Math. vol. 661 (Springer-Verlag, 1978), 141–242.

[16] M. Makkai and R. Paré, *Accessible Categories: the Foundations of Categorical Model Theory*, Contemp. Math. vol. 104 (Amer. Math. Soc., 1989).

[17] B.A. Pasynkov, 'Partial topological products', *Trans. Moscow Math. Soc.* 13 (1965), 153–272.

[18] R. Street, 'Fibrations in bicategories', *Cahiers Top. Géom. Diff.* 21 (1980), 111–160.

[19] M. Tierney, 'Forcing topologies and classifying topoi', in *Algebra, Topology and Category Theory: a Collection of Papers in Honor of Samuel Eilenberg*, Academic Press (1976), 211–219.

[20] S.J. Vickers, *Topology via Logic*, Cambridge Univ. Press (1989).

[21] S.J. Vickers, 'Geometric theories and databases', this volume.

[22] G.C. Wraith, 'Artin glueing', *J. Pure Appl. Alg.* 4 (1974), 345–348.

Printed in the United States
By Bookmasters